# RICHARD OF
# WALLINGFORD

British Museum, MS. Cotton Claud. E.iv, f. 201: Richard of Wallingford depicted dividing a metal plate, in the Abbot's study at St. Albans. See vol. ii, p. 17

# RICHARD

# OF

# WALLINGFORD

*An edition of his writings*
*with introductions, English translation*
*and commentary by*

## J. D. NORTH

## III

ILLUSTRATIONS, TABLES
APPENDIXES, GLOSSARIES
BIBLIOGRAPHY, AND
INDEXES

OXFORD
AT THE CLARENDON PRESS
1976

*Oxford University Press, Ely House, London W. 1*

OXFORD  LONDON  GLASGOW  NEW YORK
TORONTO  MELBOURNE  WELLINGTON  CAPE TOWN
IBADAN  NAIROBI  DAR ES SALAAM  LUSAKA  ADDIS ABABA
KUALA LUMPUR  SINGAPORE  JAKARTA  HONG KONG  TOKYO
DELHI  BOMBAY  CALCUTTA  MADRAS  KARACHI

ISBN 0 19 858139 4

© *Oxford University Press 1976*

MATH-STAT.

*Printed in Great Britain
at the University Press, Oxford
by Vivian Ridler
Printer to the University*

# CONTENTS

## VOLUME III

# LIST OF PLATES

# FIGURES

The following diagrams, illustrating the treatises of vol. i and the commentaries of vol. ii, are generally numbered to correspond to the chapters of the original works. The addition of an asterisk (*) to the chapter reference means that the diagram is not to be found in anything resembling its present form in any of the manuscripts.

# FIGURES TO
# QUADRIPARTITUM

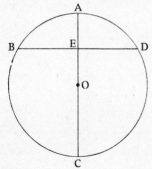

**I. Prol.**

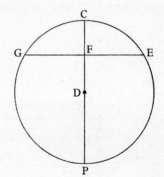

**I.1, I.2**

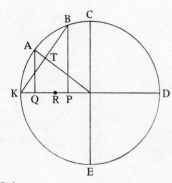

**I.3**

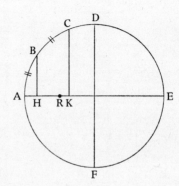

**I.4**

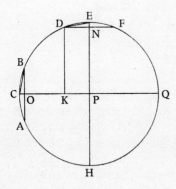

**I.5**

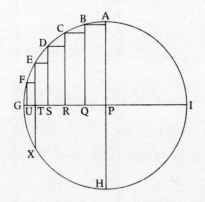

**I.6**

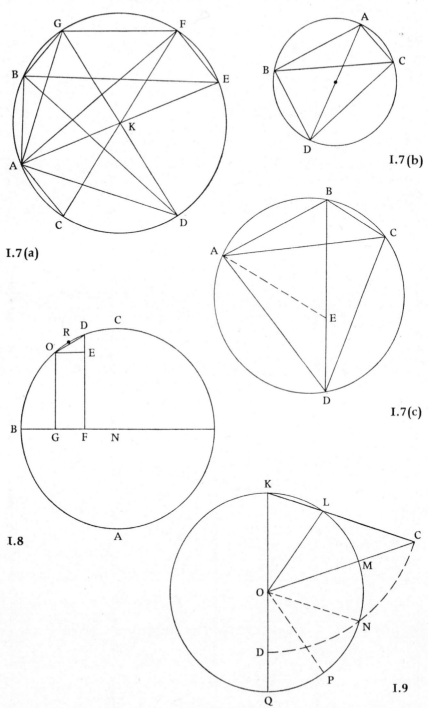

I.7(a)

I.7(b)

I.7(c)

I.8

I.9

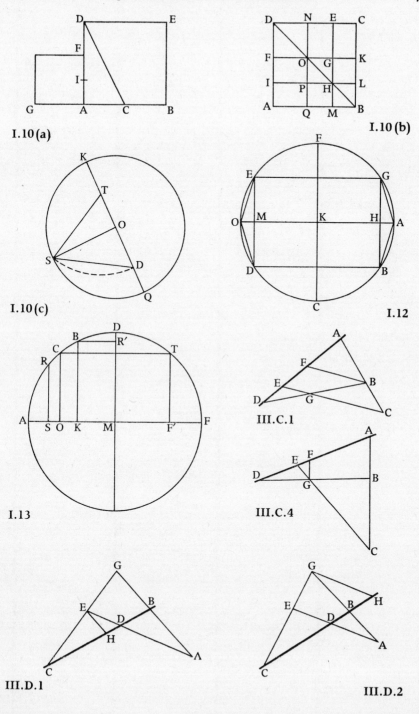

I.10 (a)

I.10 (b)

I.10 (c)

I.12

I.13

III.C.1

III.C.4

III.D.1

III.D.2

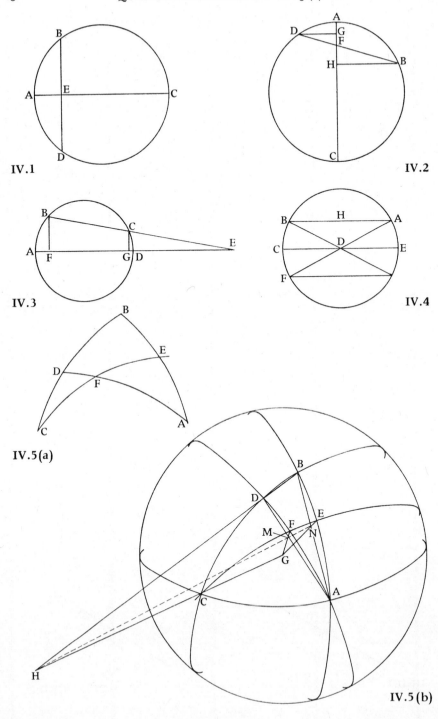

IV.1

IV.2

IV.3

IV.4

IV.5(a)

IV.5(b)

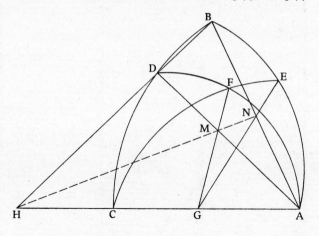

IV.5(c)

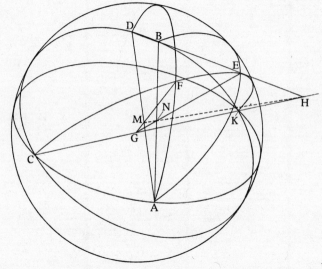

IV.5(d)

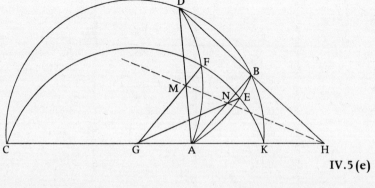

IV.5(e)

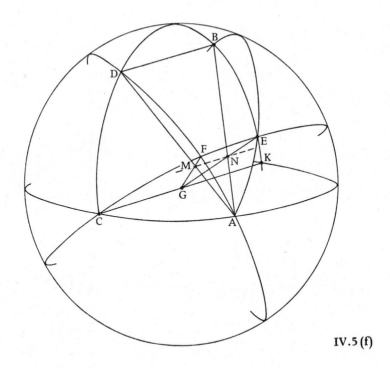

IV.5 (f)

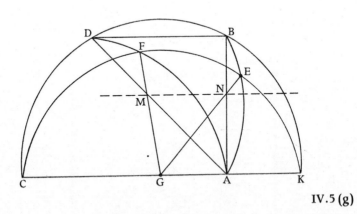

IV.5 (g)

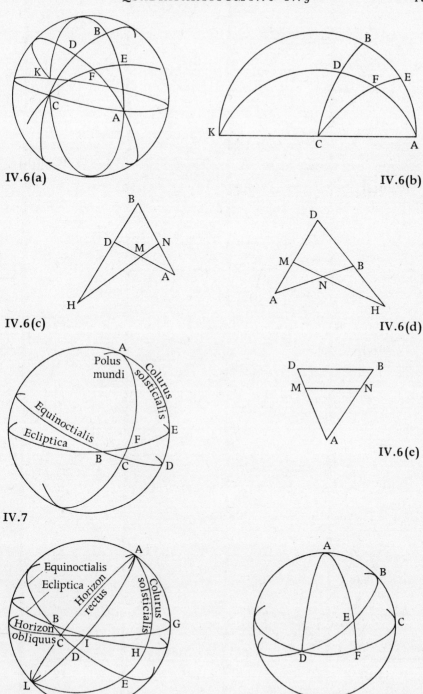

IV.6(a)

IV.6(b)

IV.6(c)

IV.6(d)

IV.6(e)

IV.7

IV.8

IV.9

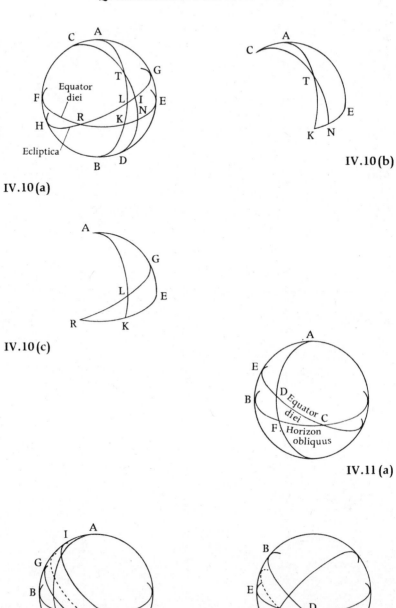

IV.10(a)

IV.10(b)

IV.10(c)

IV.11(a)

IV.11(b)

IV.11(c)

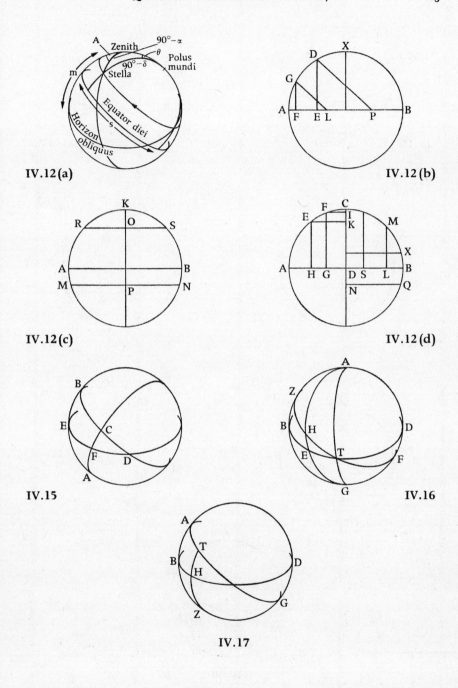

IV.12 (a)

IV.12 (b)

IV.12 (c)

IV.12 (d)

IV.15

IV.16

IV.17

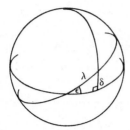

IV.18

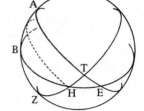

IV.22

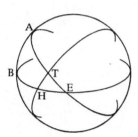

IV.23

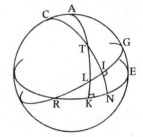

IV.26

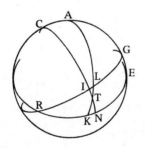

IV.27

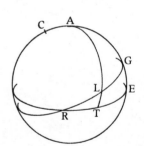

IV.29

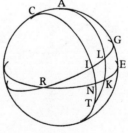

IV.30

# FIGURE TO
# EXAFRENON PRONOSTICACIONUM
## TEMPORIS

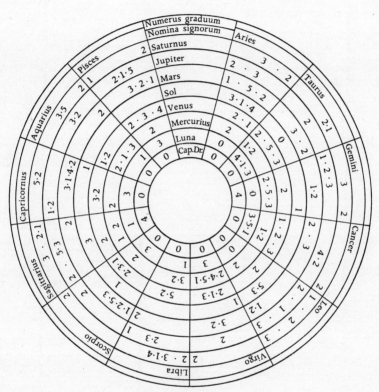

Key:  5  domicile
      4  exaltation
      3  triplicity
      2  term
      1  face

# FIGURES TO
# TRACTATUS ALBIONIS

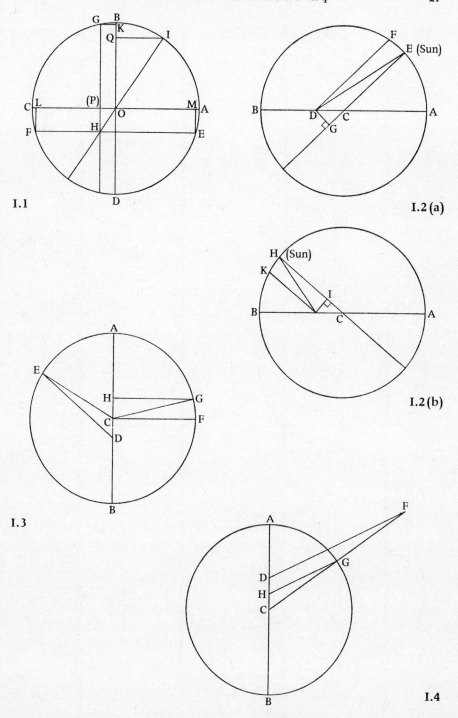

I.1

I.2 (a)

I.2 (b)

I.3

I.4

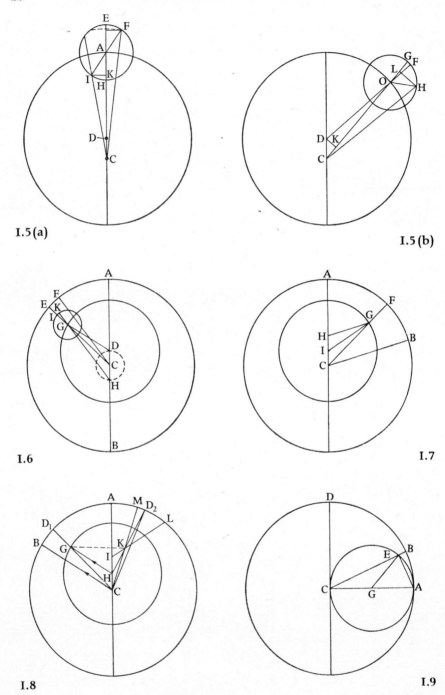

I.5(a)

I.5(b)

I.6

I.7

I.8

I.9

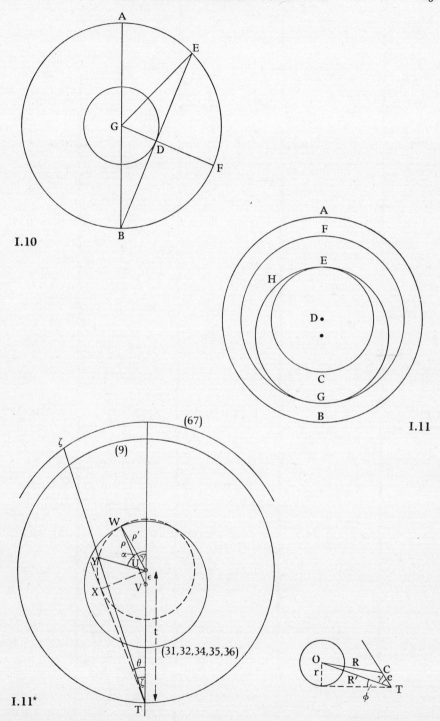

I.10

I.11

I.11*

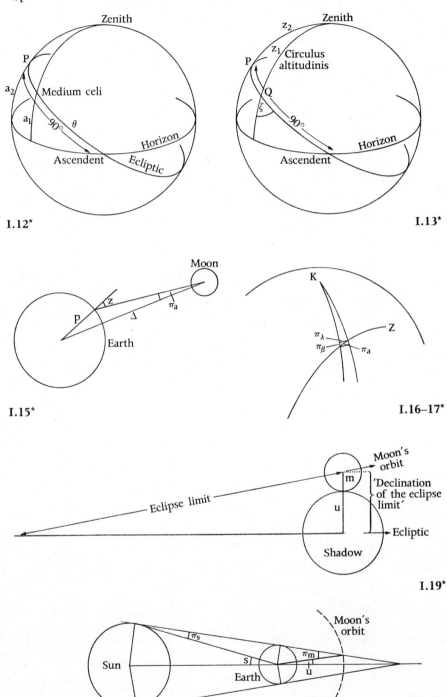

I.12*

I.13*

I.15*

I.16–17*

I.19*

I.19*

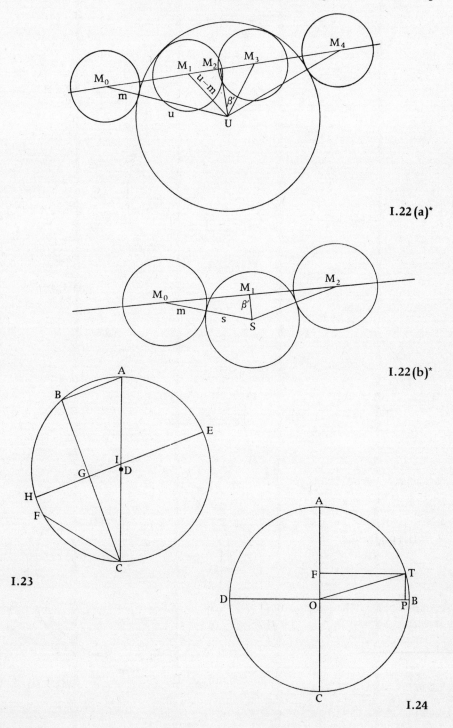

I.22 (a)*

I.22 (b)*

I.23

I.24

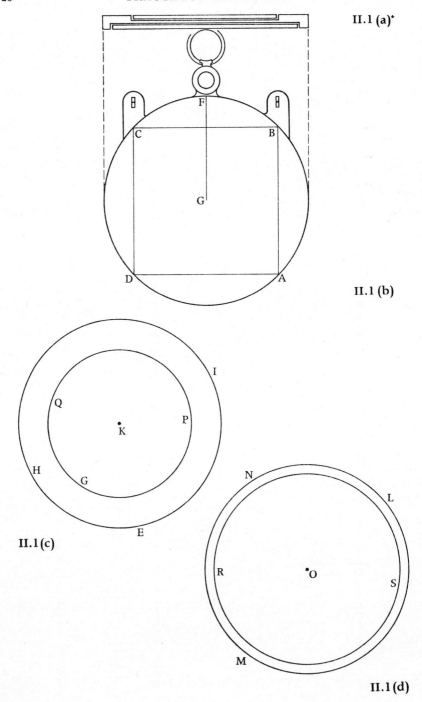

II.1 (a)*

II.1 (b)

II.1 (c)

II.1 (d)

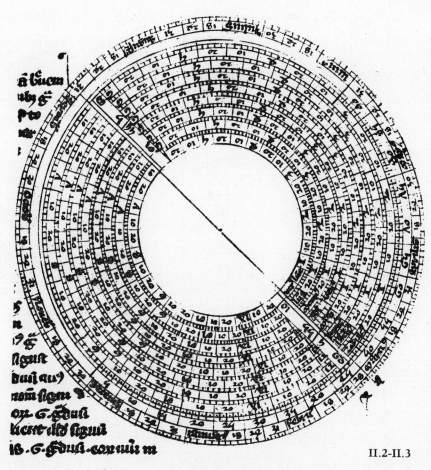

II.2-II.3

The figure is taken from MS. C.C.C. (D) 144, f. 55<sup>r</sup>

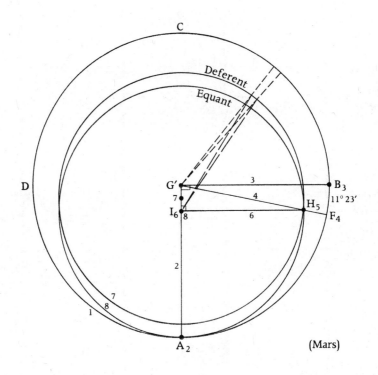

II.4*

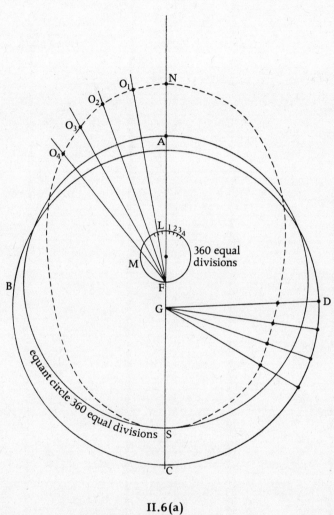

II.6(a)

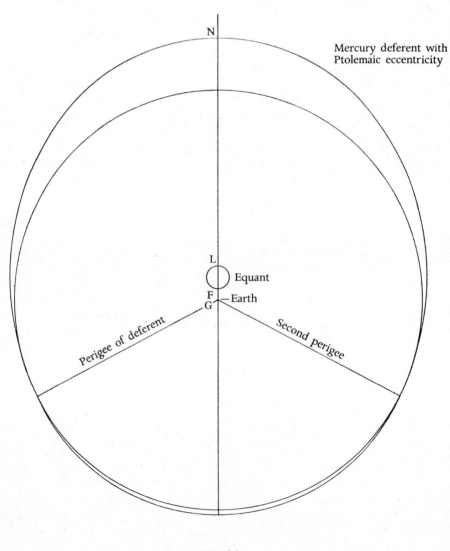

II.6 (b)

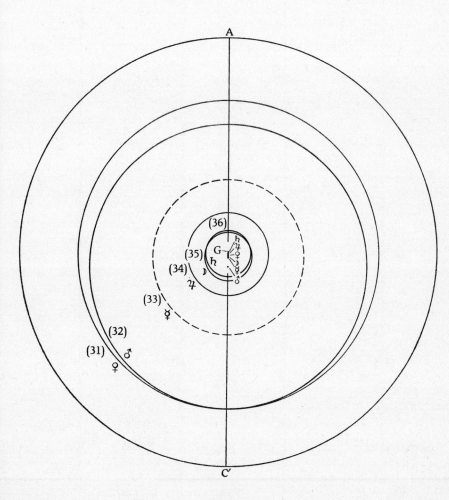

II.8*

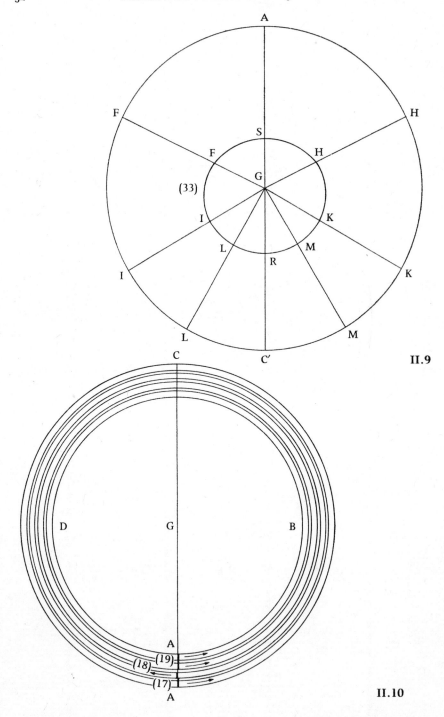

II.9

II.10

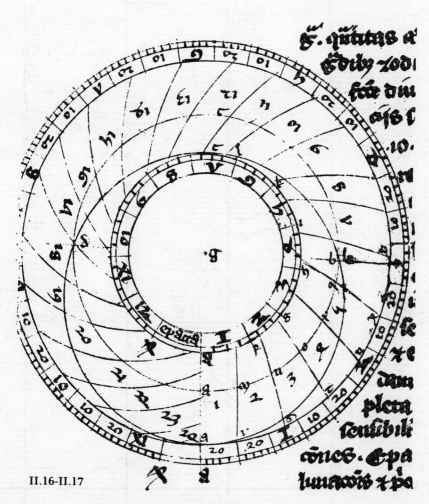

II.16-II.17

The figure is taken from MS. C.C.C. (D) 144, f. 58ᵛ.

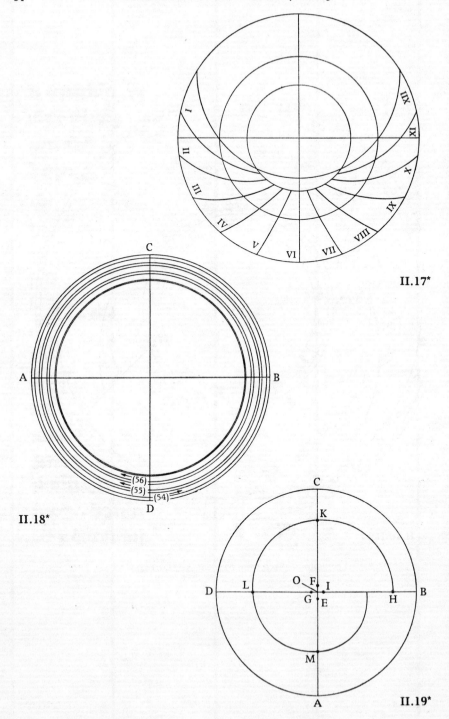

II.17*

II.18*

II.19*

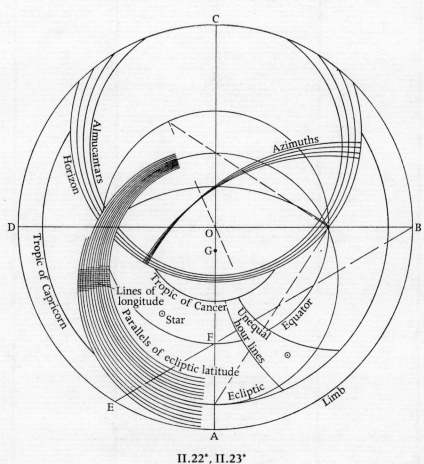

**II.22\*, II.23\***

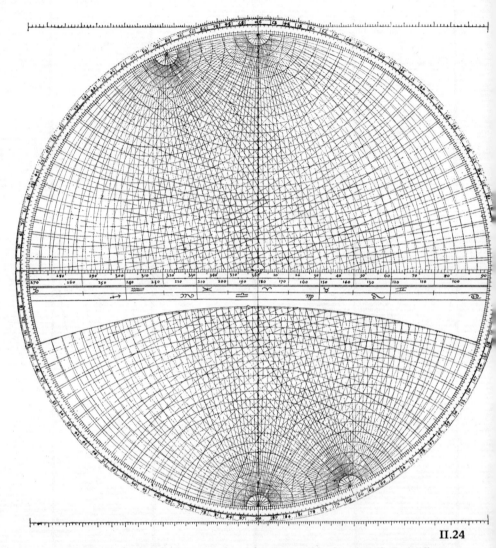

**II.24**

Double 'saphea' projection, reproduced from a movable chart in MS. Radcliffe 74.

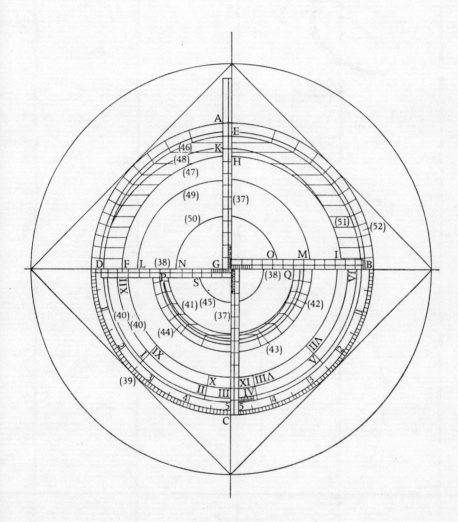

II.25* – II.28*

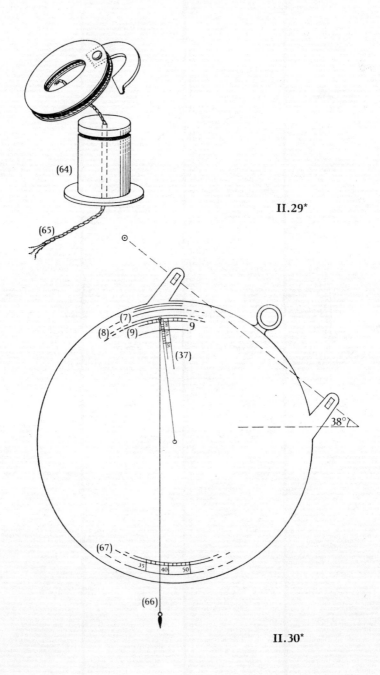

(64)

II.29*

(65)

(7)
(8)   (9)        9
(37)

38°

(67)

35   40   50

(66)

II.30*

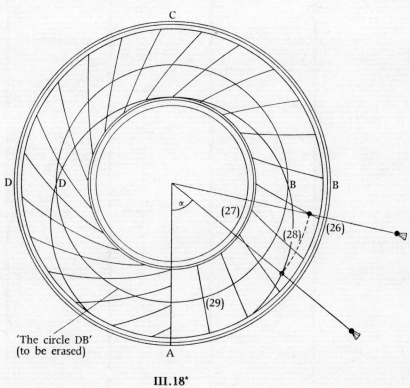

'The circle DB'
(to be erased)

III.18*

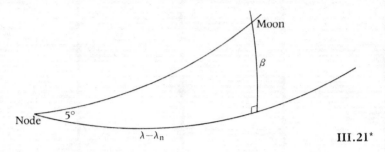

III.21*

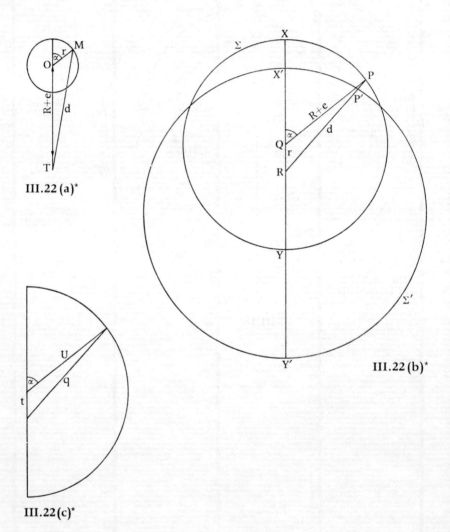

III.22 (a)*

III.22 (b)*

III.22 (c)*

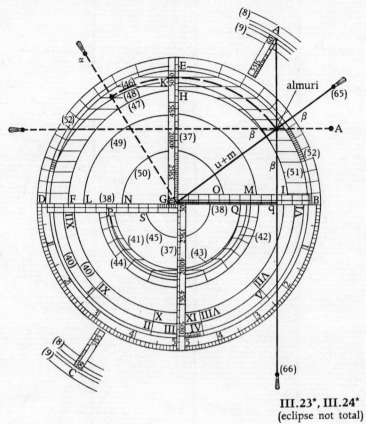

**III.23\*, III.24\***
(eclipse not total)

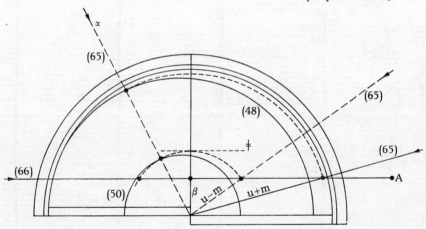

**III.24\***
(total eclipse)

‡ marks the true limiting value of β, below
which there will be a period of total eclipse.

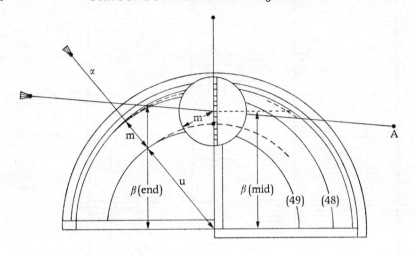

III.25*

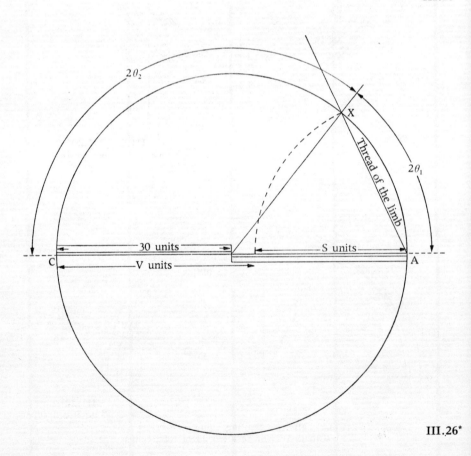

III.26*

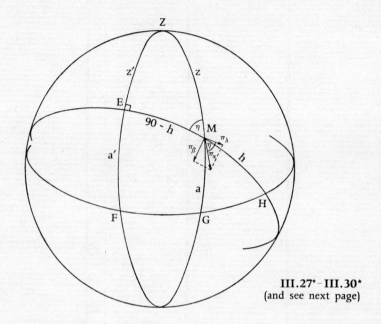

**III.27\* - III.30\***
(and see next page)

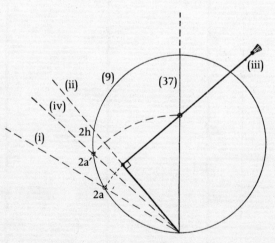

**III. 27\***
(and see next page)

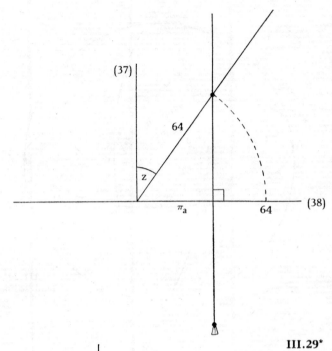

**III.29***

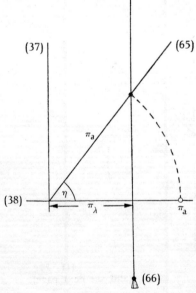

**III.30***

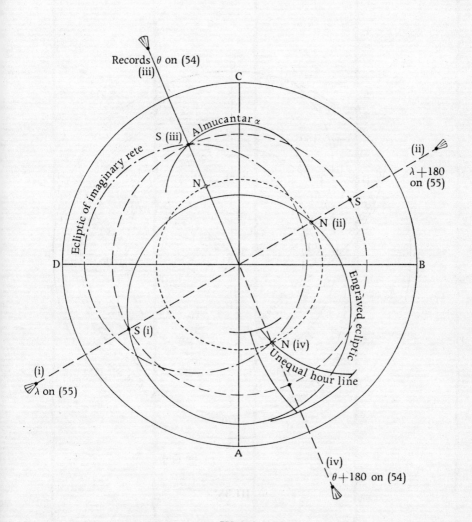

**III. 36\***

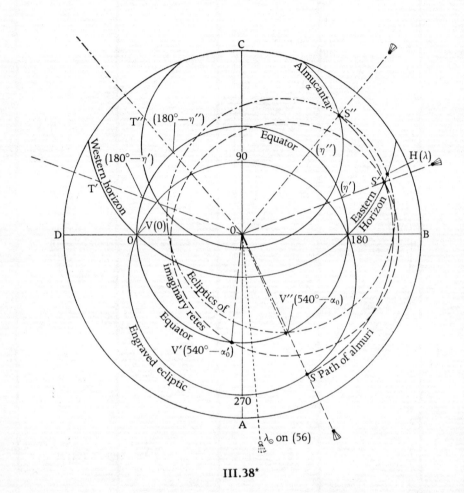

III.38*

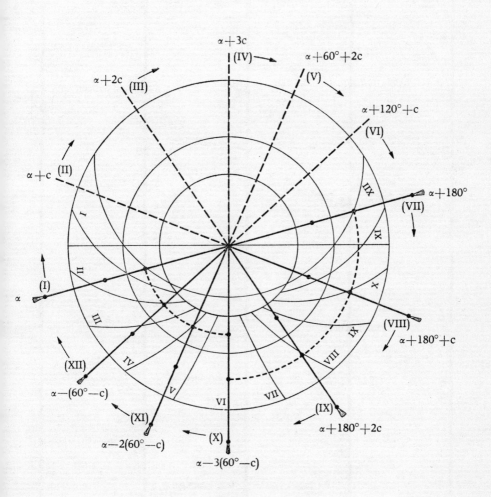

**III.39\***

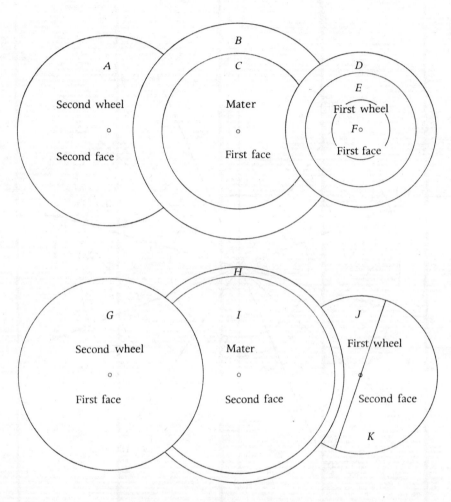

*A general view of the instrument albion**

The discs of the instrument, with their numerous scales, are here divided into eleven principal zones (*A* to *K*, in the above diagrams). The main contents of these zones, together with the principal references to those parts of the text dealing with the construction and use of the scales, will be found in the table on the opposite page.

| Zone | Scales, numbered in text | Main references to text, parts I, II, and III | General character |
|---|---|---|---|
| A | 63 | II. 22–3; III. 36–40. | Astrolabe plate |
| B | 66 | III. 9, 26, 35. | Thread of the limb—three functions |
|  | 7 | II. 2; III. 4. | Circle of the solar year |
|  | 8, 67 | II. 2; III. 9, 18, 39. | Circle of hours and of altitude |
|  | 9 | II. 2; III. 4–9. | Zodiac of the eighth sphere or zodiac of first limb |
|  | 10–16 | II. 2, 5, 6, 7; III. 5, 7, 8, 17. | Circles of the spheres of the planets, and true circle of the Moon |
| C | — | | Empty? See I below |
| D | 17, 18 | II. 10; III. 4–9. | Circle of solar year and zodiac of first disc or wheel |
|  | 19 | II. 10; III. 18. | Iomyn—equation of days |
| E | 20–4 | II. 11–15; III. 4. | Year of Saturn, Jupiter, Mars, Venus, nodes, Moon |
|  | 26–9 | II. 17; III. 19. | Equator of lunar velocity, etc. |
| F | 30 | II. 17; III. 5–7. | Circle of auges of planets and nodes |
|  | 31–6 | II. 11; II. 8, 9; III. 9, 10, 12. | Equators of the planets (the eccentric scales) |
| G | 58–61 | II. 19–21; III. 7–12. | Spiral of approx. 31 turns (approx. 1 turn—hours of mean motus Moon; approx. 3 turns—mean arg. year of Mercury; approx 13 turns—m. motus lunar year; approx. 13 turns—mean argument Moon) |
| H | 54–6 | II. 18; III. 36, 38. | Zodiac of first limb, ascensions on the direct circle, ascensions on the oblique circle |
| I | — | II. 24; III. 36. | Saphea Arzachelis? No position indicated in text. Cf. C. |
| J | 37–45 } | II. 25–28; III. 20–34. | Lunar latitude and parallax, and eclipses of Sun and Moon |
| K | 37–8, 46–52 } | | |

## AEQVATORIVM
## THEORICA ECCLIPSIS SOLARIS.

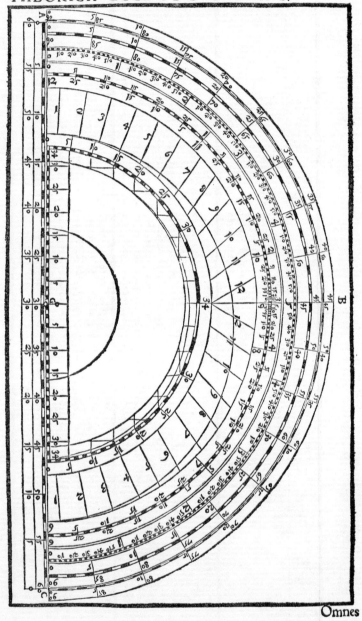

The solar eclipse instrument from John Schöner's *Opera mathematica*. See vol. ii, pp. 276–7 above, and compare the next figure.

## AEQVATORIVM
# THEORICA ECCLIPSIS LVNÆ.

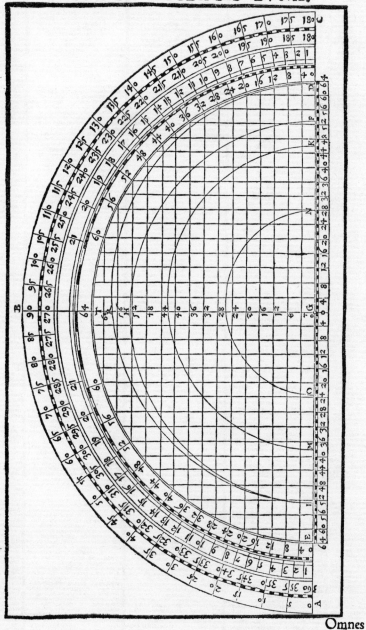

Omnes

The lunar eclipse instrument from John Schöner's *Opera mathematica*. Compare the previous figure and see vol. ii, pp. 276–7 above.

FIGURES TO

# TRACTATUS RECTANGULI
## AND
## FOUR SHORT TRACTS
## CONCERNING CELESTIAL
## COORDINATES

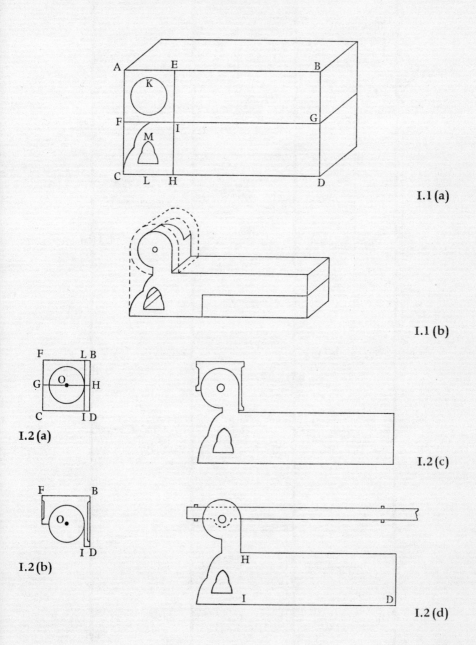

I.1 (a)

I.1 (b)

I.2 (a)

I.2 (c)

I.2 (b)

I.2 (d)

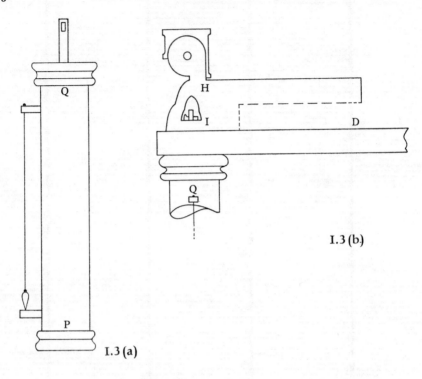

I.3 (b.)

I.3 (a)

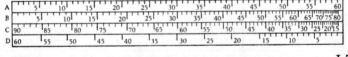

I.5

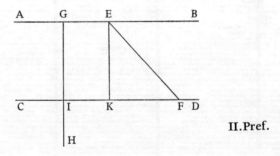

II. Pref.

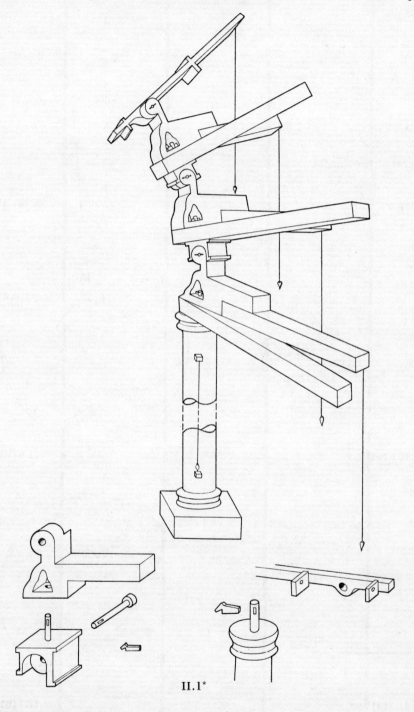

II.1*

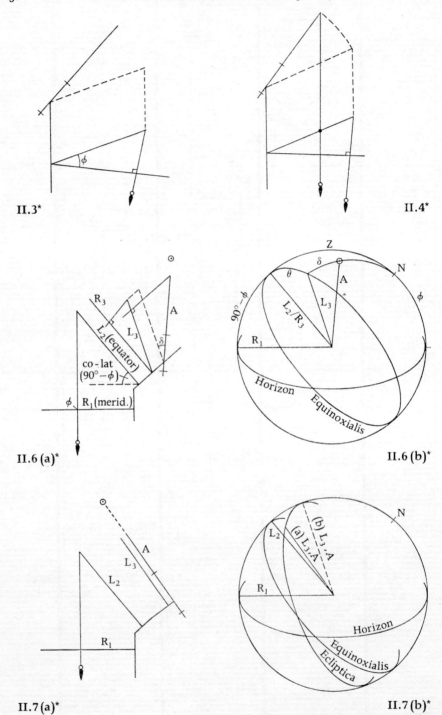

II.3*

II.4*

II.6 (a)*

II.6 (b)*

II.7 (a)*

II.7 (b)*

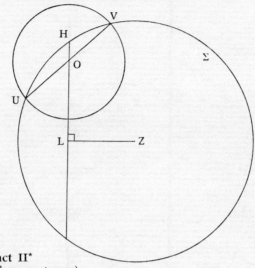

**Tract II\***
(and see next page)

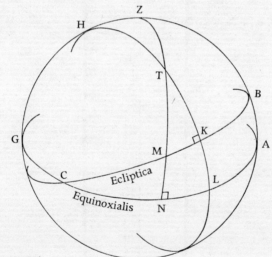

**Tract III\***

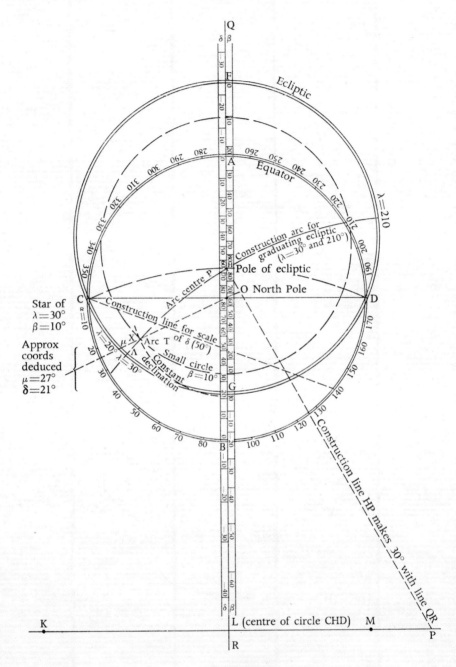

**Tract II\***

# FIGURES TO
# TRACTATUS HOROLOGII
# ASTRONOMICI

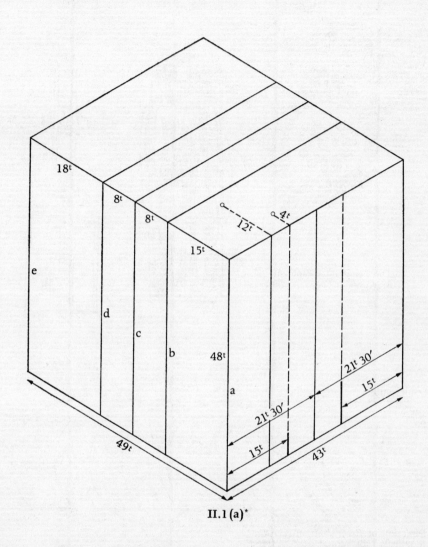

II.1 (a)*

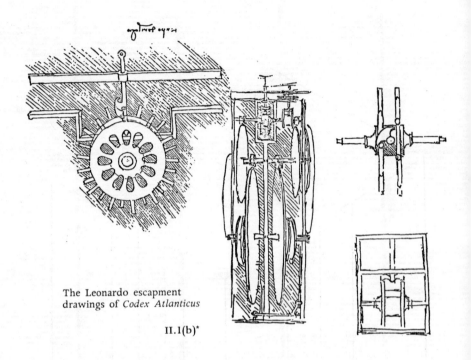

The Leonardo escapment
drawings of *Codex Atlanticus*

II.1(b)*

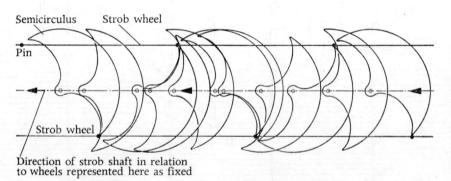

Semicirculus    Strob wheel

Pin

Strob wheel

Direction of strob shaft in relation
to wheels represented here as fixed

II.1(c)*

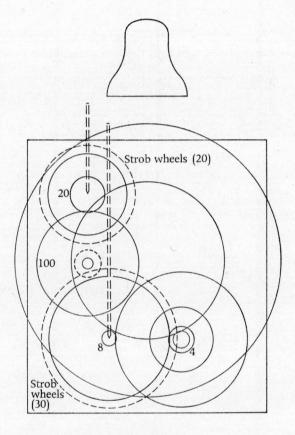

II.2(a)*

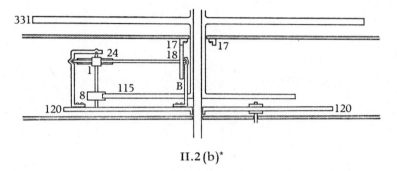

II.2 (b)*

Train for the conversion of the mean daily motion (wheel of 120) into the sidereal motion (wheel of 115) of the *caliga diurna*. The bracket *B* must span the wheel of 115. The wheel of 17 is fixed to the frame.

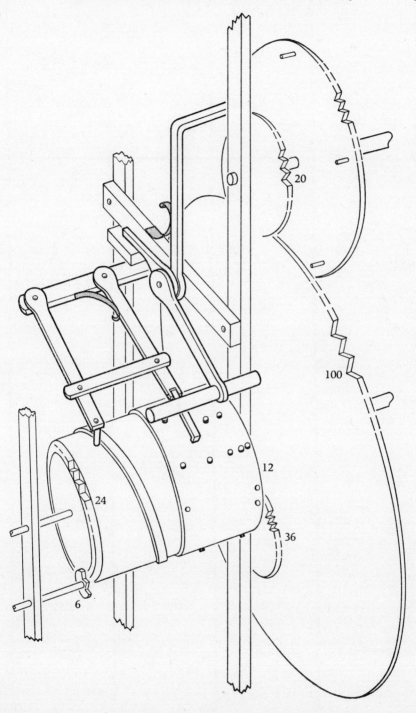

**II.3\***

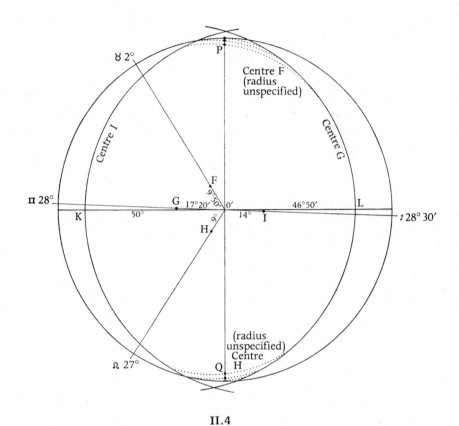

II.4

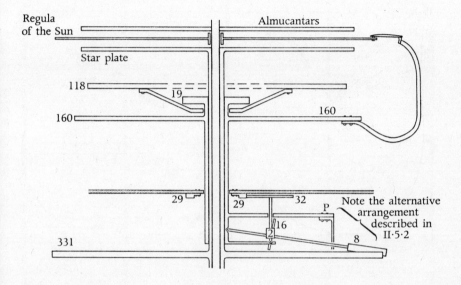

## II.4.3*, II.5.2*

The variable-velocity drive from the wheel of 331, via a transported train to the *caliga Solis*. The wheel of 29 is fixed to the frame (cf. II. 7. 1). Six distinct trains of wheels which fall within this part of the mechanism are omitted, for clarity. A wheel could have been added at the point *P* (cf. the dotted line on Plate XVII) to take a drive to peripheral planetary wheels.

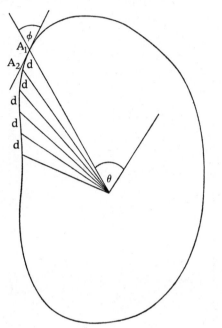

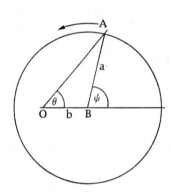

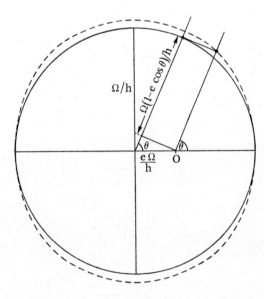

II.4.6*

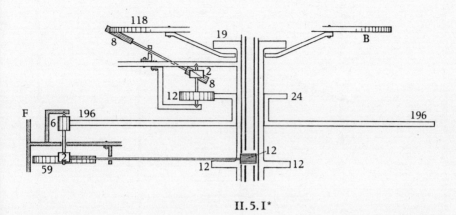

II. 5. I*

Train for the lunar motion (wheel of 118), taken from the contrate wheel of 12 (cf. fig. II. 7. 2). The point *F* is on the clock frame. *B* is the 'black wheel,' an inch broad at its rim, on which the Moon is carried.

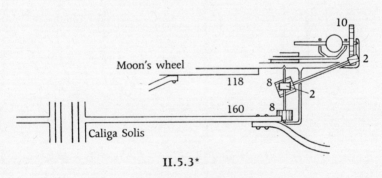

II.5.3*

Train for lunar phase.

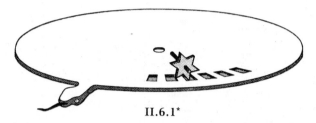

II.6.1*

Probable form of the dragon wheel, with its 177 'perforated teeth' meshing with a pinion of
six.

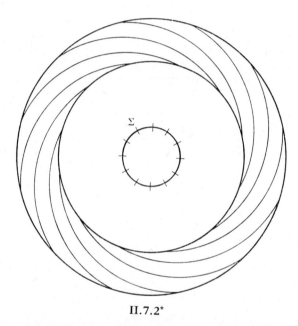

II.7.2*

Geometrical form of the contrate wheel of twelve, on the *caliga Solis*. The twelve centres for
the circular arcs of the teeth are on the construction circle Σ.

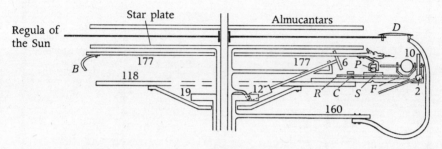

## II.7.5* (and II.6.1*)

Conjectured mechanism for lunar eclipses, in relation to the foremost wheels and plates of the mechanism. *D* is a disc representing the Earth's shadow (away from the Sun, on the rule). *R* is a rack of 18 teeth. *S* is a half-wheel of 18 teeth. *F* is a fork moving the Moon-globe along its squared shaft. *C* is a channel (*ligniculum*) for the rack *R*. *P* is a pinion of 10 teeth (conjectured), surrounded by two racks of 5. *B* is a section of the large basin (*pellvis*) 'cut off at the base'.

This diagram also illustrates the drive for the wheel carrying the dragon (*cf*. II. 6. 1).

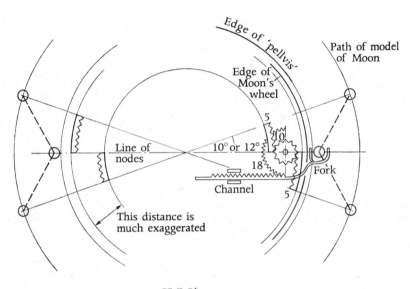

## II.7.5*

Conjectured mechanism for lunar eclipses. The size of the racking, and the movements of the ball representing the Moon, are much exaggerated.

# ALBION TABLES

## WITH SUBSIDIARY TABLES OF EMENDATIONS

| | Amended to | MSS. read | Error in mins. column of Row (degree) | Error in mins. column of Sign |
|---|---|---|---|---|
| | 39 | 40 | 3 | 0 |
| | 19 | 20 | 36 | 2 |
| | 32 | 25 | 18 | 4 |
| | 27 | 29 | 0 | 4 |
| | 49 | 46 | 26 | 7 |
| | 31 | 30 | 22 | 8 |
| | 56 | 59 | 26 | 9 |
| | 44 | 47 | 27 | 9 |
| | 41 | 31 | 9 | 9 |
| | 13 | 19 | 27 | 11 |
| | 21 | 30 | | 11 |

## IV. 1. TABULA VERI MOTUS ORBIS SATURNI

| | 0 g | 0 m | 1 g | 1 m | 2 g | 2 m | 3 g | 3 m | 4 g | 4 m | 5 g | 5 m | 6 g | 6 m | 7 g | 7 m | 8 g | 8 m | 9 g | 9 m | 10 g | 10 m | 11 g | 11 m |
|---|---|---|---|---|---|---|---|---|---|---|---|---|---|---|---|---|---|---|---|---|---|---|---|---|
| 1 | 0 | 53 | 27 | 48 | 25 | 27 | 24 | 29 | 25 | 14 | 27 | 42 | 1 | 8 | 4 | 30 | 6 | 52 | 7 | 30 | 6 | 25 | 4 | 1 |
| 2 | 1 | 46 | 28 | 42 | 26 | 23 | 25 | 29 | 26 | 17 | 28 | 48 | 2 | 15 | 5 | 36 | 7 | 55 | 8 | 30 | 7 | 22 | 4 | 55 |
| 3 | 2 | 39 | 29 | 37 | 27 | 29 | 26 | 29 | 27 | 20 | 29 | 55 | 3 | 22 | 6 | 42 | 8 | 58 | 9 | 29 | 8 | 17 | 5 | 49 |
| 4 | 3 | 33 | 1 | 16 | 28 | 16 | 27 | 29 | 28 | 24 | 5 | 1 | 4 | 29 | 7 | 48 | 10 | 4 | 10 | 28 | 9 | 13 | 6 | 43 |
| 5 | 4 | 27 | 1 | 26 | 29 | 13 | 28 | 30 | 29 | 28 | 1 | 7 | 5 | 36 | 8 | 54 | 11 | 6 | 11 | 27 | 10 | 9 | 7 | 37 |
| 6 | 5 | 21 | 2 | 31 | 2 | 7 | 29 | 30 | 4 | 32 | 2 | 14 | 6 | 43 | 10 | 0 | 12 | 8 | 12 | 25 | 11 | 5 | 8 | 31 |
| 7 | 6 | 14 | 3 | 15 | 1 | 4 | 3 | 31 | 1 | 36 | 3 | 20 | 7 | 50 | 11 | 6 | 13 | 10 | 13 | 25 | 12 | 1 | 9 | 25 |
| 8 | 7 | 8 | 4 | 10 | 2 | 58 | 1 | 31 | 2 | 40 | 4 | 26 | 8 | 57 | 12 | 12 | 14 | 12 | 14 | 23 | 13 | 57 | 10 | 19 |
| 9 | 8 | 4 | 5 | 5 | 3 | 55 | 2 | 32 | 3 | 44 | 5 | 33 | 9 | 4 | 13 | 17 | 15 | 14 | 15 | 22 | 13 | 52 | 11 | 13 |
| 10 | 8 | 55 | 6 | 0 | 4 | 50 | 3 | 33 | 4 | 48 | 6 | 39 | 11 | 11 | 14 | 23 | 16 | 66 | 16 | 21 | 14 | 48 | 12 | 7 |
| 11 | 9 | 49 | 6 | 55 | 5 | 48 | 4 | 34 | 5 | 52 | 7 | 46 | 13 | 18 | 15 | 28 | 17 | 17 | 17 | 19 | 15 | 44 | 13 | 1 |
| 12 | 11 | 36 | 7 | 50 | 6 | 44 | 5 | 35 | 6 | 57 | 8 | 53 | 14 | 25 | 16 | 33 | 18 | 19 | 18 | 18 | 16 | 39 | 13 | 55 |
| 13 | 12 | 30 | 8 | 45 | 7 | 42 | 6 | 36 | 8 | 1 | 9 | 0 | 15 | 32 | 17 | 38 | 19 | 21 | 19 | 16 | 17 | 35 | 14 | 49 |
| 14 | 13 | 24 | 9 | 40 | 8 | 41 | 7 | 38 | 9 | 7 | 11 | 7 | 16 | 39 | 18 | 43 | 20 | 24 | 20 | 14 | 18 | 30 | 15 | 43 |
| 15 | 14 | 17 | 10 | 35 | 9 | 39 | 8 | 39 | 10 | 12 | 12 | 14 | 17 | 46 | 19 | 53 | 21 | 25 | 21 | 12 | 19 | 25 | 16 | 30 |
| 16 | 15 | 11 | 11 | 30 | 10 | 38 | 9 | 41 | 11 | 17 | 13 | 21 | 18 | 53 | 20 | 59 | 22 | 26 | 22 | 10 | 20 | 15 | 16 | 24 |
| 17 | 16 | 5 | 12 | 25 | 11 | 37 | 10 | 43 | 12 | 22 | 14 | 28 | 20 | 0 | 21 | 8 | 23 | 27 | 23 | 7 | 21 | 10 | 17 | 17 |
| 18 | 16 | 59 | 13 | 16 | 12 | 35 | 11 | 44 | 13 | 27 | 15 | 35 | 21 | 7 | 23 | 12 | 24 | 28 | 24 | 5 | 22 | 5 | 18 | 11 |
| 19 | 17 | 53 | 14 | 11 | 13 | 35 | 12 | 46 | 14 | 32 | 16 | 42 | 22 | 14 | 24 | 16 | 25 | 29 | 25 | 2 | 23 | 0 | 19 | 5 |
| 20 | 18 | 47 | 15 | 8 | 14 | 33 | 13 | 48 | 15 | 37 | 17 | 49 | 23 | 21 | 25 | 20 | 27 | 29 | 26 | 59 | 24 | 55 | 20 | 56 |
| 21 | 19 | 41 | 16 | 3 | 15 | 32 | 14 | 50 | 17 | 43 | 19 | 56 | 24 | 27 | 27 | 24 | 28 | 30 | 26 | 56 | 25 | 50 | 21 | 52 |
| 22 | 20 | 35 | 17 | 59 | 16 | 31 | 15 | 52 | 18 | 48 | 21 | 3 | 25 | 34 | 28 | 28 | 29 | 30 | 27 | 53 | 26 | 45 | 22 | 46 |
| 23 | 21 | 29 | 18 | 55 | 17 | 30 | 16 | 54 | 20 | 54 | 22 | 10 | 26 | 40 | 29 | 32 | 9 | 31 | 28 | 50 | 27 | 39 | 23 | 39 |
| 24 | 22 | 23 | 19 | 51 | 18 | 29 | 17 | 56 | 21 | 0 | 23 | 17 | 27 | 46 | 8 | 36 | 1 | 30 | 29 | 47 | 28 | 34 | 24 | 33 |
| 25 | 23 | 17 | 20 | 47 | 19 | | 18 | 59 | 22 | 6 | 24 | 24 | 28 | 53 | 1 | 40 | 2 | 31 | 10 | 44 | 29 | 29 | 25 | 27 |
| 26 | 24 | 11 | 21 | 43 | 20 | | 19 | 8 | 23 | 12 | 25 | 31 | 7 | 59 | 2 | 43 | 3 | 31 | 1 | 41 | 11 | 23 | 26 | 21 |
| 27 | 25 | 5 | 22 | 38 | 21 | | 21 | 11 | 24 | 18 | 26 | 38 | 1 | 5 | 3 | 46 | 4 | 31 | 2 | 37 | 2 | 18 | 27 | 14 |
| 28 | 25 | 59 | 23 | 31 | 22 | | 22 | | 25 | 24 | 27 | 45 | 2 | 12 | 4 | 49 | 5 | 31 | 3 | 33 | 3 | 12 | 28 | 7 |
| 29 | 26 | 54 | 24 | | 23 | | 23 | | 26 | 30 | 28 | 52 | 3 | 18 | 5 | | 6 | | 4 | 29 | | 6 | 29 | 0 |
| 0 | | | | | | | 24 | | | 36 | 6 | 0 | | 24 | | | | | 5 | | | | 12 | 0 |

| Error in mins. column of | | MSS. read | Amended to |
|---|---|---|---|
| Sign | Row (degree) | | |
| 1 | 29 | 38 | 36 |
| 3 | 12 | 48 | 49 |
| 5 | 19 | 50 | 56 |
| 5 | 29 | 52 | 54 |
| 6 | 11 | 5 | 4 |
| 6 | 12 | 10 | 9 |
| 11 | 4 | 12 | 13 |
| 11 | 6 | 9 | 3 |

## IV. 2. TABULA VERI MOTUS ORBIS IOVIS

| | 0 g | 0 m | 1 g | 1 m | 2 g | 2 m | 3 g | 3 m | 4 g | 4 m | 5 g | 5 m | 6 g | 6 m | 7 g | 7 m | 8 g | 8 m | 9 g | 9 m | 10 g | 10 m | 11 g | 11 m |
|---|---|---|---|---|---|---|---|---|---|---|---|---|---|---|---|---|---|---|---|---|---|---|---|---|
| 1 | 0 | 53 | 28 | 24 | 26 | 30 | 25 | 45 | 26 | 22 | 28 | 21 | 1 | 6 | 3 | 48 | 5 | 43 | 6 | 15 | 5 | 24 | 3 | 27 |
| 2 | 1 | 48 | 29 | 19 | 27 | 27 | 26 | 45 | 27 | 26 | 29 | 26 | 2 | 12 | 4 | 53 | 6 | 45 | 7 | 14 | 6 | 21 | 4 | 22 |
| 3 | 2 | 43 | 1 | 15 | 28 | 24 | 27 | 45 | 28 | 28 | 5 | 31 | 3 | 18 | 5 | 57 | 7 | 47 | 8 | 14 | 7 | 18 | 5 | 17 |
| 4 | 3 | 38 | 2 | 10 | 29 | 21 | 28 | 45 | 29 | 31 | 1 | 36 | 4 | 24 | 7 | 2 | 8 | 49 | 9 | 14 | 8 | 15 | 6 | 13 |
| 5 | 4 | 28 | 3 | 56 | 1 | 18 | 29 | 45 | 1 | 37 | 2 | 41 | 5 | 30 | 8 | 7 | 9 | 51 | 10 | 13 | 9 | 12 | 7 | 8 |
| 6 | 5 | 28 | 4 | 52 | 2 | 16 | 3 | 45 | 2 | 40 | 3 | 46 | 6 | 35 | 9 | 11 | 10 | 53 | 11 | 13 | 10 | 8 | 8 | 3 |
| 7 | 6 | 23 | 5 | 48 | 3 | 14 | 1 | 46 | 3 | 41 | 4 | 52 | 7 | 42 | 10 | 16 | 11 | 55 | 12 | 12 | 11 | 5 | 8 | 58 |
| 8 | 7 | 18 | 6 | 44 | 4 | 10 | 2 | 46 | 4 | 48 | 5 | 56 | 8 | 47 | 11 | 21 | 12 | 57 | 13 | 11 | 11 | 59 | 9 | 53 |
| 9 | 8 | 14 | 7 | 40 | 5 | 6 | 3 | 46 | 5 | 51 | 7 | 2 | 9 | 52 | 12 | 25 | 13 | 59 | 14 | 9 | 12 | 58 | 10 | 48 |
| 10 | 9 | 7 | 8 | 36 | 6 | 4 | 4 | 47 | 6 | 55 | 8 | 7 | 10 | 58 | 13 | 30 | 15 | 1 | 15 | 8 | 13 | 55 | 11 | 46 |
| 11 | 10 | 2 | 9 | 32 | 7 | 2 | 5 | 47 | 7 | 59 | 9 | 17 | 12 | 4 | 14 | 34 | 16 | 3 | 16 | 7 | 14 | 51 | 12 | 38 |
| 12 | 10 | 57 | 10 | 28 | 8 | 0 | 6 | 49 | 9 | 6 | 10 | 24 | 13 | 9 | 15 | 38 | 17 | 4 | 17 | 6 | 15 | 47 | 13 | 33 |
| 13 | 11 | 52 | 11 | 24 | 9 | 59 | 7 | 49 | 10 | 10 | 11 | 29 | 14 | 15 | 16 | 42 | 18 | 6 | 18 | 5 | 16 | 44 | 14 | 28 |
| 14 | 12 | 47 | 12 | 20 | 10 | 57 | 8 | 50 | 11 | 14 | 12 | 35 | 15 | 20 | 17 | 46 | 19 | 8 | 19 | 1 | 17 | 40 | 15 | 23 |
| 15 | 13 | 42 | 13 | 16 | 11 | 55 | 9 | 51 | 12 | 18 | 13 | 40 | 16 | 25 | 18 | 50 | 20 | 9 | 19 | 9 | 18 | 36 | 16 | 18 |
| 16 | 14 | 37 | 14 | 13 | 12 | 54 | 10 | 52 | 13 | 22 | 14 | 45 | 17 | 31 | 19 | 54 | 21 | 11 | 20 | 58 | 19 | 32 | 17 | 13 |
| 17 | 15 | 32 | 15 | 9 | 13 | 53 | 11 | 54 | 14 | 26 | 15 | 51 | 18 | 36 | 20 | 58 | 22 | 13 | 21 | 56 | 20 | 28 | 18 | 8 |
| 18 | 16 | 27 | 16 | 5 | 14 | 52 | 12 | 56 | 15 | 30 | 16 | 56 | 19 | 43 | 21 | 1 | 23 | 13 | 22 | 54 | 21 | 24 | 19 | 3 |
| 19 | 17 | 22 | 17 | 1 | 15 | 51 | 13 | 57 | 16 | 35 | 17 | 2 | 20 | 48 | 22 | 5 | 24 | 14 | 23 | 52 | 22 | 16 | 19 | 58 |
| 20 | 18 | 14 | 17 | 55 | 16 | 49 | 14 | 59 | 17 | 39 | 19 | 8 | 21 | 53 | 23 | 9 | 25 | 14 | 24 | 50 | 23 | 12 | 20 | 53 |
| 21 | 19 | 7 | 18 | 52 | 17 | 48 | 16 | 3 | 18 | 44 | 20 | 13 | 22 | 58 | 24 | 12 | 26 | 14 | 25 | 48 | 24 | 8 | 21 | 46 |
| 22 | 20 | 2 | 19 | 49 | 18 | 47 | 17 | 5 | 19 | 49 | 21 | 18 | 24 | 4 | 25 | 16 | 27 | 15 | 26 | 46 | 25 | 4 | 22 | 32 |
| 23 | 21 | 2 | 20 | 45 | 19 | 47 | 18 | 7 | 20 | 53 | 22 | 25 | 25 | 8 | 26 | 20 | 28 | 15 | 27 | 44 | 26 | 59 | 23 | 37 |
| 24 | 21 | 57 | 21 | 42 | 20 | 46 | 19 | 9 | 21 | 58 | 23 | 30 | 26 | 14 | 27 | 23 | 29 | 15 | 28 | 42 | 26 | 55 | 24 | 32 |
| 25 | 22 | 52 | 22 | 39 | 21 | 46 | 21 | 11 | 22 | 3 | 24 | 36 | 27 | 19 | 28 | 29 | 9 | 15 | 29 | 39 | 27 | 50 | 25 | 27 |
| 26 | 23 | 47 | 23 | 36 | 22 | 46 | 23 | 13 | 24 | 7 | 25 | 42 | 28 | 24 | 8 | 32 | 1 | 15 | 10 | 36 | 28 | 45 | 26 | 22 |
| 27 | 24 | 43 | 24 | 33 | 23 | 45 | 24 | 15 | 25 | 12 | 26 | 48 | 7 | 29 | 2 | 38 | 2 | 15 | 2 | 33 | 29 | 41 | 27 | 17 |
| 28 | 25 | 38 | 25 | 30 | 24 | 45 | 25 | 17 | 26 | 17 | 27 | 54 | 1 | 34 | 3 | 40 | 3 | 15 | 3 | 30 | 1 | 36 | 28 | 12 |
| 29 | 26 | 33 | 26 | 27 | 25 | 45 | 26 | 20 | 27 | | 28 | | 2 | 39 | 4 | | 4 | | 4 | 27 | 2 | 31 | 29 | 7 |
| 0 | 27 | 29 | 27 | 24 | 26 | 45 | 27 | | | | 6 | 0 | 3 | 43 | 5 | 40 | 5 | 15 | | | | | **12** | 0 |

| g/m | Column S | Error in Row | MSS. read | Amended to |
|---|---|---|---|---|
| m | 0 | 1 | 48 | 49 |
| m | 1 | 24 | 16 | 18 |
| g | 2 | 1 | 22 | 21 |
| m | 6 | 3 | 30 | 40 |
| m | 11 | 10 | 45 | 35 |

## IV. 3. TABULA VERI MOTUS ORBIS MARTIS

| | 0 g | 0 m | 1 g | 1 m | 2 g | 2 m | 3 g | 3 m | 4 g | 4 m | 5 g | 5 m | 6 g | 6 m | 7 g | 7 m | 8 g | 8 m | 9 g | 9 m | 10 g | 10 m | 11 g | 11 m |
|---|---|---|---|---|---|---|---|---|---|---|---|---|---|---|---|---|---|---|---|---|---|---|---|---|
| 1 | 0 | 49 | 25 | 34 | 22 | 21 | 19 | 36 | 20 | 43 | 24 | 55 | 1 | 14 | 7 | 27 | 11 | 25 | 12 | 22 | 10 | 18 | 6 | 6 |
| 2 | 1 | 38 | 26 | 24 | 22 | 23 | 20 | 36 | 21 | 48 | 26 | 6 | 3 | 27 | 8 | 37 | 12 | 29 | 13 | 21 | 11 | 11 | 6 | 56 |
| 3 | 2 | 27 | 27 | 15 | 23 | 17 | 21 | 36 | 22 | 54 | 27 | 17 | 4 | 40 | 9 | 47 | 13 | 33 | 14 | 19 | 12 | 4 | 7 | 46 |
| 4 | 3 | 16 | 28 | 5 | 24 | 11 | 22 | 26 | 24 | 0 | 28 | 28 | 6 | 54 | 10 | 57 | 14 | 37 | 15 | 17 | 12 | 57 | 8 | 36 |
| 5 | 4 | 4 | 28 | 56 | 25 | 5 | 23 | 36 | 25 | 6 | 29 | 39 | 7 | 7 | 12 | 7 | 15 | 41 | 16 | 15 | 13 | 50 | 9 | 26 |
| 6 | 4 | 54 | 29 | 27 | 26 | 0 | 24 | 36 | 26 | 12 | 5 | 51 | 8 | 20 | 13 | 17 | 16 | 45 | 17 | 12 | 14 | 42 | 10 | 16 |
| 7 | 5 | 44 | 1 | 38 | 26 | 55 | 25 | 36 | 27 | 19 | 2 | 3 | 10 | 33 | 14 | 27 | 17 | 49 | 18 | 9 | 15 | 35 | 11 | 6 |
| 8 | 6 | 33 | 1 | 29 | 27 | 50 | 26 | 37 | 28 | 26 | 3 | 15 | 11 | 46 | 15 | 37 | 18 | 53 | 19 | 6 | 16 | 28 | 11 | 56 |
| 9 | 7 | 22 | 2 | 20 | 28 | 45 | 27 | 38 | 29 | 33 | 4 | 28 | 12 | 59 | 16 | 46 | 19 | 57 | 20 | 3 | 17 | 13 | 12 | 45 |
| 10 | 8 | 11 | 3 | 11 | 29 | 40 | 28 | 37 | 4 | 40 | 5 | 40 | 13 | 13 | 17 | 56 | 20 | 0 | 21 | 0 | 18 | 5 | 13 | 35 |
| 11 | 9 | 0 | 4 | 2 | 0 | 35 | 29 | 40 | 1 | 47 | 6 | 52 | 15 | 26 | 18 | 5 | 21 | 3 | 21 | 57 | 19 | 57 | 14 | 24 |
| 12 | 9 | 50 | 4 | 53 | 1 | 31 | 0 | 40 | 3 | 55 | 8 | 5 | 17 | 39 | 19 | 14 | 22 | 6 | 22 | 53 | 19 | 49 | 15 | 13 |
| 13 | 10 | 39 | 5 | 44 | 2 | 26 | 1 | 41 | 4 | 3 | 9 | 17 | 18 | 52 | 20 | 24 | 23 | 9 | 23 | 50 | 20 | 41 | 16 | 3 |
| 14 | 11 | 28 | 6 | 38 | 3 | 22 | 2 | 43 | 5 | 11 | 10 | 29 | 19 | 5 | 21 | 32 | 24 | 11 | 24 | 46 | 21 | 32 | 16 | 53 |
| 15 | 12 | 17 | 7 | 28 | 4 | 18 | 3 | 45 | 6 | 20 | 11 | 42 | 20 | 18 | 22 | 41 | 25 | 13 | 25 | 42 | 22 | 24 | 17 | 43 |
| 16 | 13 | 7 | 8 | 17 | 5 | 14 | 4 | 47 | 7 | 28 | 12 | 55 | 21 | 31 | 23 | 49 | 26 | 15 | 26 | 26 | 23 | 16 | 18 | 32 |
| 17 | 13 | 57 | 9 | 7 | 6 | 10 | 5 | 49 | 8 | 36 | 14 | 8 | 23 | 43 | 24 | 57 | 27 | 17 | 27 | 34 | 24 | 7 | 19 | 21 |
| 18 | 14 | 47 | 10 | 58 | 7 | 7 | 6 | 51 | 10 | 46 | 15 | 21 | 24 | 55 | 25 | 5 | 28 | 19 | 28 | 29 | 25 | 58 | 20 | 10 |
| 19 | 15 | 36 | 10 | 55 | 8 | 3 | 7 | 54 | 12 | 55 | 16 | 34 | 25 | 8 | 28 | 13 | 29 | 20 | 29 | 25 | 26 | 49 | 21 | 0 |
| 20 | 16 | 25 | 11 | 47 | 9 | 0 | 9 | 57 | 13 | 4 | 17 | 47 | 26 | 20 | 29 | 20 | 10 | 21 | 10 | 20 | 27 | 40 | 21 | 49 |
| 21 | 17 | 15 | 12 | 39 | 9 | 57 | 9 | 0 | 14 | 14 | 19 | 1 | 27 | 32 | 8 | 27 | 1 | 22 | 1 | 15 | 28 | 31 | 22 | 38 |
| 22 | 18 | 4 | 13 | 32 | 10 | 54 | 10 | 3 | 15 | 23 | 20 | 14 | 29 | 45 | 1 | 34 | 2 | 23 | 2 | 10 | 29 | 22 | 23 | 27 |
| 23 | 18 | 54 | 14 | 25 | 11 | 51 | 11 | 7 | 16 | 33 | 21 | 27 | 7 | 57 | 2 | 41 | 3 | 24 | 3 | 5 | 11 | 13 | 24 | 16 |
| 24 | 19 | 34 | 15 | 18 | 12 | 48 | 12 | 11 | 17 | 43 | 22 | 40 | 1 | 9 | 3 | 48 | 4 | 24 | 4 | 0 | 1 | 4 | 25 | 5 |
| 25 | 20 | 24 | 16 | 10 | 13 | 45 | 13 | 15 | 19 | 53 | 23 | 53 | 2 | 21 | 4 | 54 | 5 | 24 | 4 | 55 | 1 | 55 | 25 | 56 |
| 26 | 21 | 14 | 17 | 3 | 14 | 43 | 14 | 19 | 20 | 3 | 25 | 6 | 3 | 32 | 6 | 0 | 6 | 24 | 5 | 49 | 2 | 45 | 26 | 44 |
| 27 | 22 | 4 | 17 | 56 | 15 | 41 | 15 | 23 | 21 | 13 | 26 | 20 | 5 | 53 | 7 | 6 | 7 | 24 | 6 | 43 | 3 | 36 | 27 | 33 |
| 28 | 23 | 56 | 18 | 49 | 16 | 39 | 16 | 27 | 22 | 23 | 27 | 33 | 6 | 54 | 8 | 12 | 8 | 24 | 7 | 37 | 4 | 26 | 28 | 22 |
| 29 | 23 | 4 | 19 | 42 | 17 | 38 | 17 | 31 | 23 | 33 | 28 | 46 | 5 | 5 | 9 | 17 | 9 | 24 | 8 | 31 | 5 | 16 | 29 | 11 |
| 30 | 24 | 54 | 20 | 36 | 18 | 37 | 18 | 35 | 23 | 44 | 28 | 0 | 5 | 16 | 10 | 21 | 10 | 24 | 9 | 24 | 5 | 1 | 11 | 4 |
| 0 | 0 | 44 | 20 | 36 | 18 | 37 | 19 | 39 | 23 | 44 | 6 | 0 | 5 | 16 | 10 | 21 | 11 | 23 | 9 | 24 | 5 | 16 | 12 | 0 |

## IV. 4. TABULA VERI MOTUS ORBIS SOLIS ET VENERIS

| | 0 | | 1 | | 2 | | 3 | | 4 | | 5 | | 6 | | 7 | | 8 | | 9 | | 10 | | 11 | |
|---|---|---|---|---|---|---|---|---|---|---|---|---|---|---|---|---|---|---|---|---|---|---|---|---|
| | g | m | g | m | g | m | g | m | g | m | g | m | g | m | g | m | g | m | g | m | g | m | g | m |
| 1 | 0 | 58 | 1 | 0 | 29 | 18 | 29 | 29 | 29 | 16 | 5 | 1 | 1 | 2 | 2 | 2 | 2 | 46 | 2 | 59 | 3 | 40 | 1 | 56 |
| 2 | 1 | 56 | 0 | 59 | 2 | 17 | 3 | 3 | 4 | 17 | 1 | 2 | 2 | 4 | 3 | 5 | 3 | 47 | 3 | 59 | 4 | 39 | 2 | 54 |
| 3 | 2 | 54 | 0 | 57 | 1 | 16 | 1 | 1 | 1 | 18 | 2 | 4 | 3 | 6 | 4 | 7 | 4 | 48 | 4 | 59 | 5 | 38 | 3 | 52 |
| 4 | 3 | 52 | 1 | 55 | 2 | 15 | 1 | 1 | 2 | 19 | 3 | 6 | 4 | 9 | 5 | 9 | 5 | 49 | 5 | 58 | 6 | 37 | 4 | 51 |
| 5 | 4 | 50 | 2 | 54 | 3 | 14 | 2 | 1 | 3 | 20 | 4 | 8 | 5 | 11 | 6 | 10 | 6 | 49 | 6 | 58 | 7 | 36 | 5 | 49 |
| 6 | 5 | 48 | 3 | 52 | 4 | 13 | 3 | 1 | 4 | 22 | 5 | 10 | 6 | 13 | 7 | 12 | 7 | 50 | 7 | 58 | 8 | 34 | 6 | 47 |
| 7 | 6 | 46 | 4 | 50 | 5 | 12 | 4 | 1 | 5 | 23 | 6 | 12 | 7 | 15 | 8 | 14 | 8 | 51 | 8 | 58 | 9 | 33 | 7 | 45 |
| 8 | 7 | 44 | 5 | 49 | 6 | 11 | 5 | 1 | 6 | 24 | 7 | 14 | 8 | 17 | 9 | 15 | 9 | 52 | 9 | 57 | 10 | 32 | 8 | 43 |
| 9 | 8 | 42 | 6 | 47 | 7 | 10 | 6 | 2 | 7 | 25 | 8 | 16 | 9 | 19 | 10 | 17 | 10 | 53 | 10 | 57 | 11 | 31 | 9 | 41 |
| 10 | 9 | 40 | 7 | 45 | 8 | 9 | 7 | 2 | 8 | 27 | 9 | 18 | 10 | 21 | 11 | 19 | 11 | 53 | 11 | 57 | 12 | 29 | 10 | 39 |
| 11 | 10 | 38 | 8 | 44 | 9 | 9 | 8 | 2 | 9 | 28 | 10 | 20 | 11 | 24 | 12 | 20 | 12 | 54 | 12 | 56 | 13 | 28 | 11 | 38 |
| 12 | 11 | 36 | 9 | 42 | 10 | 8 | 9 | 3 | 10 | 29 | 11 | 22 | 12 | 26 | 13 | 22 | 13 | 54 | 13 | 56 | 14 | 26 | 12 | 36 |
| 13 | 12 | 34 | 10 | 41 | 11 | 7 | 10 | 3 | 11 | 31 | 12 | 24 | 13 | 28 | 14 | 23 | 14 | 55 | 14 | 55 | 15 | 25 | 13 | 34 |
| 14 | 13 | 32 | 11 | 39 | 12 | 6 | 11 | 4 | 12 | 32 | 13 | 26 | 14 | 30 | 15 | 26 | 15 | 55 | 15 | 55 | 16 | 24 | 14 | 32 |
| 15 | 14 | 30 | 12 | 38 | 13 | 5 | 12 | 4 | 13 | 34 | 14 | 28 | 15 | 32 | 16 | 26 | 16 | 56 | 16 | 54 | 17 | 22 | 15 | 30 |
| 16 | 15 | 28 | 13 | 36 | 14 | 5 | 13 | 5 | 14 | 34 | 15 | 30 | 16 | 34 | 17 | 28 | 17 | 57 | 17 | 53 | 18 | 21 | 16 | 28 |
| 17 | 16 | 26 | 14 | 34 | 15 | 4 | 14 | 6 | 15 | 37 | 16 | 32 | 17 | 36 | 18 | 29 | 18 | 57 | 18 | 53 | 19 | 20 | 17 | 26 |
| 18 | 17 | 24 | 15 | 32 | 16 | 4 | 15 | 6 | 16 | 38 | 17 | 34 | 18 | 38 | 19 | 31 | 19 | 57 | 19 | 52 | 20 | 18 | 18 | 24 |
| 19 | 18 | 22 | 16 | 31 | 17 | 3 | 16 | 7 | 17 | 40 | 18 | 36 | 19 | 40 | 20 | 32 | 20 | 58 | 20 | 51 | 21 | 16 | 19 | 22 |
| 20 | 19 | 20 | 17 | 29 | 18 | 3 | 17 | 7 | 18 | 41 | 19 | 39 | 20 | 42 | 21 | 33 | 21 | 58 | 21 | 51 | 22 | 15 | 20 | 20 |
| 21 | 20 | 19 | 18 | 28 | 19 | 2 | 18 | 8 | 19 | 43 | 20 | 41 | 21 | 44 | 22 | 35 | 22 | 58 | 22 | 50 | 23 | 13 | 21 | 18 |
| 22 | 21 | 17 | 19 | 26 | 20 | 2 | 19 | 9 | 20 | 45 | 21 | 43 | 22 | 46 | 23 | 36 | 23 | 58 | 23 | 49 | 24 | 10 | 22 | 16 |
| 23 | 22 | 15 | 20 | 24 | 21 | 2 | 20 | 10 | 21 | 46 | 22 | 45 | 23 | 48 | 24 | 38 | 24 | 59 | 24 | 48 | 25 | 8 | 23 | 14 |
| 24 | 23 | 13 | 21 | 24 | 22 | 1 | 21 | 11 | 22 | 48 | 23 | 47 | 24 | 50 | 25 | 40 | 25 | 59 | 25 | 47 | 26 | 6 | 24 | 12 |
| 25 | 24 | 11 | 22 | 23 | 23 | 1 | 22 | 11 | 23 | 50 | 24 | 49 | 25 | 52 | 26 | 41 | 26 | 59 | 26 | 46 | 27 | 5 | 25 | 10 |
| 26 | 25 | 9 | 23 | 21 | 24 | 1 | 23 | 12 | 24 | 51 | 25 | 51 | 26 | 54 | 27 | 42 | 27 | 59 | 27 | 45 | 28 | 3 | 26 | 8 |
| 27 | 26 | 8 | 24 | 20 | 25 | | 24 | 13 | 25 | 53 | 26 | 54 | 27 | 56 | 28 | 43 | 28 | 59 | 28 | 44 | 29 | 1 | 27 | 6 |
| 28 | 27 | 6 | 25 | 19 | 26 | | 25 | 13 | 26 | 55 | 27 | 56 | 28 | 58 | 29 | 44 | 29 | 59 | 29 | 43 | 11 | 0 | 28 | 4 |
| 29 | 28 | 4 | 26 | 18 | 27 | | 26 | 14 | 27 | 58 | 28 | 58 | 29 | 59 | 8 | 44 | 9 | 59 | 10 | 42 | 0 | 58 | 29 | 2 |
| 0 | 29 | 2 | 27 | 17 | 28 | | 28 | 15 | 28 | 59 | 6 | 0 | 7 | 1 | 1 | 45 | 1 | 59 | 1 | 41 | | | 12 | 0 |

## IV. 5. TABULA VERI MOTUS ORBIS MERCURII

| | 0 | | 1 | | 2 | | 3 | | 4 | | 5 | | 6 | | 7 | | 8 | | 9 | | 10 | | 11 | |
|---|---|---|---|---|---|---|---|---|---|---|---|---|---|---|---|---|---|---|---|---|---|---|---|---|
| | g | m | g | m | g | m | g | m | g | m | g | m | g | m | g | m | g | m | g | m | g | m | g | m |
| 1 | 0 | 57 | 29 | 40 | 28 | 33 | 27 | 59 | 28 | 21 | 29 | 30 | 1 | 3 | 2 | 35 | 3 | 43 | 4 | 0 | 3 | 23 | 2 | 15 |
| 2 | 1 | 54 | 1 | 37 | 29 | 31 | 28 | 59 | 29 | 23 | 5 | 33 | 2 | 6 | 3 | 38 | 4 | 45 | 5 | 0 | 4 | 21 | 3 | 13 |
| 3 | 2 | 51 | 1 | 35 | 2 | 29 | 29 | 58 | 4 | 25 | 1 | 36 | 3 | 9 | 4 | 41 | 5 | 46 | 5 | 59 | 5 | 19 | 4 | 10 |
| 4 | 3 | 48 | 2 | 32 | 3 | 27 | 3 | 58 | 1 | 26 | 2 | 39 | 4 | 13 | 5 | 44 | 6 | 48 | 6 | 59 | 6 | 18 | 5 | 8 |
| 5 | 4 | 45 | 3 | 29 | 4 | 26 | 1 | 58 | 2 | 28 | 3 | 42 | 5 | 16 | 6 | 47 | 7 | 49 | 7 | 58 | 7 | 16 | 6 | 5 |
| 6 | 5 | 43 | 4 | 27 | 5 | 24 | 2 | 58 | 3 | 30 | 4 | 45 | 6 | 19 | 7 | 49 | 8 | 50 | 8 | 57 | 8 | 14 | 7 | 2 |
| 7 | 6 | 40 | 5 | 24 | 6 | 22 | 3 | 58 | 4 | 32 | 5 | 48 | 7 | 22 | 8 | 52 | 9 | 51 | 9 | 57 | 9 | 12 | 8 | 0 |
| 8 | 7 | 37 | 6 | 22 | 7 | 20 | 4 | 58 | 5 | 34 | 6 | 51 | 8 | 25 | 9 | 55 | 10 | 52 | 10 | 56 | 10 | 10 | 8 | 58 |
| 9 | 8 | 35 | 7 | 20 | 8 | 19 | 5 | 59 | 6 | 36 | 7 | 54 | 9 | 28 | 10 | 57 | 11 | 53 | 11 | 55 | 11 | 8 | 9 | 55 |
| 10 | 9 | 32 | 8 | 17 | 9 | 17 | 6 | 59 | 7 | 38 | 8 | 57 | 10 | 32 | 11 | 0 | 12 | 54 | 12 | 54 | 12 | 6 | 10 | 53 |
| 11 | 10 | 30 | 9 | 15 | 10 | 16 | 8 | 0 | 8 | 40 | 10 | 0 | 11 | 35 | 12 | 2 | 13 | 55 | 13 | 53 | 13 | 4 | 11 | 50 |
| 12 | 11 | 28 | 10 | 13 | 11 | 15 | 9 | 1 | 9 | 42 | 11 | 1 | 12 | 39 | 13 | 4 | 14 | 56 | 14 | 52 | 14 | 1 | 12 | 47 |
| 13 | 12 | 25 | 11 | 10 | 12 | 13 | 10 | 1 | 11 | 43 | 12 | 6 | 13 | 42 | 14 | 7 | 15 | 57 | 15 | 51 | 15 | 59 | 13 | 45 |
| 14 | 13 | 22 | 12 | 8 | 13 | 12 | 11 | 1 | 12 | 44 | 13 | 9 | 14 | 45 | 15 | 9 | 16 | 57 | 16 | 50 | 16 | 57 | 14 | 43 |
| 15 | 14 | 20 | 13 | 6 | 14 | 11 | 12 | 2 | 13 | 49 | 14 | 12 | 15 | 48 | 16 | 11 | 17 | 58 | 17 | 49 | 17 | 54 | 15 | 40 |
| 16 | 15 | 17 | 14 | 3 | 15 | 10 | 13 | 2 | 14 | 51 | 15 | 15 | 16 | 51 | 17 | 16 | 18 | 58 | 18 | 48 | 18 | 52 | 16 | 38 |
| 17 | 16 | 15 | 15 | 1 | 16 | 9 | 14 | 3 | 15 | 53 | 16 | 18 | 17 | 54 | 18 | 17 | 19 | 59 | 19 | 47 | 19 | 50 | 17 | 35 |
| 18 | 17 | 13 | 15 | 59 | 17 | 8 | 15 | 4 | 16 | 56 | 17 | 21 | 18 | 59 | 19 | 18 | 20 | 45 | 20 | 43 | 20 | 40 | 18 | 32 |
| 19 | 18 | 10 | 16 | 56 | 18 | 7 | 16 | 5 | 17 | 58 | 18 | 25 | 19 | 0 | 20 | 19 | 21 | 44 | 21 | 43 | 21 | 38 | 19 | 30 |
| 20 | 19 | 7 | 17 | 54 | 19 | 6 | 17 | 6 | 18 | 0 | 19 | 28 | 20 | 20 | 21 | 0 | 22 | 43 | 22 | 40 | 22 | 36 | 20 | 28 |
| 21 | 20 | 5 | 18 | 52 | 20 | 5 | 18 | 7 | 19 | 3 | 20 | 32 | 21 | 22 | 22 | 1 | 23 | 41 | 23 | 38 | 23 | 33 | 21 | 25 |
| 22 | 21 | 2 | 19 | 50 | 21 | 4 | 19 | 8 | 20 | 5 | 21 | 35 | 22 | 24 | 23 | 20 | 24 | 40 | 24 | 36 | 24 | 28 | 22 | 23 |
| 23 | 22 | 0 | 20 | 48 | 22 | 3 | 20 | 9 | 21 | 8 | 22 | 38 | 23 | 26 | 24 | 1 | 25 | 38 | 25 | 33 | 25 | 25 | 23 | 20 |
| 24 | 22 | 58 | 21 | 46 | 23 | 2 | 21 | 10 | 22 | 11 | 23 | 41 | 24 | 28 | 25 | 2 | 26 | 36 | 26 | 34 | 26 | 28 | 24 | 17 |
| 25 | 23 | 55 | 22 | 44 | 24 | 1 | 22 | 12 | 23 | 13 | 24 | 44 | 25 | 30 | 26 | 2 | 27 | 34 | 27 | 31 | 27 | 23 | 25 | 15 |
| 26 | 24 | 52 | 23 | 42 | 25 | 0 | 23 | 14 | 24 | 16 | 25 | 47 | 26 | 32 | 28 | 2 | 28 | 33 | 28 | 29 | 28 | 20 | 26 | 12 |
| 27 | 25 | 50 | 24 | 41 | 26 | 0 | 24 | 15 | 25 | 19 | 26 | 51 | 27 | 34 | 29 | 2 | 29 | 31 | 29 | 31 | 29 | 25 | 27 | 9 |
| 28 | 26 | 47 | 25 | 39 | 26 | 59 | 25 | 17 | 26 | 22 | 27 | 54 | 28 | 35 | 9 | 1 | 10 | 29 | 10 | 29 | 11 | 23 | 28 | 6 |
| 29 | 27 | 45 | 26 | 37 | 26 | 59 | 26 | 19 | 27 | 25 | 28 | 57 | 29 | 39 | 1 | 1 | 1 | 27 | 1 | 20 | 1 | 20 | 29 | 3 |
| 0 | 28 | 43 | 27 | 35 | 26 | 59 | 27 | 19 | 28 | 28 | 6 | 0 | 1 | 32 | 2 | 41 | 3 | 1 | 2 | 25 | 1 | 17 | 12 | 0 |

Error in last entry under col. 2ˢ: MSS. read 27 gra. Amended to 26 gra. (The amendment has been made by a glossator in MS. C.)

| Error in mins col. of | | MSS. read | Amended to |
|---|---|---|---|
| Sign | Row | | |
| 4 | 7 | 12 or 13 | 14 |
| 11 | 7 | 32 | 23 |
| 11 | 26 | 38 | 36 |

## IV. 6. TABULA VERI MOTUS ORBIS LUNE

| | 0 | | 1 | | 2 | | 3 | | 4 | | 5 | | 6 | | 7 | | 8 | | 9 | | 10 | | 11 | |
|---|---|---|---|---|---|---|---|---|---|---|---|---|---|---|---|---|---|---|---|---|---|---|---|---|
| | g | m | g | m | g | m | g | m | g | m | g | m | g | m | g | m | g | m | g | m | g | m | g | m |
| 1 | 0 | 51 | 26 | 28 | 22 | 16 | 18 | 55 | 17 | 57 | 21 | 52 | 1 | 23 | 10 | 35 | 14 | 5 | 12 | 55 | 9 | 28 | 5 | 15 |
| 2 | 1 | 42 | 27 | 19 | 23 | 8 | 19 | 50 | 18 | 59 | 23 | 7 | 3 | 45 | 11 | 48 | 15 | 6 | 13 | 50 | 10 | 20 | 6 | 6 |
| 3 | 2 | 33 | 28 | 11 | 24 | 1 | 20 | 45 | 20 | 1 | 24 | 22 | 4 | 7 | 13 | 0 | 16 | 7 | 14 | 44 | 11 | 12 | 6 | 57 |
| 4 | 3 | 24 | 29 | 2 | 24 | 53 | 21 | 40 | 21 | 4 | 25 | 38 | 5 | 29 | 14 | 11 | 17 | 7 | 15 | 39 | 12 | 4 | 7 | 49 |
| 5 | 4 | 15 | 29 | 53 | 25 | 55 | 22 | 36 | 22 | 10 | 26 | 55 | 6 | 50 | 15 | 22 | 18 | 8 | 16 | 33 | 12 | 56 | 8 | 40 |
| 6 | 5 | 7 | 1 | 45 | 26 | 38 | 23 | 32 | 23 | 10 | 28 | 29 | 8 | 32 | 17 | 33 | 19 | 9 | 17 | 27 | 13 | 48 | 9 | 31 |
| 7 | 5 | 58 | 1 | 36 | 27 | 30 | 24 | 28 | 24 | 14 | 29 | 46 | 9 | 12 | 18 | 43 | 20 | 8 | 18 | 21 | 14 | 40 | 10 | 23 |
| 8 | 6 | 49 | 2 | 27 | 28 | 23 | 25 | 24 | 25 | 19 | 2 | 4 | 10 | 52 | 18 | 53 | 21 | 7 | 19 | 15 | 15 | 32 | 11 | 14 |
| 9 | 7 | 40 | 3 | 18 | 29 | 16 | 26 | 20 | 26 | 30 | 2 | 21 | 12 | 12 | 19 | 11 | 22 | 6 | 20 | 2 | 16 | 23 | 12 | 5 |
| 10 | 8 | 31 | 4 | 10 | 2 | 8 | 27 | 18 | 27 | 37 | 3 | 39 | 13 | 32 | 20 | 11 | 23 | 5 | 21 | 55 | 17 | 15 | 12 | 57 |
| 11 | 9 | 22 | 5 | 1 | 1 | 1 | 28 | 16 | 28 | 30 | 4 | 57 | 14 | 52 | 21 | 20 | 24 | 4 | 22 | 48 | 18 | 7 | 13 | 48 |
| 12 | 10 | 13 | 6 | 52 | 1 | 54 | 29 | 12 | 29 | 44 | 5 | 15 | 16 | 11 | 22 | 29 | 25 | 2 | 23 | 41 | 18 | 58 | 14 | 40 |
| 13 | 11 | 5 | 7 | 44 | 3 | 47 | 3 | 9 | 4 | 51 | 7 | 33 | 17 | 30 | 24 | 38 | 26 | 0 | 24 | 34 | 19 | 50 | 15 | 31 |
| 14 | 11 | 56 | 8 | 35 | 4 | 40 | 3 | 6 | 5 | 58 | 8 | 52 | 18 | 49 | 25 | 46 | 26 | 58 | 25 | 27 | 20 | 42 | 16 | 22 |
| 15 | 12 | 47 | 9 | 27 | 4 | 33 | 4 | 4 | 6 | 14 | 9 | 11 | 20 | 27 | 26 | 17 | 27 | 54 | 26 | 20 | 21 | 33 | 17 | 13 |
| 16 | 13 | 38 | 10 | 18 | 5 | 26 | 4 | 2 | 7 | 19 | 12 | 30 | 21 | 45 | 28 | 9 | 28 | 51 | 27 | 13 | 22 | 25 | 18 | 4 |
| 17 | 14 | 29 | 11 | 10 | 6 | 19 | 4 | 0 | 8 | 31 | 13 | 49 | 22 | 3 | 29 | 16 | 29 | 48 | 28 | 6 | 23 | 16 | 18 | 55 |
| 18 | 15 | 20 | 12 | 2 | 7 | 12 | 5 | 58 | 9 | 40 | 16 | 8 | 24 | 21 | 1 | 23 | 1 | 44 | 28 | 59 | 24 | 8 | 19 | 47 |
| 19 | 16 | 12 | 13 | 53 | 8 | 5 | 6 | 56 | 10 | 49 | 16 | 28 | 25 | 39 | 2 | 36 | 2 | 42 | 28 | 51 | 24 | 59 | 20 | 38 |
| 20 | 17 | 3 | 14 | 45 | 8 | 58 | 7 | 55 | 11 | 58 | 17 | 48 | 26 | 56 | 3 | 36 | 3 | 39 | 10 | 44 | 25 | 50 | 21 | 29 |
| 21 | 18 | 55 | 15 | 37 | 9 | 56 | 9 | 54 | 12 | 7 | 19 | 8 | 27 | 14 | 4 | 41 | 4 | 36 | 2 | 37 | 27 | 33 | 22 | 20 |
| 22 | 18 | 46 | 16 | 28 | 10 | 55 | 10 | 53 | 13 | 17 | 20 | 28 | 29 | 31 | 5 | 46 | 5 | 32 | 3 | 30 | 28 | 24 | 23 | 11 |
| 23 | 19 | 37 | 17 | 20 | 11 | 52 | 11 | 52 | 14 | 27 | 21 | 49 | 7 | 48 | 6 | 50 | 6 | 28 | 4 | 22 | 29 | 15 | 24 | 2 |
| 24 | 20 | 29 | 17 | 12 | 12 | 45 | 12 | 51 | 15 | 38 | 23 | 10 | 1 | 5 | 8 | 53 | 7 | 24 | 5 | 15 | 11 | 7 | 24 | 53 |
| 25 | 21 | 20 | 18 | 4 | 13 | 39 | 13 | 52 | 17 | 49 | 24 | 31 | 3 | 22 | 9 | 59 | 8 | 20 | 6 | 7 | 0 | 58 | 25 | 45 |
| 26 | 22 | 11 | 19 | 56 | 14 | 33 | 14 | 53 | 18 | 0 | 25 | 53 | 5 | 38 | 11 | 3 | 9 | 15 | 7 | 59 | 1 | 49 | 26 | 36 |
| 27 | 23 | 3 | 19 | 48 | 15 | 27 | 15 | 53 | 18 | 12 | 27 | 15 | 6 | 53 | 12 | 4 | 10 | 10 | 7 | 52 | 2 | 41 | 27 | 27 |
| 28 | 23 | 54 | 20 | 40 | 16 | 21 | 15 | 54 | 19 | 25 | 28 | 15 | 8 | 8 | 12 | 3 | 11 | 5 | 8 | 44 | 3 | 32 | 28 | 18 |
| 29 | 24 | 45 | 21 | 32 | 17 | 20 | 16 | 55 | 20 | 38 | 28 | 37 | 8 | 22 | 13 | 4 | 12 | 0 | 9 | 36 | 4 | 23 | 29 | 9 |
| 30 | 25 | 37 | 21 | 24 | 18 | 0 | 16 | 56 | 20 | 38 | 6 | 0 | 9 | 0 | 13 | 0 | 12 | 0 | 8 | 0 | 4 | 0 | 12 | 0 |

## IV. 7. TABULA VERI MOTUS LUNE ET EQUATIONIS ARGUMENTI PRO HORA CONIUNCTIONIS

| | 0 | | | 1 | | | 2 | | | 3 | | | 4 | | | 5 | | | 6 | | |
|---|---|---|---|---|---|---|---|---|---|---|---|---|---|---|---|---|---|---|---|---|---|
| | g | m | s | g | m | s | g | m | s | g | m | s | g | m | s | g | m | s | g | m | s |
| 1 | 0 | 55 | 1 | 28 | 35 | 59 | 26 | 48 | 7 | 25 | 59 | 34 | 26 | 30 | 34 | 28 | 22 | 11 | 1 | 5 | 45 |
| 2 | 0 | 50 | 20 | 29 | 31 | 44 | 27 | 45 | 23 | 26 | 59 | 16 | 27 | 33 | 4 | 29 | 27 | 3 | 2 | 11 | 30 |
| 3 | 1 | 45 | 31 | 1 | 27 | 30 | 28 | 42 | 42 | 27 | 59 | 8 | 28 | 35 | 16 | 5 | 31 | 59 | 3 | 17 | 14 |
| 4 | 2 | 40 | 42 | 1 | 23 | 18 | 29 | 40 | 6 | 28 | 59 | 1 | 29 | 38 | 2 | 1 | 36 | 58 | 4 | 22 | 58 |
| 5 | 3 | 35 | 53 | 2 | 19 | 8 | 2 | 37 | 35 | 29 | 59 | 1 | 4 | 41 | 55 | 2 | 42 | 0 | 5 | 28 | 42 |
| 6 | 4 | 31 | 4 | 3 | 15 | 2 | 1 | 35 | 9 | 3 | 59 | 3 | 1 | 43 | 54 | 3 | 47 | 5 | 6 | 34 | 36 |
| 7 | 5 | 26 | 16 | 4 | 10 | 58 | 2 | 32 | 48 | 1 | 59 | 11 | 2 | 46 | 57 | 4 | 52 | 12 | 7 | 40 | 9 |
| 8 | 6 | 21 | 28 | 5 | 6 | 55 | 3 | 30 | 32 | 2 | 59 | 23 | 3 | 49 | 3 | 5 | 57 | 24 | 8 | 45 | 51 |
| 9 | 7 | 16 | 41 | 6 | 2 | 54 | 4 | 28 | 19 | 3 | 59 | 39 | 4 | 53 | 13 | 7 | 2 | 38 | 9 | 51 | 32 |
| 10 | 8 | 11 | 55 | 6 | 58 | 57 | 5 | 26 | 9 | 4 | 59 | 59 | 5 | 56 | 28 | 8 | 7 | 56 | 10 | 57 | 11 |
| 11 | 9 | 7 | 9 | 7 | 55 | 3 | 6 | 24 | 3 | 6 | 0 | 28 | 6 | 59 | 49 | 9 | 13 | 16 | 12 | 2 | 48 |
| 12 | 10 | 2 | 24 | 8 | 51 | 12 | 7 | 22 | 8 | 7 | 1 | 5 | 8 | 2 | 16 | 10 | 18 | 39 | 13 | 8 | 23 |
| 13 | 11 | 57 | 40 | 9 | 47 | 23 | 8 | 20 | 17 | 8 | 2 | 48 | 9 | 6 | 51 | 11 | 24 | 4 | 14 | 13 | 58 |
| 14 | 11 | 53 | 56 | 10 | 43 | 37 | 9 | 18 | 31 | 9 | 3 | 36 | 10 | 9 | 33 | 12 | 29 | 31 | 15 | 19 | 29 |
| 15 | 12 | 48 | 13 | 11 | 39 | 54 | 11 | 16 | 50 | 11 | 4 | 27 | 11 | 13 | 22 | 13 | 35 | 0 | 16 | 20 | 0 |
| 16 | 13 | 43 | 32 | 12 | 36 | 38 | 12 | 14 | 14 | 12 | 5 | 22 | 12 | 17 | 15 | 14 | 40 | 31 | 17 | 30 | 29 |
| 17 | 14 | 38 | 52 | 13 | 32 | 5 | 13 | 13 | 43 | 13 | 6 | 40 | 13 | 21 | 10 | 15 | 46 | 2 | 18 | 35 | 56 |
| 18 | 15 | 34 | 13 | 14 | 29 | 35 | 14 | 11 | 18 | 14 | 8 | 0 | 14 | 25 | 8 | 16 | 52 | 37 | 19 | 41 | 21 |
| 19 | 16 | 29 | 35 | 15 | 25 | 8 | 15 | 10 | 55 | 15 | 9 | 27 | 15 | 29 | 11 | 17 | 57 | 12 | 20 | 46 | 44 |
| 20 | 17 | 24 | 57 | 16 | 22 | 44 | 16 | 8 | 37 | 16 | 10 | 58 | 16 | 33 | 19 | 19 | 2 | 49 | 21 | 52 | 4 |
| 21 | 18 | 20 | 21 | 17 | 18 | 23 | 17 | 7 | 24 | 17 | 12 | 34 | 17 | 37 | 34 | 20 | 8 | 28 | 22 | 57 | 22 |
| 22 | 19 | 15 | 47 | 18 | 15 | 6 | 18 | 6 | 17 | 18 | 14 | 13 | 18 | 41 | 19 | 21 | 14 | 9 | 24 | 2 | 36 |
| 23 | 20 | 11 | 15 | 19 | 12 | 53 | 19 | 5 | 17 | 19 | 15 | 54 | 19 | 45 | 34 | 22 | 19 | 51 | 25 | 7 | 48 |
| 24 | 21 | 7 | 43 | 20 | 8 | 43 | 20 | 4 | 17 | 20 | 17 | 37 | 20 | 49 | 55 | 23 | 25 | 24 | 26 | 12 | 55 |
| 25 | 22 | 3 | 12 | 21 | 5 | 37 | 21 | 3 | 21 | 21 | 19 | 29 | 21 | 54 | 19 | 24 | 31 | 18 | 27 | 18 | 0 |
| 26 | 23 | 57 | 43 | 22 | 2 | 34 | 22 | 2 | 30 | 22 | 21 | 37 | 22 | 58 | 47 | 25 | 37 | 9 | 28 | 23 | 2 |
| 27 | 23 | 53 | 17 | 22 | 59 | 38 | 23 | 1 | 44 | 23 | 23 | 51 | 25 | 13 | 10 | 26 | 42 | 46 | 29 | 18 | 1 |
| 28 | 24 | 48 | 55 | 23 | 56 | 44 | 24 | 0 | 29 | 24 | 25 | 37 | 26 | 7 | 40 | 27 | 48 | 30 | 7 | 32 | 57 |
| 29 | 25 | 44 | 35 | 24 | 53 | 54 | 24 | 59 | 58 | 25 | 28 | 10 | 27 | 12 | 24 | 28 | 54 | 15 | 1 | 37 | 49 |
| 0 | 27 | 40 | 16 | 25 | 50 | 54 | 24 | 59 | 58 | 25 | 28 | 10 | 27 | 17 | 24 | 6 | 0 | 0 | 2 | 42 | 36 |

| Error in | | | MSS. read | Amended to |
|---|---|---|---|---|
| Column g, m, or s | Column of sign | Row | | |
| g | 1 | 27 | 23 | 22 |
| m | 1 | 27 | 0 | 59 |
| m | 3 | 1 | 50 | 59 |
| s | 3 | 22 | 32 | 34 |
| m | 3 | 29 | 26 | 25 |
| s | 4 | 15 | 53 | 33 |
| s | 5 | 20 | 39 | 49 |
| m | 6 | 19 | 14 | 44 |
| m | 7 | 3 | 58 | 56 |
| m | 7 | 15 | 40 | 46 |
| m | 7 | 19 | 32 | 0‡ |
| s | 7 | 19 | 33 | 32 |
| m | 8 | 20 | 5 | 3 |
| s | 9 | 9 | 13 | 23 |
| m | 9 | 1 | 50 | 59 |
| s | 9 | 16 | 49 | 43 |
| s | 10 | 27 | 16 | 18 |
| s | 10 | 14 | 24 | 46 |
| s | | 18 | 27 | 37 |

‡ This and the following error are amended by a contemporaneous scribe in C.

| Row | 7 g | 7 m | 7 s | 8 g | 8 m | 8 s | 9 g | 9 m | 9 s | 10 g | 10 m | 10 s | 11 g | 11 m | 11 s |
|---|---|---|---|---|---|---|---|---|---|---|---|---|---|---|---|
| 1 | 3 | 47 | 20 | 5 | 34 | 9 | 5 | 59 | 31 | 5 | 6 | 16 | 3 | 15 | 25 |
| 2 | 4 | 52 | 50 | 6 | 36 | 23 | 6 | 58 | 56 | 6 | 3 | 22 | 4 | 11 | 5 |
| 3 | 5 | 56 | 39 | 7 | 38 | 31 | 7 | 58 | 16 | 7 | 0 | 24 | 5 | 6 | 43 |
| 4 | 7 | 1 | 13 | 8 | 40 | 31 | 8 | 57 | 30 | 8 | 57 | 23 | 6 | 2 | 17 |
| 5 | 8 | 5 | 41 | 9 | 42 | 23 | 9 | 56 | 39 | 9 | 54 | 17 | 6 | 57 | 48 |
| 6 | 9 | 10 | 5 | 10 | 44 | 6 | 10 | 55 | 43 | 10 | 51 | 7 | 7 | 53 | 17 |
| 7 | 10 | 14 | 26 | 11 | 45 | 47 | 11 | 54 | 43 | 11 | 47 | 54 | 8 | 48 | 45 |
| 8 | 11 | 18 | 41 | 12 | 47 | 26 | 12 | 53 | 36 | 12 | 44 | 37 | 9 | 44 | 13 |
| 9 | 12 | 22 | 49 | 13 | 49 | 2 | 13 | 52 | 23 | 13 | 41 | 16 | 10 | 39 | 39 |
| 10 | 13 | 26 | 52 | 14 | 50 | 33 | 14 | 51 | 5 | 14 | 37 | 52 | 11 | 35 | 25 |
| 11 | 14 | 30 | 52 | 15 | 52 | 0 | 15 | 49 | 42 | 15 | 34 | 25 | 12 | 30 | 47 |
| 12 | 15 | 34 | 50 | 16 | 53 | 20 | 16 | 48 | 17 | 16 | 30 | 55 | 13 | 25 | 8 |
| 13 | 16 | 38 | 45 | 17 | 54 | 33 | 17 | 46 | 46 | 17 | 27 | 22 | 14 | 21 | 28 |
| 14 | 17 | 42 | 38 | 18 | 55 | 38 | 18 | 45 | 10 | 18 | 23 | 46 | 15 | 16 | 47 |
| 15 | 18 | 46 | 27 | 19 | 56 | 33 | 19 | 43 | 29 | 19 | 20 | 6 | 16 | 11 | 4 |
| 16 | 19 | 50 | 9 | 20 | 57 | 24 | 20 | 41 | 43 | 20 | 16 | 23 | 16 | 7 | 20 |
| 17 | 20 | 53 | 44 | 21 | 58 | 12 | 21 | 39 | 52 | 21 | 12 | 27 | 17 | 2 | 36 |
| 18 | 21 | 57 | 11 | 22 | 58 | 55 | 22 | 37 | 57 | 22 | 8 | 37 | 18 | 57 | 51 |
| 19 | 23 | 0 | 32 | 23 | 59 | 32 | 23 | 35 | 57 | 23 | 4 | 57 | 18 | 52 | 5 |
| 20 | 24 | 3 | 47 | 25 | 0 | 1 | 24 | 33 | 51 | 23 | 1 | 3 | 19 | 48 | 19 |
| 21 | 25 | 6 | 57 | 26 | 0 | 21 | 26 | 31 | 41 | 24 | 57 | 6 | 20 | 43 | 32 |
| 22 | 26 | 10 | 6 | 27 | 0 | 37 | 27 | 29 | 28 | 25 | 53 | 5 | 21 | 38 | 44 |
| 23 | 27 | 13 | 3 | 28 | 0 | 49 | 28 | 27 | 12 | 26 | 49 | 2 | 22 | 33 | 56 |
| 24 | 28 | 16 | 6 | 29 | 0 | 57 | 29 | 24 | 51 | 27 | 44 | 58 | 23 | 28 | 7 |
| 25 | 29 | 18 | 58 | 9 | 0 | 59 | 10 | 22 | 25 | 28 | 40 | 52 | 24 | 24 | 18 |
| 26 | 8 | 21 | 44 | 1 | 0 | 52 | 1 | 19 | 54 | 29 | 36 | 42 | 25 | 19 | 29 |
| 27 | 1 | 24 | 23 | 2 | 0 | 59 | 2 | 17 | 37 | 11 | 32 | 30 | 26 | 14 | 40 |
| 28 | 2 | 26 | 56 | 3 | 0 | 44 | 3 | 14 | 53 | 1 | 28 | 1 | 27 | 9 | 59 |
| 29 | 3 | 29 | 26 | 4 | 0 | 26 | 4 | 11 | 6 | 2 | 24 | 44 | 28 | 4 | 40 |
| 0 | 4 | 31 | 50 | 5 | 0 | 2 | | 9 | | | 19 | | 29 | 0 | 59 |
| | | | | | | | | | | | | | **12** | 0 | 0 |

## IV. 8. TABULA IOMYN SIVE CIRCULI EQUANTIS MOTUM DIURNUM

| Row | 0 g | 0 m | 1 g | 1 m | 2 g | 2 m | 3 g | 3 m | 4 g | 4 m | 5 g | 5 m | 6 g | 6 m | 7 g | 7 m | 8 g | 8 m | 9 g | 9 m | 10 g | 10 m | 11 g | 11 m |
|---|---|---|---|---|---|---|---|---|---|---|---|---|---|---|---|---|---|---|---|---|---|---|---|---|
| 0  | 2  | 10 | 4  | 38 | 5  | 33 | 4  | 20 | 3  | 5  | 3  | 45 | 6  | 4  | 7  | 48 | 7  | 3  | 3  | 49 | 10 | 41 | 11 | 10 |
| 1  | 3  | 15 | 5  | 41 | 6  | 33 | 5  | 17 | 4  | 4  | 4  | 48 | 7  | 9  | 8  | 49 | 7  | 59 | 4  | 41 | 1  | 37 | 1  | 12 |
| 2  | 4  | 20 | 6  | 45 | 7  | 33 | 6  | 14 | 5  | 4  | 5  | 51 | 8  | 13 | 9  | 50 | 8  | 55 | 5  | 33 | 2  | 33 | 2  | 15 |
| 3  | 5  | 25 | 7  | 48 | 8  | 33 | 7  | 10 | 6  | 4  | 6  | 54 | 9  | 18 | 10 | 51 | 9  | 50 | 6  | 25 | 3  | 30 | 3  | 17 |
| 4  | 6  | 30 | 8  | 52 | 9  | 33 | 8  | 7  | 7  | 4  | 7  | 57 | 10 | 22 | 11 | 52 | 10 | 44 | 7  | 18 | 4  | 26 | 4  | 20 |
| 5  | 7  | 35 | 9  | 55 | 10 | 33 | 9  | 3  | 8  | 4  | 9  | 0  | 11 | 27 | 12 | 52 | 11 | 38 | 8  | 11 | 5  | 23 | 5  | 22 |
| 6  | 8  | 41 | 10 | 59 | 11 | 32 | 9  | 0  | 9  | 4  | 10 | 6  | 12 | 32 | 13 | 53 | 12 | 32 | 9  | 4  | 6  | 20 | 6  | 25 |
| 7  | 9  | 46 | 12 | 2  | 12 | 31 | 10 | 57 | 10 | 4  | 11 | 11 | 13 | 36 | 14 | 53 | 13 | 27 | 9  | 57 | 7  | 18 | 7  | 28 |
| 8  | 10 | 52 | 13 | 5  | 13 | 29 | 11 | 54 | 11 | 4  | 12 | 16 | 14 | 41 | 15 | 54 | 14 | 22 | 10 | 50 | 8  | 15 | 8  | 32 |
| 9  | 11 | 57 | 14 | 8  | 14 | 26 | 12 | 51 | 12 | 4  | 13 | 21 | 15 | 45 | 16 | 54 | 15 | 17 | 11 | 43 | 9  | 13 | 9  | 35 |
| 10 | 13 | 3  | 15 | 10 | 15 | 23 | 13 | 48 | 13 | 5  | 14 | 26 | 16 | 49 | 17 | 53 | 16 | 12 | 12 | 37 | 10 | 10 | 10 | 38 |
| 11 | 14 | 8  | 16 | 13 | 16 | 20 | 14 | 45 | 14 | 6  | 15 | 31 | 17 | 53 | 18 | 53 | 17 | 6  | 13 | 31 | 11 | 8  | 11 | 42 |
| 12 | 15 | 14 | 17 | 16 | 17 | 17 | 15 | 42 | 15 | 7  | 16 | 36 | 18 | 57 | 19 | 51 | 17 | 59 | 14 | 24 | 12 | 6  | 12 | 45 |
| 13 | 16 | 19 | 18 | 18 | 18 | 15 | 16 | 39 | 16 | 8  | 17 | 41 | 20 | 1  | 20 | 50 | 18 | 53 | 15 | 17 | 13 | 5  | 13 | 49 |
| 14 | 17 | 24 | 19 | 20 | 19 | 12 | 17 | 36 | 17 | 10 | 18 | 46 | 21 | 6  | 21 | 49 | 19 | 46 | 16 | 10 | 14 | 4  | 14 | 54 |
| 15 | 18 | 29 | 20 | 22 | 20 | 10 | 18 | 33 | 18 | 11 | 19 | 51 | 22 | 10 | 22 | 47 | 20 | 39 | 17 | 3  | 15 | 3  | 15 | 58 |
| 16 | 19 | 34 | 21 | 23 | 21 | 8  | 19 | 30 | 19 | 13 | 20 | 56 | 23 | 15 | 23 | 45 | 21 | 32 | 17 | 56 | 16 | 2  | 17 | 3  |
| 17 | 20 | 39 | 22 | 25 | 23 | 5  | 20 | 27 | 20 | 14 | 22 | 1  | 24 | 19 | 24 | 43 | 22 | 18 | 18 | 49 | 17 | 1  | 18 | 7  |
| 18 | 21 | 44 | 23 | 27 | 23 | 3  | 21 | 25 | 21 | 16 | 23 | 6  | 25 | 23 | 25 | 41 | 23 | 11 | 19 | 43 | 18 | 0  | 19 | 12 |
| 19 | 22 | 49 | 24 | 28 | 24 | 0  | 23 | 23 | 22 | 19 | 24 | 11 | 26 | 27 | 26 | 39 | 24 | 4  | 20 | 37 | 19 | 0  | 20 | 16 |
| 20 | 23 | 54 | 25 | 30 | 25 | 56 | 24 | 21 | 23 | 21 | 25 | 16 | 27 | 31 | 27 | 37 | 25 | 57 | 21 | 31 | 20 | 1  | 21 | 25 |
| 21 | 24 | 0  | 26 | 31 | 26 | 53 | 25 | 19 | 24 | 24 | 26 | 21 | 28 | 34 | 28 | 34 | 26 | 49 | 22 | 25 | 21 | 1  | 22 | 30 |
| 22 | 25 | 5  | 27 | 31 | 27 | 49 | 26 | 17 | 25 | 27 | 27 | 26 | 29 | 38 | 29 | 31 | 27 | 42 | 23 | 19 | 22 | 2  | 23 | 34 |
| 23 | 26 | 10 | 28 | 32 | 28 | 46 | 27 | 15 | 26 | 30 | 28 | 31 | 7  | 39 | 8  | 27 | 28 | 34 | 24 | 14 | 23 | 3  | 24 | 39 |
| 24 | 27 | 16 | 29 | 32 | 29 | 43 | 28 | 13 | 27 | 33 | 29 | 36 | 1  | 41 | 1  | 23 | 29 | 27 | 25 | 9  | 24 | 4  | 25 | 44 |
| 25 | 28 | 21 | 2  | 32 | 1  | 40 | 29 | 11 | 28 | 36 | 6  | 41 | 2  | 42 | 2  | 20 | 9  | 19 | 26 | 59 | 25 | 6  | 26 | 49 |
| 26 | 29 | 27 | 1  | 1  | 3  | 36 | 4  | 9  | 29 | 39 | 1  | 45 | 3  | 44 | 3  | 16 | 1  | 12 | 27 | 54 | 26 | 7  | 27 | 54 |
| 27 | 1  | 32 | 2  | 32 | 4  | 33 | 1  | 8  | 5  | 42 | 2  | 50 | 4  | 44 | 4  | 12 | 2  | 4  | 28 | 49 | 27 | 8  | 12 | 0  |
| 28 | 2  | 31 | 3  | 33 | 1  | 30 | 2  | 7  | 1  | 39 | 3  | 55 | 5  | 45 | 5  | 8  | 2  | 56 | 29 | 44 | 28 |    | 1  | 5  |
| 29 | 3  | 34 | 4  | 33 | 2  | 27 |    | 6  | 2  | 42 | 4  | 59 | 6  | 47 | 6  | 5  |    |    |    |    | 29 |    |    |    |

Error in mins col. of:

| Sign | Row | MSS. read | Nallino ed. |
|---|---|---|---|
| 0  | 14 | 20 | 24 |
| 1  | 2  | 44 | 45 |
| 2  | 29 | 34 | 24 |
| 3  | 1  | 16 | 17 |
| 3  | 8  | 54 | 55 |
| 4  | 22 | 20 | 21 |
| 4  | 23 | 21 | 24 |
| 5  | 7  | 17 | 16 |
| 5  | 27 | 49 | 16 |
| 6  | 7  | 14 | 50 |
| 6  | 8  | 25 | 31 |
| 8  | 9  | 19 | 27 |
| 8  | 10 | 12 | 22 |
| 8  | 12 | 6  | 17 |
| 8  | 13 | 59 | 12 |
| 8  | 14 | 53 | 6  |
| 8  | 11 | 48 | 59 |
| 9  | 0  | 43 | 53 |
| 11 | 8  | 43 | 46 |
| 11 |    | 13 | 31 |
| 11 |    | 16 | 10 |
|    |    | 31 | 32 |

| Sign | Row | MSS. read | Nallino ed. |
|---|---|---|---|
| 1  | 23 | 31 | 32 |
| 2  | 17 | 3  | 0  |
| 2  | 18 | 0  | 58 |
| 3  | 8  | 54 | 55 |
| 7  | 8  | 53 | 54 |
| 7  | 10 | 52 | 53 |
| 7  | 11 | 51 | 52 |
| 7  | 12 | 50 | 51 |
| 7  | 13 | 49 | 50 |
| 7  | 14 | 47 | 49 |
| 7  | 15 | 45 | 49 |
| 7  | 16 | 43 | 47 |
| 7  | 17 | 41 | 45 |
| 7  | 18 | 39 | 43 |
| 7  | 19 | 37 | 41 |
| 7  | 20 | 34 | 39 |
| 7  | 21 | 31 | 37 |
| 7  | 22 | 27 | 34 |
| 7  | 23 | 23 | 31 |
| 7  | 24 | 20 | 27 |
| 7  | 25 | 16 | 23 |
| 7  | 26 | 16 | 20 |
| 7  | 27 | 8  | 16 |
| 8  | 28 | 8  | 2  |
| 8  | 29 | 5  | 7  |
| 8  | 4  | 44 | 43 |

# IV. 9. TABULA LATITUDINIS LUNE IN ECLIPTICA IN SEPTENTRIONEM ET MERIDIEM

| | | 0/11 | | | 1/10 | | | 2/9 | | | Table of errors (in seconds) (subtract from opposite table to give correct reading) | | |
|---|---|---|---|---|---|---|---|---|---|---|---|---|---|
| | | g | m | s | g | m | s | g | m | s | 0/11 | 1/10 | 2/9 |
| 1 | 29 | 0 | 5 | 13 | 2 | 34 | 24 | 4 | 22 | 20 | −1 | +2 | +1 |
| 2 | 28 | 0 | 10 | 27 | 2 | 38 | 52 | 4 | 24 | 49 | 0 | +2 | 0 |
| 3 | 27 | 0 | 15 | 40 | 2 | 43 | 17 | 4 | 27 | 14 | −1 | +2 | 0 |
| 4 | 26 | 0 | 20 | 53 | 2 | 47 | 39 | 4 | 29 | 34 | −1 | +3 | 0 |
| 5 | 25 | 0 | 26 | 7 | 2 | 51 | 56 | 4 | 31 | 49 | 0 | +1 | −1 |
| 6 | 24 | 0 | 31 | 19 | 2 | 56 | 11 | 4 | 33 | 59 | +1 | 0 | −1 |
| 7 | 23 | 0 | 36 | 31 | 3 | 0 | 21 | 4 | 36 | 4 | 0 | −3 | −2 |
| 8 | 22 | 0 | 41 | 42 | 3 | 4 | 29 | 4 | 38 | 4 | +2 | −4 | −2 |
| 9 | 21 | 0 | 46 | 52 | 3 | 8 | 35 | 4 | 40 | 0 | 0 | −4 | −2 |
| 10 | 20 | 0 | 52 | 1 | 3 | 12 | 39 | 4 | 41 | 52 | −1 | −2 | 0 |
| 11 | 19 | 0 | 57 | 9 | 3 | 16 | 39 | 4 | 43 | 38 | −2 | −2 | +1 |
| 12 | 18 | 1 | 2 | 16 | 3 | 20 | 35 | 4 | 45 | 18 | −2 | −1 | +1 |
| 13 | 17 | 1 | 7 | 23 | 3 | 24 | 27 | 4 | 46 | 52 | −1 | −1 | 0 |
| 14 | 16 | 1 | 12 | 30 | 3 | 28 | 15 | 4 | 48 | 20 | +1 | −1 | −1 |
| 15 | 15 | 1 | 17 | 33 | 3 | 32 | 0 | 4 | 49 | 44 | 0 | 0 | −1 |
| 16 | 14 | 1 | 22 | 35 | 3 | 35 | 41 | 4 | 51 | 3 | −1 | +1 | −1 |
| 17 | 13 | 1 | 27 | 33 | 3 | 39 | 17 | 4 | 52 | 17 | −4 | 0 | −1 |
| 18 | 12 | 1 | 32 | 31 | 3 | 42 | 49 | 4 | 53 | 25 | −5 | 0 | −1 |
| 19 | 11 | 1 | 37 | 29 | 3 | 46 | 17 | 4 | 54 | 38 | −5 | 0 | 0 |
| 20 | 10 | 1 | 42 | 27 | 3 | 49 | 40 | 4 | 55 | 25 | −3 | −2 | −1 |
| 21 | 9 | 1 | 47 | 23 | 3 | 53 | 0 | 4 | 56 | 17 | 0 | −2 | −1 |
| 22 | 8 | 1 | 52 | 17 | 3 | 56 | 16 | 4 | 57 | 4 | +1 | −2 | 0 |
| 23 | 7 | 1 | 57 | 8 | 3 | 59 | 28 | 4 | 57 | 45 | +2 | −1 | 4 |
| 24 | 6 | 2 | 1 | 56 | 4 | 2 | 35 | 4 | 58 | 21 | +3 | −1 | 0 |
| 25 | 5 | 2 | 6 | 40 | 4 | 5 | 38 | 4 | 58 | 51 | +1 | 0 | 0 |
| 26 | 4 | 2 | 11 | 22 | 4 | 8 | 37 | 4 | 59 | 15 | −1 | 0 | −1 |
| 27 | 3 | 2 | 16 | 2 | 4 | 11 | 30 | 4 | 59 | 35 | −2 | 0 | 0 |
| 28 | 2 | 2 | 20 | 40 | 4 | 14 | 22 | 4 | 59 | 50 | −2 | +3 | +1 |
| 29 | 1 | 2 | 25 | 17 | 4 | 17 | 7 | 4 | 59 | 56 | −1 | +3 | −1 |
| 30 | 0 | 2 | 29 | 52 | 4 | 19 | 47 | 5 | 0 | 0 | +1 | −3 | 0 |
| | | | 6/5 | | | 7/4 | | | 8/3 | | | | |

| Column (sign) | Row | MSS. read (seconds) | Nallino ed. |
|---|---|---|---|
| 0 | 15 | 36 | 33 |
| 0 | 25 | 40 | (37) |
| 0 | 28 | 40 | (39) |
| 1 | 3 | 57 | 17 |
| 1 | 5 | 57 | 56 |
| 1 | 6 | 10 | 11 |
| 1 | 13 | 26 | 27 |
| 1 | 16 | 42 | 41 |
| 1 | 24 | 37 | 35 |
| 1 | 27 | 34 | 30 |
| 2 | 1 | 22 | 20 |
| 2 | 2 | 51 | 49 |
| 2 | 5 | 43 | 49 |
| 2 | 18 | 35 | 25 |
| 2 | 20 | 35 | 25 |
| 2 | 28 | 50 | (47) |
| 2 | 29 | 58 | 56 |

Note an error in Nallino's minutes column:

| 2 | 3 | 27$^m$ | (26$^m$) |

## IV. 10. TABULA LONGITUDINIS CUM SUA PARTE DUODECIMA, ET TABULA CONIUNCTIONUM DUODECIM

| | s | g | m | s | g | m |
|---|---|---|---|---|---|---|
| 1 | 1 | 2 | 30 | 0 | 29 | 6 |
| 2 | 2 | 5 | 0 | 1 | 28 | 13 |
| 3 | 3 | 7 | 30 | 2 | 27 | 19 |
| 4 | 4 | 10 | 0 | 3 | 26 | 26 |
| 5 | 5 | 12 | 30 | 4 | 25 | 32 |
| 6 | 6 | 15 | 0 | 5 | 24 | 38 |
| 7 | 7 | 17 | 30 | 6 | 23 | 45 |
| 8 | 8 | 20 | 0 | 7 | 22 | 51 |
| 9 | 9 | 22 | 30 | 8 | 21 | 58 |
| 10 | 10 | 25 | 0 | 9 | 21 | 4 |
| 11 | 11 | 27 | 30 | 10 | 20 | 10 |
| 12 | 13 | 0 | 0 | 11 | 19 | 17 |

| Row | MSS. | Correct |
|---|---|---|
| 3 | 17 | 7 |
| 2 | 12 | 13 |

## IV. 11. TABULA MOTUS LUNE IN UNA HORA AD AUGEM† ET AD LONGITUDINEM MEDIAM‡ ET AD EIUS OPPOSITUM𝜙

| | † | | | ‡ | | | 𝜙 | | |
|---|---|---|---|---|---|---|---|---|---|
| | g | m | s | g | m | s | g | m | s |
| 1 | 0 | 30 | 18 | 0 | 32 | 42 | 0 | 36 | 4 |
| 2 | 1 | 0 | 36 | 1 | 5 | 24 | 1 | 12 | 8 |
| 3 | 1 | 30 | 54 | 1 | 38 | 6 | 1 | 48 | 12 |
| 4 | 2 | 0 | 12 | 2 | 10 | 48 | 2 | 24 | 16 |
| 5 | 2 | 31 | 30 | 2 | 43 | 30 | 3 | 0 | 20 |
| 6 | 3 | 1 | 48 | 3 | 16 | 12 | 3 | 36 | 24 |
| 7 | 3 | 32 | 6 | 3 | 48 | 54 | 4 | 12 | 28 |
| 8 | 4 | 2 | 24 | 4 | 21 | 36 | 4 | 48 | 32 |
| 9 | 4 | 32 | 42 | 4 | 54 | 18 | 5 | 24 | 36 |
| 10 | 5 | 3 | 0 | 5 | 27 | 0 | 6 | 0 | 40 |
| 11 | 5 | 33 | 18 | 5 | 59 | 42 | 6 | 36 | 44 |
| 12 | 6 | 3 | 36 | 6 | 32 | 24 | 7 | 12 | 48 |
| 13 | 6 | 33 | 54 | 7 | 5 | 6 | 7 | 48 | 52 |
| 14 | 7 | 4 | 12 | 7 | 37 | 48 | 8 | 24 | 56 |
| 15 | 7 | 34 | 30 | 8 | 10 | 30 | 9 | 1 | 0 |
| 16 | 8 | 4 | 48 | 8 | 43 | 12 | 9 | 37 | 4 |
| 17 | 8 | 35 | 6 | 9 | 15 | 54 | 10 | 13 | 8 |
| 18 | 9 | 5 | 24 | 9 | 48 | 36 | 10 | 49 | 12 |
| 19 | 9 | 35 | 42 | 10 | 21 | 18 | 11 | 25 | 16 |
| 20 | 10 | 6 | 0 | 10 | 54 | 0 | 12 | 1 | 20 |
| 21 | 10 | 36 | 58 | 11 | 26 | 42 | 12 | 37 | 24 |
| 22 | 11 | 6 | 36 | 11 | 59 | 24 | 13 | 13 | 28 |
| 23 | 11 | 36 | 54 | 12 | 32 | 6 | 13 | 49 | 32 |
| 24 | 12 | 7 | 12 | 13 | 4 | 48 | 14 | 25 | 36 |

(No emendation)

## IV. 12. TABULA STELLARUM FIXARUM

| [No.] | Nomina stellarum | Verus gradus stelle in ecliptica | | | Latitudo ab ecliptica | | Pars mer. | Gradus cum quo stella mediat celum | | | Declinacio stelle ab ecliptica | | | | Magn. stelle | [Modern designation] |
|---|---|---|---|---|---|---|---|---|---|---|---|---|---|---|---|---|
| | | s | g | m | g | m | | s | g | m | s | g | m | | | |
| 1 | Alayoch | 2 | 12 | 0 | 22 | 30 | S | 2 | 6 | 0 | 1 | 15 | 0 | S | 1 | α Aur |
| 2 | Caput Orionis | 2 | 14 | (20) | (24) | 50 | M | 2 | 15 | 10 | 2 | 9 | 33 | S | (2) | ε Ori |
| 3 | Rigil | 2 | 6 | 50 | 30 | 30 | M | 2 | 11 | 12 | 0 | 10 | 30 | M | 1 | β Ori |
| 4 | Alabor | 3 | 4 | 40 | 39 | 10 | M | 3 | 3 | 0 | 0 | 15 | 30 | M | 1 | α CMa |
| 5 | Aldebaran | 1 | 29 | 40 | 5 | 10 | M | 1 | 29 | 0 | 0 | 14 | 30 | M | 1 | α Tau |
| 6 | Alramech | 6 | 13 | 0 | 31 | 30 | S | 6 | 29 | 0 | 0 | 24 | 0 | S | 1 | α Boo |
| 7 | Cor leonis | 4 | 19 | 30 | 0 | 10 | (S) | 4 | 20 | 0 | 0 | 15 | 0 | S | 1 | α Leo |
| 8 | Vega | 9 | 4 | 20 | 62 | 0 | S | 9 | 3 | 0 | 1 | 7 | 59 | S | 1 | α Lyr |
| 9 | Algomesa | 3 | 16 | (30) | 16 | 10 | S | 3 | 13 | 0 | 0 | 7 | 30 | S | 1 | α CMi |
| 10 | Altayir | (9) | 20 | 10) | (29) | 10) | (S) | 9 | 16 | 0 | 0 | 7 | 0 | M | 1 | α Aql |
| 11 | Cor Scorpionis | 7 | 29 | 40 | 4 | 0 | M | 7 | 27 | 0 | 0 | 29 | 30 | S | 1 | α Sco |
| 12 | Caput Geminorum | 3 | 10 | (40) | 9 | 40 | S | 3 | 9 | 0 | 1 | 3 | 30 | | 1 | α Gem |
| 13 | Effeta | 7 | 1 | 40 | 44 | 30 | S | 7 | 18 | 0 | 1 | 2 | 30 | | (2) | α CrB |
| 14 | Suel | 3 | 4 | 10 | 29 | 0 | M | | | | | | | | 1 | α Car |
| 15 | Centaurus | 6 | 25 | 20 | 41 | 10 | M | | | | | | | | (2) | α Cen |

(For an explanation of emendations, see Commentary.)

## IV. 13. TABULA MEDII MOTUS [ARGUMENTI] MERCURII

| | Marcius | | | Aprilis | | | Maius | | | Iunius | | | Iulius | | | Augustus | | |
|---|---|---|---|---|---|---|---|---|---|---|---|---|---|---|---|---|---|---|
| | s | g | m | s | g | m | s | g | m | s | g | m | s | g | m | s | g | m |
| 1 | 0 | 3 | 6 | 3 | 9 | 25 | 6 | 12 | 37 | 9 | 18 | 56 | 0 | 22 | 8 | 3 | 28 | 26 |
| 2 | 0 | 6 | 13 | 3 | 12 | 31 | 6 | 15 | 43 | 9 | 22 | 2 | 0 | 25 | 14 | 4 | 1 | 33 |
| 3 | 0 | 9 | 19 | 3 | 15 | 38 | 6 | 18 | 50 | 9 | 25 | 8 | 0 | 28 | 20 | 4 | 4 | 39 |
| 4 | 0 | 12 | 26 | 3 | 18 | 44 | 6 | 21 | 56 | 9 | 28 | 15 | 1 | 1 | 27 | 4 | 7 | 45 |
| 5 | 0 | 15 | 32 | 3 | 21 | 51 | 6 | 25 | 3 | 10 | 1 | 21 | 1 | 4 | 33 | 4 | 10 | 52 |
| 6 | 0 | 18 | 38 | 3 | 24 | 57 | 6 | 28 | 9 | 10 | 4 | 28 | 1 | 7 | 41 | 4 | 13 | 58 |
| 7 | 0 | 21 | 45 | 3 | 28 | 3 | 7 | 1 | 15 | 10 | 7 | 34 | 1 | 10 | 46 | 4 | 17 | 5 |
| 8 | 0 | 24 | 51 | 4 | 1 | 10 | 7 | 4 | 22 | 10 | 10 | 40 | 1 | 13 | 52 | 4 | 20 | 11 |
| 9 | 0 | 27 | 58 | 4 | 4 | 16 | 7 | 7 | 28 | 10 | 13 | 47 | 1 | 16 | 59 | 4 | 23 | 17 |
| 10 | 1 | 1 | 4 | 4 | 7 | 23 | 7 | 10 | 35 | 10 | 16 | 53 | 1 | 20 | 5 | 4 | 26 | 24 |
| 11 | 1 | 4 | 10 | 4 | 10 | 29 | 7 | 13 | 41 | 10 | 20 | 0 | 1 | 23 | 12 | 4 | 29 | 30 |
| 12 | 1 | 7 | 17 | 4 | 13 | 35 | 7 | 16 | 47 | 10 | 23 | 6 | 1 | 26 | 18 | 5 | 2 | 37 |
| 13 | 1 | 10 | 23 | 4 | 16 | 42 | 7 | 19 | 54 | 10 | 26 | 12 | 1 | 29 | 25 | 5 | 5 | 43 |
| 14 | 1 | 13 | 30 | 4 | 19 | 48 | 7 | 23 | 0 | 10 | 29 | 19 | 2 | 2 | 31 | 5 | 8 | 49 |
| 15 | 1 | 16 | 36 | 4 | 22 | 55 | 7 | 26 | 7 | 11 | 2 | 25 | 2 | 5 | 37 | 5 | 11 | 56 |
| 16 | 1 | 19 | 42 | 4 | 26 | 1 | 8 | 29 | 13 | 11 | 5 | 32 | 2 | 8 | 44 | 5 | 15 | 2 |
| 17 | 1 | 22 | 49 | 4 | 29 | 7 | 8 | 2 | 19 | 11 | 8 | 38 | 2 | 11 | 50 | 5 | 18 | 9 |
| 18 | 1 | 25 | 55 | 5 | 2 | 14 | 8 | 5 | 26 | 11 | 11 | 44 | 2 | 14 | 57 | 5 | 21 | 15 |
| 19 | 1 | 29 | 2 | 5 | 5 | 20 | 8 | 8 | 32 | 11 | 14 | 51 | 2 | 18 | 3 | 5 | 24 | 21 |
| 20 | 2 | 2 | 8 | 5 | 8 | 27 | 8 | 11 | 39 | 11 | 17 | 57 | 2 | 21 | 9 | 5 | 27 | 28 |
| 21 | 2 | 5 | 14 | 5 | 11 | 33 | 8 | 14 | 45 | 11 | 21 | 4 | 2 | 24 | 16 | 6 | 0 | 34 |
| 22 | 2 | 8 | 21 | 5 | 14 | 39 | 8 | 17 | 52 | 11 | 24 | 10 | 2 | 27 | 22 | 6 | 3 | 41 |
| 23 | 2 | 11 | 27 | 5 | 17 | 46 | 8 | 20 | 58 | 11 | 27 | 16 | 3 | 0 | 29 | 6 | 6 | 47 |
| 24 | 2 | 14 | 34 | 5 | 20 | 52 | 8 | 24 | 4 | 0 | 0 | 23 | 3 | 3 | 35 | 6 | 9 | 53 |
| 25 | 2 | 17 | 40 | 5 | 23 | 59 | 8 | 27 | 11 | 0 | 3 | 29 | 3 | 6 | 41 | 6 | 13 | 0 |
| 26 | 2 | 20 | 46 | 5 | 27 | 5 | 9 | 0 | 17 | 0 | 6 | 36 | 3 | 9 | 48 | 6 | 16 | 6 |
| 27 | 2 | 23 | 53 | 6 | 0 | 11 | 9 | 3 | 24 | 0 | 9 | 42 | 3 | 12 | 54 | 6 | 19 | 13 |
| 28 | 2 | 26 | 59 | 6 | 3 | 18 | 9 | 6 | 30 | 0 | 12 | 48 | 3 | 16 | 1 | 6 | 22 | 19 |
| 29 | 3 | 0 | 6 | 6 | 6 | 24 | 9 | 9 | 36 | 0 | 15 | 55 | 3 | 19 | 7 | 6 | 25 | 25 |
| 30 | 3 | 3 | 12 | 6 | 9 | 31 | 9 | 12 | 42 | 0 | 19 | 1 | 3 | 22 | 13 | 6 | 28 | 32 |
| 31 | 3 | 6 | 19 | 6 | : | : | 9 | 15 | 49 | 0 | : | : | 3 | 25 | 20 | 7 | 1 | 38 |

(MSS)

| | Marcius | | | Aprilis | | | Maius | | | Iunius | | | Iulius | | | Augustus | | |
|---|---|---|---|---|---|---|---|---|---|---|---|---|---|---|---|---|---|---|
| 30 | 3 | 6 | 18 | 6 | 9 | 28 | 9 | 15 | 46 | 0 | 18 | 57 | 3 | 25 | 15 | 7 | 1 | 34 |
| 31 | | | | | | | | | | | | | | | | | | |

**Error accumulated during the month named (see foot of main table)**

| Month | Error |
|---|---|
| Mar. | −1′ |
| Apr. | −2′ |
| May | 0 |
| Iun. | −1′ |
| Iul. | −1′ |
| Aug. | +1′ |
| Sept. | 0 |
| Oct. | −2′ |
| Nov. | 0 |
| Dec. | −2′ |
| Ian. | 0 |
| Feb. | 0 |
| Total | −8′ |

| Day | September s | g | m | October s | g | m | November s | g | m | December s | g | m | Ianuarius s | g | m | Februarius s | g | m |
|---|---|---|---|---|---|---|---|---|---|---|---|---|---|---|---|---|---|---|
| 1 | 7 | 4 | 45 | 10 | 7 | 57 | 1 | 14 | 15 | 4 | 17 | 27 | 7 | 23 | 46 | 11 | 0 | 4 |
| 2 | 7 | 7 | 51 | 10 | 11 | 3 | 1 | 17 | 22 | 4 | 20 | 34 | 7 | 26 | 52 | 11 | 3 | 11 |
| 3 | 7 | 10 | 58 | 10 | 14 | 10 | 1 | 20 | 28 | 4 | 23 | 40 | 7 | 29 | 59 | 11 | 6 | 17 |
| 4 | 7 | 14 | 4 | 10 | 17 | 16 | 1 | 23 | 35 | 4 | 26 | 47 | 8 | 3 | 5 | 11 | 9 | 24 |
| 5 | 7 | 17 | 10 | 10 | 20 | 22 | 1 | 26 | 41 | 4 | 29 | 53 | 8 | 6 | 12 | 11 | 12 | 30 |
| 6 | 7 | 20 | 17 | 10 | 23 | 29 | 1 | 29 | 47 | 5 | 2 | 59 | 8 | 9 | 18 | 11 | 15 | 37 |
| 7 | 7 | 23 | 23 | 10 | 26 | 35 | 2 | 2 | 54 | 5 | 6 | 6 | 8 | 12 | 24 | 11 | 18 | 43 |
| 8 | 7 | 26 | 30 | 10 | 29 | 42 | 2 | 6 | 0 | 5 | 9 | 12 | 8 | 15 | 31 | 11 | 21 | 49 |
| 9 | 7 | 29 | 36 | 11 | 2 | 48 | 2 | 9 | 7 | 5 | 12 | 19 | 8 | 18 | 37 | 11 | 24 | 56 |
| 10 | 8 | 2 | 42 | 11 | 5 | 54 | 2 | 12 | 13 | 5 | 15 | 25 | 8 | 21 | 44 | 11 | 28 | 2 |
| 11 | 8 | 5 | 49 | 11 | 9 | 1 | 2 | 15 | 19 | 5 | 18 | 31 | 8 | 24 | 50 | 0 | 1 | 9 |
| 12 | 8 | 8 | 55 | 11 | 12 | 7 | 2 | 18 | 26 | 5 | 21 | 38 | 8 | 27 | 56 | 0 | 4 | 15 |
| 13 | 8 | 12 | 2 | 11 | 15 | 14 | 2 | 21 | 32 | 5 | 24 | 44 | 9 | 1 | 3 | 0 | 7 | 21 |
| 14 | 8 | 15 | 8 | 11 | 18 | 20 | 2 | 24 | 39 | 5 | 27 | 51 | 9 | 4 | 9 | 0 | 10 | 28 |
| 15 | 8 | 18 | 14 | 11 | 21 | 26 | 2 | 27 | 45 | 6 | 0 | 57 | 9 | 7 | 16 | 0 | 13 | 34 |
| 16 | 8 | 21 | 21 | 11 | 24 | 33 | 3 | 0 | 51 | 6 | 4 | 4 | 9 | 10 | 22 | 0 | 16 | 41 |
| 17 | 8 | 24 | 27 | 11 | 27 | 39 | 3 | 3 | 58 | 6 | 7 | 10 | 9 | 13 | 28 | 0 | 19 | 47 |
| 18 | 8 | 27 | 34 | 0 | 0 | 46 | 3 | 7 | 4 | 6 | 10 | 16 | 9 | 16 | 35 | 0 | 22 | 53 |
| 19 | 9 | 0 | 40 | 0 | 3 | 52 | 3 | 10 | 11 | 6 | 13 | 23 | 9 | 19 | 41 | 0 | 25 | 0 |
| 20 | 9 | 3 | 46 | 0 | 6 | 58 | 3 | 13 | 17 | 6 | 16 | 29 | 9 | 22 | 48 | 0 | 28 | 6 |
| 21 | 9 | 6 | 53 | 0 | 10 | 5 | 3 | 16 | 23 | 6 | 19 | 36 | 9 | 25 | 54 | 1 | 2 | 13 |
| 22 | 9 | 9 | 59 | 0 | 13 | 11 | 3 | 19 | 30 | 6 | 22 | 42 | 9 | 29 | 0 | 1 | 5 | 19 |
| 23 | 9 | 13 | 6 | 0 | 16 | 18 | 3 | 22 | 36 | 6 | 25 | 48 | 10 | 2 | 7 | 1 | 8 | 25 |
| 24 | 9 | 16 | 12 | 0 | 19 | 24 | 3 | 25 | 43 | 6 | 28 | 55 | 10 | 5 | 13 | 1 | 11 | 32 |
| 25 | 9 | 19 | 18 | 0 | 22 | 31 | 3 | 28 | 49 | 7 | 2 | 1 | 10 | 8 | 20 | 1 | 14 | 38 |
| 26 | 9 | 22 | 25 | 0 | 25 | 37 | 4 | 1 | 55 | 7 | 5 | 8 | 10 | 11 | 26 | 1 | 17 | 45 |
| 27 | 9 | 25 | 31 | 0 | 28 | 43 | 4 | 5 | 2 | 7 | 8 | 14 | 10 | 14 | 32 | 1 | 20 | 51 |
| 28 | 9 | 28 | 38 | 1 | 1 | 50 | 4 | 8 | 8 | 7 | 11 | 20 | 10 | 17 | 39 | 1 | 23 | 57 |
| 29 | 10 | 1 | 44 | 1 | 4 | 56 | 4 | 11 | 15 | 7 | 14 | 27 | 10 | 20 | 45 | Addicio motus | | |
| 30 | 10 | 4 | 50 | 1 | 7 | 57 | 4 | 14 | 21 | 7 | 17 | 33 | 10 | 23 | 52 | 6 horarum | | |
| 31 | .. | .. | .. | 1 | 11 | 9 | .. | .. | .. | 7 | 20 | 40 | 10 | 26 | 58 | 1 | 24 | 44 |
| (MSS) 30 | 10 | 4 | 46 | 1 | 11 | 3 | 4 | 14 | 15 | 7 | 20 | 32 | 10 | 26 | 50 | 1 | 23 | 49 |
| 31 | | | | | | | | | | | | | | | | and 1 | 24 | 36 |

## IV. 14. TABULA MEDII MOTUS LUNE

| | Marcius | | | Aprilis | | | Mayus | | | Iunius | | | Iulius | | | Augustus | | |
|---|---|---|---|---|---|---|---|---|---|---|---|---|---|---|---|---|---|---|
| | s | g | m | s | g | m | s | g | m | s | g | m | s | g | m | s | g | m |
| 1 | 0 | 13 | 11 | 2 | 1 | 39 | 3 | 6 | 56 | 4 | 25 | 24 | 6 | 0 | 41 | 7 | 19 | 9 |
| 2 | 0 | 26 | 21 | 2 | 14 | 49 | 3 | 20 | 7 | 5 | 8 | 35 | 6 | 13 | 52 | 8 | 2 | 20 |
| 3 | 1 | 9 | 32 | 2 | 28 | 0 | 4 | 3 | 17 | 5 | 21 | 45 | 6 | 27 | 3 | 8 | 15 | 31 |
| 4 | 1 | 22 | 42 | 3 | 11 | 10 | 4 | 16 | 28 | 6 | 4 | 56 | 7 | 10 | 13 | 8 | 28 | 41 |
| 5 | 2 | 5 | 53 | 3 | 24 | 21 | 4 | 29 | 38 | 6 | 18 | 6 | 7 | 23 | 24 | 9 | 11 | 52 |
| 6 | 2 | 19 | 3 | 4 | 7 | 31 | 5 | 12 | 49 | 7 | 1 | 17 | 8 | 6 | 34 | 9 | 25 | 2 |
| 7 | 3 | 2 | 14 | 4 | 20 | 42 | 5 | 25 | 59 | 7 | 14 | 27 | 8 | 19 | 45 | 10 | 8 | 13 |
| 8 | 3 | 15 | 25 | 5 | 3 | 53 | 6 | 9 | 10 | 7 | 27 | 38 | 9 | 2 | 56 | 10 | 21 | 24 |
| 9 | 3 | 28 | 35 | 5 | 17 | 3 | 6 | 22 | 21 | 8 | 10 | 49 | 9 | 16 | 6 | 11 | 4 | 34 |
| 10 | 4 | 11 | 46 | 6 | 0 | 14 | 7 | 5 | 31 | 8 | 23 | 59 | 9 | 29 | 17 | 11 | 17 | 45 |
| 11 | 4 | 24 | 56 | 6 | 13 | 24 | 7 | 18 | 42 | 9 | 7 | 10 | 10 | 12 | 27 | 0 | 0 | 55 |
| 12 | 5 | 8 | 7 | 6 | 26 | 35 | 8 | 1 | 52 | 9 | 20 | 20 | 10 | 25 | 38 | 0 | 14 | 6 |
| 13 | 5 | 21 | 18 | 7 | 9 | 46 | 8 | 15 | 3 | 10 | 3 | 31 | 11 | 8 | 48 | 0 | 27 | 16 |
| 14 | 6 | 4 | 28 | 7 | 22 | 56 | 8 | 28 | 14 | 10 | 16 | 42 | 11 | 21 | 59 | 1 | 10 | 27 |
| 15 | 6 | 17 | 39 | 8 | 6 | 7 | 9 | 11 | 24 | 10 | 29 | 52 | 0 | 5 | 10 | 1 | 23 | 38 |
| 16 | 7 | 0 | 49 | 8 | 19 | 17 | 9 | 24 | 35 | 11 | 13 | 3 | 0 | 18 | 20 | 2 | 6 | 48 |
| 17 | 7 | 14 | 0 | 9 | 2 | 28 | 10 | 7 | 45 | 11 | 26 | 13 | 1 | 1 | 31 | 2 | 19 | 59 |
| 18 | 7 | 27 | 10 | 9 | 15 | 38 | 10 | 20 | 56 | 0 | 9 | 24 | 1 | 14 | 41 | 3 | 3 | 9 |
| 19 | 8 | 10 | 21 | 9 | 28 | 49 | 11 | 4 | 6 | 0 | 22 | 34 | 1 | 27 | 52 | 3 | 16 | 20 |
| 20 | 8 | 23 | 32 | 10 | 12 | 0 | 11 | 17 | 17 | 1 | 5 | 45 | 2 | 11 | 2 | 3 | 29 | 30 |
| 21 | 9 | 6 | 42 | 10 | 25 | 10 | 0 | 0 | 28 | 1 | 18 | 56 | 2 | 24 | 13 | 4 | 12 | 41 |
| 22 | 9 | 19 | 53 | 11 | 8 | 21 | 0 | 13 | 38 | 2 | 2 | 6 | 3 | 7 | 24 | 4 | 25 | 52 |
| 23 | 10 | 3 | 3 | 11 | 21 | 31 | 0 | 26 | 49 | 2 | 15 | 17 | 3 | 20 | 34 | 5 | 9 | 2 |
| 24 | 10 | 16 | 14 | 0 | 4 | 42 | 1 | 9 | 59 | 2 | 28 | 27 | 4 | 3 | 45 | 5 | 22 | 13 |
| 25 | 10 | 29 | 25 | 0 | 17 | 53 | 1 | 23 | 10 | 3 | 11 | 38 | 4 | 16 | 55 | 6 | 5 | 23 |
| 26 | 11 | 12 | 35 | 1 | 1 | 3 | 2 | 6 | 21 | 3 | 24 | 49 | 5 | 0 | 6 | 6 | 18 | 34 |
| 27 | 11 | 25 | 46 | 1 | 14 | 14 | 2 | 19 | 31 | 4 | 7 | 59 | 5 | 13 | 17 | 7 | 1 | 45 |
| 28 | 0 | 8 | 56 | 1 | 27 | 24 | 3 | 2 | 42 | 4 | 21 | 10 | 5 | 26 | 27 | 7 | 14 | 55 |
| 29 | 0 | 22 | 7 | 2 | 10 | 35 | 3 | 15 | 52 | 5 | 4 | 20 | 6 | 9 | 38 | 7 | 28 | 6 |
| 30 | 1 | 5 | 17 | 2 | 23 | 45 | 3 | 29 | 3 | 5 | 17 | 31 | 6 | 22 | 48 | 8 | 11 | 16 |
| 31 | 1 | 18 | 28 | : | : | : | 4 | 12 | 13 | : | : | : | 7 | 5 | 59 | 8 | 24 | 27 |

| Entry | MSS. (minutes) | Amended to |
|---|---|---|
| Mar. 21 | 4 | 2 |
| May 20 | 10 | 17 |
| Iul. 14 | 54 | 59 |
| Aug. 12 | 2 | 6 |
| Aug. 16 | 46 | 48 |
| Sep. 2 | 40 | 48 |
| Sep. 3 | 5 | 59 |
| Sep. 13 | 46 | 44 |
| Sep. 25 | 53 | 51 |
| Nov. 13 | 13 | 30 |

*See also commentary to IV, 14, 15*

| | September | | | October | | | November | | | December | | | Ianuarius | | | Februarius | | |
|---|---|---|---|---|---|---|---|---|---|---|---|---|---|---|---|---|---|---|
| | s | g | m | s | g | m | s | g | m | s | g | m | s | g | m | s | g | m |
| 1 | 9 | 7 | 37 | 10 | 12 | 55 | 0 | 1 | 23 | 1 | 6 | 40 | 2 | 25 | 8 | 4 | 13 | 36 |
| 2 | 9 | 20 | 48 | 10 | 26 | 5 | 0 | 14 | 33 | 1 | 19 | 51 | 3 | 8 | 19 | 4 | 26 | 47 |
| 3 | 10 | 3 | 59 | 11 | 9 | 16 | 0 | 27 | 44 | 2 | 3 | 12 | 3 | 21 | 29 | 5 | 9 | 57 |
| 4 | 10 | 17 | 9 | 11 | 22 | 27 | 1 | 10 | 55 | 2 | 16 | 23 | 4 | 4 | 40 | 5 | 23 | 8 |
| 5 | 11 | 0 | 20 | 0 | 5 | 37 | 1 | 24 | 5 | 2 | 29 | 33 | 4 | 17 | 51 | 6 | 6 | 19 |
| 6 | 11 | 13 | 30 | 0 | 18 | 48 | 2 | 7 | 16 | 3 | 12 | 44 | 5 | 1 | 1 | 6 | 19 | 29 |
| 7 | 11 | 26 | 41 | 1 | 1 | 58 | 2 | 20 | 26 | 3 | 25 | 54 | 5 | 14 | 12 | 7 | 2 | 40 |
| 8 | 0 | 9 | 52 | 1 | 15 | 9 | 3 | 3 | 37 | 4 | 8 | 5 | 5 | 27 | 22 | 7 | 15 | 50 |
| 9 | 0 | 23 | 2 | 1 | 28 | 20 | 3 | 16 | 48 | 4 | 22 | 16 | 6 | 10 | 33 | 7 | 29 | 1 |
| 10 | 1 | 6 | 13 | 2 | 11 | 30 | 3 | 29 | 58 | 5 | 5 | 26 | 6 | 23 | 44 | 8 | 12 | 12 |
| 11 | 1 | 19 | 23 | 2 | 24 | 41 | 4 | 13 | 9 | 5 | 18 | 37 | 7 | 6 | 54 | 8 | 25 | 22 |
| 12 | 2 | 2 | 34 | 3 | 7 | 51 | 4 | 26 | 19 | 6 | 1 | 47 | 7 | 20 | 5 | 9 | 8 | 33 |
| 13 | 2 | 15 | 44 | 3 | 21 | 2 | 5 | 9 | 30 | 6 | 14 | 58 | 8 | 3 | 15 | 9 | 21 | 43 |
| 14 | 2 | 28 | 55 | 4 | 4 | 12 | 5 | 22 | 40 | 6 | 27 | 8 | 8 | 16 | 26 | 9 | 4 | 54 |
| 15 | 3 | 12 | 6 | 4 | 17 | 23 | 6 | 5 | 51 | 7 | 11 | 19 | 8 | 29 | 36 | 10 | 18 | 4 |
| 16 | 3 | 25 | 16 | 4 | 0 | 34 | 6 | 19 | 2 | 7 | 24 | 30 | 9 | 12 | 47 | 10 | 1 | 15 |
| 17 | 4 | 8 | 27 | 5 | 13 | 44 | 7 | 2 | 12 | 8 | 7 | 40 | 9 | 25 | 58 | 11 | 14 | 26 |
| 18 | 4 | 21 | 37 | 5 | 26 | 55 | 7 | 15 | 23 | 8 | 20 | 51 | 10 | 9 | 8 | 11 | 27 | 36 |
| 19 | 5 | 4 | 48 | 6 | 10 | 5 | 7 | 28 | 33 | 9 | 3 | 12 | 11 | 22 | 19 | 0 | 10 | 47 |
| 20 | 5 | 17 | 58 | 6 | 23 | 16 | 8 | 11 | 44 | 9 | 17 | 22 | 11 | 5 | 29 | 0 | 23 | 57 |
| 21 | 6 | 0 | 9 | 7 | 6 | 26 | 8 | 24 | 54 | 10 | 0 | 33 | 0 | 18 | 40 | 1 | 7 | 8 |
| 22 | 6 | 14 | 20 | 7 | 19 | 37 | 9 | 8 | 5 | 10 | 13 | 44 | 0 | 1 | 51 | 1 | 20 | 19 |
| 23 | 6 | 27 | 30 | 8 | 2 | 48 | 9 | 21 | 16 | 10 | 26 | 54 | 1 | 15 | 1 | 2 | 3 | 29 |
| 24 | 7 | 10 | 41 | 8 | 15 | 58 | 10 | 4 | 26 | 11 | 10 | 5 | 1 | 28 | 12 | 2 | 16 | 40 |
| 25 | 7 | 23 | 51 | 8 | 29 | 9 | 10 | 17 | 37 | 11 | 22 | 15 | 1 | 11 | 22 | 2 | 29 | 50 |
| 26 | 8 | 7 | 2 | 9 | 12 | 19 | 11 | 0 | 47 | 0 | 6 | 26 | 2 | 24 | 33 | 3 | 13 | 1 |
| 27 | 8 | 20 | 13 | 9 | 25 | 30 | 11 | 13 | 58 | 0 | 19 | 37 | 2 | 20 | 43 | 3 | 26 | 11 |
| 28 | 9 | 3 | 23 | 10 | 8 | 41 | 11 | 27 | 9 | 1 | 1 | 47 | 3 | 4 | 54 | 4 | 9 | 22 |
| 29 | 9 | 16 | 34 | 10 | 21 | 51 | 0 | 10 | 19 | 1 | 15 | 37 | 3 | 17 | 5 | Addicio motus |  |  |
| 30 | 9 | 29 | 44 | 11 | 5 | 2 | 0 | 23 | 30 | 1 | 28 | 47 | 4 | 0 | 15 | 6 horarum |  |  |
| 31 |  |  |  | 11 | 18 | 12 |  |  |  | 2 | 11 | 58 | 4 | 17 | 26 | 4 | 12 | 40 |

## IV. 15. TABULA MEDII MOTUS ARGUMENTI LUNE

| | Martius | | | Aprilis | | | Mayus | | | Iunius | | | Iulius | | | Augustus | | | September | | |
|---|---|---|---|---|---|---|---|---|---|---|---|---|---|---|---|---|---|---|---|---|---|
| | s | g | m | s | g | m | s | g | m | s | g | m | s | g | m | s | g | m | s | g | m |
| 1 | 0 | 13 | 4 | 1 | 28 | 5 | 3 | 0 | 2 | 4 | 15 | 3 | 5 | 17 | 0 | 7 | 2 | 0 | 8 | 17 | 1 |
| 2 | 0 | 26 | 8 | 2 | 11 | 9 | 3 | 13 | 6 | 4 | 28 | 6 | 6 | 0 | 3 | 7 | 15 | 4 | 9 | 0 | 5 |
| 3 | 1 | 9 | 12 | 2 | 24 | 13 | 3 | 26 | 10 | 5 | 11 | 10 | 6 | 13 | 7 | 7 | 28 | 8 | 9 | 13 | 9 |
| 4 | 1 | 22 | 16 | 3 | 7 | 16 | 4 | 9 | 13 | 5 | 24 | 14 | 6 | 26 | 11 | 8 | 11 | 12 | 9 | 26 | 13 |
| 5 | 2 | 5 | 19 | 3 | 20 | 20 | 4 | 22 | 17 | 6 | 7 | 18 | 7 | 9 | 15 | 8 | 24 | 16 | 10 | 9 | 17 |
| 6 | 2 | 18 | 23 | 4 | 3 | 24 | 5 | 5 | 21 | 6 | 20 | 22 | 7 | 22 | 19 | 9 | 7 | 20 | 10 | 22 | 21 |
| 7 | 3 | 1 | 27 | 4 | 16 | 28 | 5 | 18 | 25 | 7 | 3 | 26 | 8 | 5 | 23 | 9 | 20 | 24 | 11 | 5 | 25 |
| 8 | 3 | 14 | 31 | 4 | 29 | 32 | 6 | 1 | 29 | 7 | 16 | 30 | 8 | 18 | 27 | 10 | 3 | 28 | 11 | 18 | 29 |
| 9 | 3 | 27 | 35 | 5 | 12 | 36 | 6 | 14 | 33 | 7 | 29 | 34 | 9 | 1 | 31 | 10 | 16 | 32 | 0 | 1 | 32 |
| 10 | 4 | 10 | 39 | 5 | 25 | 40 | 6 | 27 | 37 | 8 | 12 | 38 | 9 | 14 | 35 | 10 | 29 | 36 | 0 | 14 | 36 |
| 11 | 4 | 23 | 43 | 6 | 8 | 44 | 7 | 10 | 41 | 8 | 25 | 42 | 9 | 27 | 39 | 11 | 12 | 39 | 0 | 27 | 40 |
| 12 | 5 | 6 | 47 | 6 | 21 | 48 | 7 | 23 | 45 | 9 | 8 | 45 | 10 | 10 | 42 | 11 | 25 | 43 | 1 | 10 | 44 |
| 13 | 5 | 19 | 51 | 7 | 4 | 52 | 8 | 6 | 48 | 9 | 21 | 49 | 10 | 23 | 46 | 0 | 8 | 47 | 1 | 23 | 48 |
| 14 | 6 | 2 | 55 | 7 | 17 | 55 | 8 | 19 | 52 | 10 | 4 | 53 | 11 | 6 | 50 | 0 | 21 | 51 | 2 | 6 | 52 |
| 15 | 6 | 15 | 58 | 7 | 29 | 59 | 8 | 2 | 56 | 10 | 17 | 57 | 11 | 19 | 54 | 1 | 4 | 55 | 2 | 19 | 56 |
| 16 | 6 | 29 | 2 | 8 | 14 | 3 | 9 | 16 | 0 | 11 | 0 | 0 | 0 | 2 | 58 | 1 | 18 | 59 | 3 | 3 | 0 |
| 17 | 7 | 12 | 6 | 8 | 27 | 7 | 9 | 29 | 4 | 11 | 14 | 5 | 0 | 16 | 2 | 2 | 1 | 3 | 3 | 16 | 4 |
| 18 | 7 | 25 | 10 | 8 | 10 | 11 | 9 | 12 | 8 | 11 | 27 | 9 | 0 | 29 | 6 | 2 | 14 | 7 | 3 | 29 | 8 |
| 19 | 8 | 8 | 14 | 9 | 23 | 15 | 10 | 25 | 12 | 0 | 10 | 13 | 1 | 12 | 10 | 2 | 27 | 11 | 3 | 12 | 11 |
| 20 | 8 | 21 | 18 | 9 | 6 | 19 | 10 | 8 | 16 | 0 | 23 | 17 | 1 | 25 | 14 | 3 | 10 | 14 | 4 | 25 | 15 |
| 21 | 9 | 4 | 22 | 10 | 19 | 23 | 11 | 21 | 20 | 1 | 6 | 21 | 2 | 8 | 18 | 3 | 23 | 18 | 5 | 8 | 19 |
| 22 | 9 | 17 | 26 | 11 | 2 | 27 | 0 | 4 | 24 | 1 | 19 | 24 | 2 | 21 | 21 | 4 | 6 | 22 | 5 | 21 | 23 |
| 23 | 10 | 0 | 30 | 11 | 15 | 31 | 0 | 17 | 27 | 2 | 2 | 28 | 3 | 4 | 25 | 4 | 19 | 26 | 6 | 4 | 27 |
| 24 | 10 | 13 | 34 | 0 | 28 | 34 | 1 | 0 | 31 | 2 | 15 | 32 | 3 | 17 | 29 | 5 | 2 | 30 | 6 | 17 | 31 |
| 25 | 10 | 26 | 37 | 1 | 11 | 38 | 1 | 13 | 35 | 2 | 28 | 36 | 4 | 0 | 33 | 5 | 15 | 34 | 7 | 0 | 35 |
| 26 | 11 | 9 | 41 | 1 | 24 | 42 | 1 | 26 | 39 | 3 | 11 | 40 | 4 | 13 | 37 | 5 | 28 | 38 | 7 | 13 | 39 |
| 27 | 11 | 22 | 45 | 2 | 7 | 46 | 2 | 9 | 43 | 3 | 24 | 44 | 4 | 26 | 41 | 6 | 11 | 42 | 7 | 26 | 43 |
| 28 | 0 | 5 | 49 | 2 | 20 | 50 | 2 | 22 | 47 | 4 | 7 | 48 | 5 | 9 | 45 | 6 | 24 | 46 | 8 | 9 | 47 |
| 29 | 0 | 18 | 53 | 2 | 3 | 54 | 3 | 5 | 51 | 4 | 20 | 52 | 5 | 22 | 49 | 7 | 7 | 50 | 8 | 22 | 50 |
| 30 | 1 | 1 | 57 | 2 | 16 | 58 | 3 | 18 | 55 | 5 | 3 | 56 | 6 | 5 | 53 | 7 | 20 | 53 | 9 | 5 | 54 |
| 31 | 1 | 15 | 0 | 0 | 0 | 0 | 4 | 1 | 59 | 5 | 0 | 0 | 6 | 18 | 57 | 8 | 3 | 57 | 0 | 0 | 0 |

| | October | | | November | | | December | | | Ianuarius | | | Februarius | | |
|---|---|---|---|---|---|---|---|---|---|---|---|---|---|---|---|
| | s | g | m | s | g | m | s | g | m | s | g | m | s | g | m |
| 1 | 9 | 18 | 58 | 11 | 3 | 59 | 0 | 5 | 26 | 1 | 20 | 57 | 3 | 5 | 58 |
| 2 | 10 | 2 | 2 | 11 | 17 | 3 | 0 | 18 | 30 | 2 | 4 | 1 | 3 | 19 | 2 |
| 3 | 10 | 15 | 6 | 0 | 0 | 7 | 1 | 1 | 34 | 2 | 17 | 5 | 4 | 2 | 6 |
| 4 | 10 | 28 | 10 | 0 | 13 | 11 | 1 | 14 | 38 | 3 | 0 | 9 | 4 | 15 | 10 |
| 5 | 11 | 11 | 14 | 0 | 26 | 15 | 1 | 27 | 42 | 3 | 13 | 13 | 4 | 28 | 14 |
| 6 | 11 | 24 | 18 | 1 | 9 | 19 | 2 | 10 | 46 | 3 | 26 | 17 | 5 | 11 | 18 |
| 7 | 0 | 7 | 21 | 1 | 22 | 22 | 2 | 23 | 49 | 4 | 9 | 20 | 5 | 24 | 21 |
| 8 | 0 | 20 | 25 | 2 | 5 | 26 | 3 | 6 | 53 | 4 | 22 | 24 | 6 | 7 | 25 |
| 9 | 1 | 3 | 29 | 2 | 18 | 30 | 3 | 19 | 57 | 5 | 5 | 28 | 6 | 20 | 29 |
| 10 | 1 | 16 | 33 | 3 | 1 | 34 | 4 | 3 | 1 | 5 | 18 | 32 | 7 | 3 | 33 |
| 11 | 1 | 29 | 37 | 3 | 14 | 38 | 4 | 16 | 5 | 6 | 1 | 36 | 7 | 16 | 37 |
| 12 | 2 | 12 | 41 | 3 | 27 | 42 | 4 | 29 | 9 | 6 | 14 | 40 | 7 | 29 | 41 |
| 13 | 2 | 25 | 45 | 4 | 10 | 46 | 5 | 12 | 13 | 6 | 27 | 44 | 8 | 12 | 45 |
| 14 | 3 | 8 | 49 | 4 | 23 | 50 | 5 | 25 | 17 | 7 | 10 | 48 | 8 | 25 | 49 |
| 15 | 3 | 21 | 53 | 5 | 6 | 54 | 6 | 8 | 21 | 7 | 23 | 52 | 9 | 8 | 53 |
| 16 | 4 | 4 | 57 | 5 | 19 | 58 | 6 | 21 | 25 | 8 | 6 | 56 | 9 | 21 | 57 |
| 17 | 4 | 18 | 0 | 6 | 3 | 1 | 7 | 4 | 28 | 8 | 19 | 59 | 10 | 5 | 0 |
| 18 | 5 | 1 | 4 | 6 | 16 | 5 | 7 | 17 | 32 | 9 | 3 | 3 | 10 | 18 | 4 |
| 19 | 5 | 14 | 8 | 6 | 29 | 9 | 8 | 0 | 36 | 9 | 16 | 7 | 11 | 1 | 8 |
| 20 | 5 | 27 | 12 | 7 | 12 | 13 | 8 | 13 | 40 | 9 | 29 | 11 | 11 | 14 | 12 |
| 21 | 6 | 10 | 16 | 7 | 25 | 17 | 8 | 26 | 44 | 10 | 12 | 15 | 11 | 27 | 16 |
| 22 | 6 | 23 | 20 | 8 | 8 | 21 | 9 | 9 | 48 | 10 | 25 | 19 | 0 | 10 | 20 |
| 23 | 7 | 6 | 24 | 8 | 21 | 25 | 9 | 22 | 52 | 11 | 8 | 23 | 0 | 23 | 24 |
| 24 | 7 | 19 | 28 | 9 | 4 | 29 | 10 | 5 | 56 | 11 | 21 | 27 | 1 | 6 | 28 |
| 25 | 8 | 2 | 32 | 9 | 17 | 33 | 10 | 19 | 0 | 0 | 4 | 31 | 1 | 19 | 32 |
| 26 | 8 | 15 | 36 | 10 | 0 | 37 | 11 | 2 | 4 | 0 | 17 | 35 | 2 | 2 | 36 |
| 27 | 8 | 28 | 40 | 10 | 13 | 41 | 11 | 15 | 7 | 1 | 0 | 38 | 2 | 15 | 39 |
| 28 | 9 | 11 | 43 | 10 | 26 | 44 | 11 | 28 | 11 | 1 | 13 | 42 | 2 | 28 | 43 |
| 29 | 9 | 24 | 47 | 11 | 9 | 48 | 0 | 11 | 15 | 1 | 26 | 46 | addicio motus | | |
| 30 | 10 | 7 | 51 | 11 | 22 | 52 | 0 | 24 | 19 | 2 | 9 | 50 | 6 horarum | | |
| 31 | 10 | 20 | 55 | 0 | 5 | 56 | 1 | 7 | 23 | 2 | 22 | 54 | 3 | 1 | 59 |

| Entry | MSS. | Amended to |
|---|---|---|
| Mar. 29 | 56 | 53 |
| Iul. 21 | 16 | 18 |
| Aug. 21 | 16 | 18 |
| Nov. 19 | 5 | 9 |
| Nov. 26 | 35 | 37 |
| Ian. 15 | 54 | 51 |
| Ian. 16 | 58 | 55 |

See also Commentary to iv. 14, 15.

## IV. 16. TABULA ASCENSIONUM SIGNORUM

| | 9 | | 10 | | 11 | | 0 | | 1 | | 2 | | 3 | |
|---|---|---|---|---|---|---|---|---|---|---|---|---|---|---|
| | g | m | g | m | g | m | g | m | g | m | g | m | g | m |
| 1 | 1 | 5 | 3 | 15 | 3 | 4 | 0 | 54 | 28 | 50 | 28 | 50 | 1 | 5 |
| 2 | 2 | 11 | 4 | 17 | 4 | 1 | 1 | 50 | 29 | 48 | 29 | 53 | 2 | 11 |
| 3 | 3 | 16 | 5 | 19 | 4 | 58 | 2 | 45 | 1 | 46 | 2 | 56 | 3 | 16 |
| 4 | 4 | 22 | 6 | 21 | 5 | 55 | 3 | 40 | 1 | 43 | 1 | 59 | 4 | 22 |
| 5 | 5 | 27 | 7 | 23 | 6 | 52 | 4 | 35 | 2 | 41 | 3 | 2 | 5 | 27 |
| 6 | 6 | 33 | 8 | 24 | 7 | 48 | 5 | 30 | 3 | 39 | 4 | 5 | 6 | 33 |
| 7 | 7 | 38 | 9 | 26 | 8 | 45 | 6 | 25 | 4 | 38 | 5 | 9 | 7 | 38 |
| 8 | 8 | 43 | 10 | 27 | 9 | 41 | 7 | 20 | 5 | 36 | 6 | 13 | 8 | 43 |
| 9 | 9 | 48 | 11 | 28 | 10 | 37 | 8 | 16 | 6 | 35 | 7 | 16 | 9 | 48 |
| 10 | 10 | 53 | 12 | 29 | 11 | 33 | 9 | 11 | 7 | 34 | 8 | 20 | 10 | 53 |
| 11 | 11 | 58 | 13 | 29 | 12 | 29 | 10 | 6 | 8 | 33 | 9 | 25 | 11 | 58 |
| 12 | 13 | 3 | 14 | 29 | 13 | 25 | 11 | 1 | 9 | 32 | 10 | 29 | 13 | 3 |
| 13 | 14 | 8 | 15 | 30 | 14 | 21 | 11 | 57 | 10 | 31 | 11 | 33 | 14 | 8 |
| 14 | 15 | 13 | 16 | 30 | 15 | 17 | 12 | 52 | 11 | 31 | 12 | 38 | 15 | 13 |
| 15 | 16 | 18 | 17 | 30 | 16 | 12 | 13 | 48 | 12 | 30 | 13 | 42 | 16 | 18 |
| 16 | 17 | 22 | 18 | 29 | 17 | 8 | 14 | 43 | 13 | 30 | 14 | 47 | 17 | 22 |
| 17 | 18 | 27 | 19 | 29 | 18 | 3 | 15 | 39 | 14 | 30 | 15 | 52 | 18 | 27 |
| 18 | 19 | 31 | 20 | 28 | 18 | 59 | 16 | 35 | 15 | 31 | 16 | 57 | 19 | 31 |
| 19 | 20 | 35 | 21 | 27 | 19 | 54 | 17 | 31 | 16 | 31 | 18 | 2 | 20 | 35 |
| 20 | 21 | 40 | 22 | 26 | 20 | 49 | 18 | 27 | 17 | 31 | 19 | 7 | 21 | 40 |
| 21 | 22 | 44 | 23 | 25 | 21 | 44 | 19 | 23 | 18 | 32 | 20 | 12 | 22 | 44 |
| 22 | 23 | 47 | 24 | 24 | 22 | 40 | 20 | 19 | 19 | 33 | 21 | 17 | 23 | 47 |
| 23 | 24 | 51 | 25 | 22 | 23 | 35 | 21 | 15 | 20 | 34 | 22 | 22 | 24 | 51 |
| 24 | 25 | 55 | 26 | 21 | 24 | 30 | 22 | 12 | 21 | 36 | 23 | 27 | 25 | 55 |
| 25 | 26 | 58 | 27 | 19 | 25 | 25 | 23 | 8 | 22 | 37 | 24 | 33 | 26 | 58 |
| 26 | 28 | 1 | 28 | 17 | 26 | 20 | 24 | 5 | 23 | 39 | 25 | 38 | 28 | 1 |
| 27 | 29 | 4 | 29 | 14 | 27 | 15 | 25 | 2 | 24 | 41 | 26 | 44 | 29 | 4 |
| 28 | 10 | 7 | 11 | 12 | 28 | 10 | 25 | 59 | 25 | 43 | 27 | 49 | 4 | 7 |
| 29 | 1 | 10 | 1 | 10 | 29 | 5 | 26 | 56 | 26 | 45 | 28 | 55 | 1 | 10 |
| 30 | 2 | 13 | 2 | 7 | 12 | 0 | 27 | 53 | 27 | 47 | 3 | 0 | 2 | 13 |

# IN CIRCULO DIRECTO

| 4 | | 5 | | 6 | | 7 | | 8 | |
|---|---|---|---|---|---|---|---|---|---|
| g | m | g | m | g | m | g | m | g | m |
| 3 | 15 | 3 | 4 | 0 | 54 | 28 | 50 | 28 | 50 |
| 4 | 17 | 4 | 1 | 1 | 50 | 29 | 48 | 29 | 53 |
| 5 | 19 | 4 | 58 | 2 | 45 | 7 | 46 | 8 | 56 |
| 6 | 21 | 5 | 55 | 3 | 40 | 1 | 43 | 1 | 59 |
| 7 | 23 | 6 | 52 | 4 | 35 | 2 | 42 | 3 | 2 |
| 8 | 24 | 7 | 48 | 5 | 30 | 3 | 40 | 4 | 5 |
| 9 | 26 | 8 | 45 | 6 | 25 | 4 | 38 | 5 | 9 |
| 10 | 27 | 9 | 41 | 7 | 20 | 5 | 36 | 6 | 13 |
| 11 | 28 | 10 | 37 | 8 | 16 | 6 | 35 | 7 | 17 |
| 12 | 29 | 11 | 33 | 9 | 11 | 7 | 34 | 8 | 21 |
| 13 | 29 | 12 | 29 | 10 | 6 | 8 | 33 | 9 | 25 |
| 14 | 29 | 13 | 25 | 11 | 1 | 9 | 32 | 10 | 29 |
| 15 | 30 | 14 | 21 | 11 | 57 | 10 | 31 | 11 | 33 |
| 16 | 30 | 15 | 17 | 12 | 52 | 11 | 30 | 12 | 37 |
| 17 | 30 | 16 | 12 | 13 | 48 | 12 | 30 | 13 | 42 |
| 18 | 29 | 17 | 8 | 14 | 43 | 13 | 30 | 14 | 47 |
| 19 | 29 | 18 | 3 | 15 | 39 | 14 | 30 | 15 | 52 |
| 20 | 28 | 18 | 59 | 16 | 35 | 15 | 30 | 16 | 59 |
| 21 | 27 | 19 | 54 | 17 | 31 | 16 | 31 | 18 | 2 |
| 22 | 26 | 20 | 49 | 18 | 27 | 17 | 31 | 19 | 7 |
| 23 | 25 | 21 | 44 | 19 | 23 | 18 | 32 | 20 | 12 |
| 24 | 24 | 22 | 40 | 20 | 19 | 19 | 33 | 21 | 17 |
| 25 | 22 | 23 | 35 | 21 | 15 | 20 | 34 | 22 | 22 |
| 26 | 21 | 24 | 30 | 22 | 12 | 21 | 36 | 23 | 27 |
| 27 | 19 | 25 | 25 | 23 | 8 | 22 | 37 | 24 | 33 |
| 28 | 17 | 26 | 20 | 24 | 5 | 23 | 39 | 25 | 38 |
| 29 | 14 | 27 | 15 | 25 | 2 | 24 | 41 | 26 | 43 |
| 5 | 12 | 28 | 10 | 25 | 59 | 25 | 43 | 27 | 49 |
| 1 | 10 | 29 | 5 | 26 | 56 | 26 | 45 | 28 | 54 |
| 2 | 7 | 6 | 0 | 27 | 53 | 27 | 47 | 9 | 0 |

| Column (sign) | Row | MSS. read (m) | Nallino ed. | Amended to |
|---|---|---|---|---|
| 9 | 1 | 6 | 5 | 5 |
| 9 | 3 | 16 | 17 | — |
| 9 | 4 | 21 | 22 | 22 |
| 9 | 5 | 28 | 27 | 27 |
| 9 | 16 | 23 | 23 | 22 |
| 9 | 20 | 39 | 39 | 40 |
| 9 | 21 | 43 | 43 | 44 |
| 10 | 6 | 24 | 25 | — |
| 10 | 12 | 30 | 30 | 29 |
| 10 | 16 | 30 | 30 | 29 |
| 10 | 24 | 20 | 20 | 21 |
| 0 | 1 | 55 | 55 | 54 |
| 0 | 15 | 47 | 48 | 48 |
| 1 | 4 | 44 | 43 | 43 |
| 1 | 5 | 42 | 41 | 41 |
| 1 | 8 | 40 | 40 | 39 |
| 1 | 8 | 36 | 37 | — |
| 1 | 14 | 30 | 30 | 31 |
| 1 | 18 | 30 | 30 | 31 |
| 2 | 9 | 17 | 17 | 16 |
| 2 | 10 | 21 | 21 | 20 |
| 2 | 14 | 37 | 37 | 38 |
| 2 | 25 | 33 | 32 | — |
| 2 | 27 | 43 | 43 | 44 |
| 2 | 29 | 54 | 54 | 55 |
| 3 | 1 | 6 | 6 | 5 |
| 3 | 3 | 17 | 16 | 16 |
| 3 | 5 | 28 | 28 | 27 |
| 3 | 16 | 23 | 23 | 22 |
| 3 | 20 | 39 | 39 | 40 |
| 3 | 21 | 43 | 43 | 44 |
| 4 | 6 | 24 | 25 | — |
| 4 | 12 | 30 | 30 | 29 |
| 4 | 16 | 30 | 30 | 29 |
| 4 | 24 | 20 | 20 | 21 |
| 4 | 25 | 18 | 18 | 19 |
| 4 | 26 | 16 | 16 | 17 |
| 5 | 8 | 31 | 41 | 41 |
| 6 | 1 | 55 | 55 | 54 |
| 7 | 4 | 44 | 43 | 43 |
| 8 | 11 | 24 | 24 | 25 |
| 8 | 25 | 32 | 32 | 33 |

## IV. 17. TABULA ASCENSIONUM SIGNORUM IN

| | 0 | | 1 | | 2 | | 3 | | 4 | | 5 | | 6 | | 7 | |
|---|---|---|---|---|---|---|---|---|---|---|---|---|---|---|---|---|
| | g | m | g | m | g | m | g | m | g | m | g | m | g | m | g | m |
| 1 | 0 | 25 | 13 | 19 | 1 | 30 | 27 | 25 | 5 | 37 | 18 | 31 | 1 | 25 | 14 | 21 |
| 2 | 0 | 49 | 13 | 48 | 1 | 13 | 28 | 31 | 7 | 0 | 19 | 58 | 2 | 51 | 15 | 48 |
| 3 | 1 | 14 | 14 | 17 | 1 | 56 | 29 | 39 | 8 | 24 | 21 | 24 | 4 | 16 | 17 | 15 |
| 4 | 1 | 38 | 14 | 46 | 2 | 40 | 2 | 47 | 9 | 49 | 22 | 50 | 5 | 42 | 18 | 42 |
| 5 | 2 | 3 | 15 | 16 | 3 | 24 | 1 | 56 | 11 | 14 | 24 | 17 | 7 | 7 | 20 | 8 |
| 6 | 2 | 27 | 15 | 46 | 4 | 9 | 3 | 7 | 12 | 38 | 25 | 42 | 8 | 33 | 21 | 34 |
| 7 | 2 | 52 | 16 | 16 | 4 | 55 | 4 | 17 | 14 | 3 | 27 | 9 | 9 | 58 | 23 | 0 |
| 8 | 3 | 16 | 16 | 46 | 5 | 43 | 5 | 28 | 15 | 29 | 28 | 35 | 11 | 24 | 24 | 26 |
| 9 | 3 | 42 | 17 | 17 | 6 | 30 | 6 | 40 | 16 | 54 | 5 | 1 | 12 | 50 | 25 | 53 |
| 10 | 4 | 6 | 17 | 48 | 7 | 19 | 7 | 52 | 18 | 20 | 1 | 26 | 14 | 16 | 27 | 20 |
| 11 | 4 | 31 | 18 | 20 | 8 | 8 | 9 | 6 | 19 | 46 | 2 | 53 | 15 | 41 | 28 | 46 |
| 12 | 4 | 56 | 18 | 52 | 8 | 58 | 10 | 20 | 21 | 12 | 4 | 19 | 17 | 6 | 8 | 12 |
| 13 | 5 | 22 | 19 | 24 | 9 | 48 | 11 | 35 | 22 | 38 | 5 | 45 | 18 | 32 | 1 | 38 |
| 14 | 5 | 47 | 19 | 56 | 10 | 39 | 12 | 51 | 24 | 4 | 7 | 11 | 19 | 57 | 3 | 4 |
| 15 | 6 | 12 | 20 | 29 | 11 | 32 | 14 | 8 | 25 | 29 | 8 | 36 | 21 | 24 | 4 | 31 |
| 16 | 6 | 37 | 21 | 4 | 12 | 25 | 15 | 25 | 26 | 56 | 10 | 3 | 22 | 49 | 5 | 56 |
| 17 | 7 | 3 | 21 | 38 | 13 | 19 | 16 | 42 | 28 | 22 | 11 | 28 | 24 | 15 | 7 | 22 |
| 18 | 7 | 29 | 22 | 12 | 14 | 14 | 18 | 0 | 29 | 48 | 12 | 54 | 25 | 41 | 8 | 48 |
| 19 | 7 | 55 | 22 | 48 | 15 | 10 | 19 | 18 | 4 | 14 | 14 | 19 | 27 | 7 | 10 | 14 |
| 20 | 8 | 20 | 23 | 22 | 16 | 6 | 20 | 37 | 2 | 40 | 15 | 44 | 28 | 34 | 11 | 40 |
| 21 | 8 | 47 | 23 | 58 | 17 | 4 | 21 | 56 | 4 | 7 | 17 | 10 | 29 | 59 | 13 | 6 |
| 22 | 9 | 13 | 24 | 35 | 18 | 2 | 23 | 17 | 5 | 34 | 18 | 36 | 7 | 25 | 14 | 31 |
| 23 | 9 | 39 | 25 | 11 | 19 | 1 | 24 | 37 | 7 | 0 | 20 | 2 | 2 | 51 | 15 | 57 |
| 24 | 10 | 6 | 25 | 50 | 20 | 1 | 25 | 59 | 8 | 26 | 21 | 27 | 4 | 18 | 17 | 22 |
| 25 | 10 | 33 | 26 | 28 | 21 | 2 | 27 | 20 | 9 | 52 | 22 | 53 | 5 | 43 | 18 | 46 |
| 26 | 11 | 0 | 27 | 7 | 22 | 3 | 28 | 42 | 11 | 18 | 24 | 18 | 7 | 10 | 20 | 11 |
| 27 | 11 | 28 | 27 | 46 | 23 | 5 | 3 | 4 | 12 | 45 | 25 | 44 | 8 | 36 | 21 | 36 |
| 28 | 11 | 56 | 28 | 26 | 24 | 9 | 1 | 27 | 14 | 12 | 27 | 9 | 10 | 2 | 23 | 0 |
| 29 | 12 | 23 | 29 | 7 | 25 | 13 | 2 | 50 | 15 | 39 | 28 | 35 | 11 | 29 | 24 | 23 |
| 30 | 12 | 51 | 29 | 48 | 26 | 18 | 4 | 14 | 17 | 5 | 6 | 0 | 12 | 55 | 25 | 46 |

# CIRCULO OBLIQUO [in latitudine 51 gra. 50 min.]

| 8 | | 9 | | 10 | | 11 | |
|---|---|---|---|---|---|---|---|
| g | m | g | m | g | m | g | m |
| 27 | 10 | 4 | 47 | 0 | 53 | 17 | 37 |
| 28 | 33 | 5 | 51 | 1 | 34 | 18 | 4 |
| 29 | 56 | 6 | 55 | 2 | 14 | 18 | 32 |
| 9 | 18 | 7 | 57 | 2 | 53 | 19 | 0 |
| 2 | 40 | 8 | 58 | 3 | 32 | 19 | 27 |
| 4 | 1 | 9 | 59 | 4 | 10 | 19 | 54 |
| 5 | 23 | 10 | 59 | 4 | 49 | 20 | 21 |
| 6 | 43 | 11 | 58 | 5 | 25 | 20 | 47 |
| 8 | 4 | 12 | 56 | 6 | 2 | 21 | 13 |
| 9 | 23 | 13 | 54 | 6 | 38 | 21 | 40 |
| 10 | 42 | 14 | 50 | 7 | 12 | 22 | 5 |
| 12 | 0 | 15 | 46 | 7 | 48 | 22 | 31 |
| 13 | 18 | 16 | 41 | 8 | 22 | 22 | 57 |
| 14 | 35 | 17 | 35 | 8 | 56 | 23 | 23 |
| 16 | 52 | 18 | 28 | 9 | 31 | 23 | 48 |
| 17 | 9 | 19 | 21 | 10 | 4 | 24 | 13 |
| 18 | 25 | 20 | 12 | 10 | 36 | 24 | 38 |
| 19 | 40 | 21 | 2 | 11 | 8 | 25 | 4 |
| 20 | 54 | 21 | 52 | 11 | 40 | 25 | 29 |
| 22 | 8 | 22 | 41 | 12 | 12 | 25 | 54 |
| 23 | 20 | 23 | 30 | 12 | 43 | 26 | 18 |
| 24 | 32 | 24 | 17 | 13 | 14 | 26 | 44 |
| 25 | 43 | 25 | 5 | 13 | 44 | 27 | 8 |
| 26 | 53 | 25 | 51 | 14 | 14 | 27 | 33 |
| 28 | 4 | 26 | 36 | 14 | 44 | 27 | 57 |
| 29 | 13 | 27 | 20 | 15 | 14 | 28 | 22 |
| 10 | 21 | 28 | 4 | 15 | 43 | 28 | 46 |
| 1 | 29 | 28 | 47 | 16 | 12 | 29 | 11 |
| 2 | 35 | 29 | 30 | 16 | 41 | 29 | 35 |
| 3 | 42 | 11 | 12 | 17 | 9 | 12 | 0 |

| Column (sign) | Row | MSS. read (m) | MS Land misc. 674 | Amended to |
|---|---|---|---|---|
| 0 | 19 | 59 | 55 | 55 |
| 2 | 8 | 45 | 43 | 43 |
| 2 | 29 | 14 | 13 | 13 |
| 3 | 3 | 29 | 39 | 39 |
| 4 | 18 | 38 | 48 | 48 |
| 6 | 2 | 21 | 51 | 51 |
| 7 | 3 | 18 | 15 | 15 |
| 7 | 6 | 32 | 34 | 34 |
| 7 | 29 | 22 | 23 | 23 |
| 9 | 12 | 44 | 46 | 46 |
| 10 | 27 | 44 | — | 43 |
| 10 | 29 | 42 | 41 | 41 |
| 11 | 12 | 21 | 39 | 31 |
| 11 | 28 | 2 | 3 | 11 |

# ALFONSINE TABLES FOR THE ALBION

The following tables are an abstract of the supplementary tables in MS. A, to which manuscript the folio references at the head of each table apply. Only those tables are included which could not have been found in the original work.

## I (f. 107r)

Tabula limbi primi et dividens circulum mensium in 365d. cum quarta parte fere

| | Ian. 0 | | Feb. 1 | | Mar. 2 | | Apr. 3 | | May. 4 | | Iun. 5 | | Iul. 6 | | Aug. 7 | | Sept. 8 | | Oct. 9 | | Nov. 10 | | Dec. 11 | |
|---|---|---|---|---|---|---|---|---|---|---|---|---|---|---|---|---|---|---|---|---|---|---|---|---|
| | g | m | g | m | g | m | g | m | g | m | g | m | g | m | g | m | g | m | g | m | g | m | g | m |
| 5 | 4 | 56 | 5 | 29 | 3 | 5 | 3 | 38 | 3 | 12 | 3 | 46 | 3 | 20 | 3 | 53 | 4 | 26 | 4 | 1 | 4 | 34 | 4 | 8 |
| 10 | 9 | 51 | 10 | 25 | 8 | 1 | 8 | 34 | 8 | 8 | 8 | 41 | 8 | 16 | 8 | 49 | 9 | 22 | 8 | 56 | 9 | 30 | 9 | 4 |
| 15 | 14 | 47 | 15 | 20 | 12 | 56 | 13 | 30 | 13 | 37 | 13 | 4 | 13 | 11 | 13 | 45 | 14 | 18 | 13 | 52 | 14 | 26 | 13 | 59 |
| 20 | 19 | 43 | 20 | 16 | 17 | 52 | 18 | 25 | 17 | 59 | 18 | 33 | 18 | 7 | 18 | 40 | 29 | 14 | 18 | 48 | 19 | 21 | 18 | 55 |
| 25 | 24 | 38 | 25 | 12 | 22 | 48 | 23 | 21 | 22 | 55 | 23 | 28 | 23 | 3 | 23 | 36 | 24 | 9 | 23 | 43 | 24 | 17 | 23 | 51 |
| 31 | 0 | 33 | 28 | 9 | 28 | 42 | 28 | 17 | 28 | 50 | 28 | 24 | 28 | 57 | 29 | 31 | 29 | 5 | 29 | 38 | 29 | 12 | 29 | 46 |

## II (f. (108ᵛ))

Tabula equacio dierum in minuta horarum resoluta

| Gradus | 0 | 1 | 2 | 3 | 4 | 5 | 6 | 7 | 8 | 9 | 10 | 11 |
|---|---|---|---|---|---|---|---|---|---|---|---|---|
| | m | m | m | m | m | m | m | m | m | m | m | m |
| 1 | 9 | 19 | 22 | 17 | 12 | 15 | 25 | 31 | 28 | 15 | 2 | 1 |
| 2 | 9 | 19 | 22 | 17 | 12 | 15 | 25 | 31 | 28 | 14 | 2 | 1 |
| 3 | 10 | 19 | 22 | 17 | 12 | 16 | 25 | 31 | 27 | 14 | 2 | 1 |
| 4 | 10 | 19 | 22 | 16 | 12 | 16 | 25 | 31 | 27 | 13 | 2 | 1 |
| 5 | 10 | 20 | 22 | 16 | 12 | 16 | 26 | 32 | 27 | 23 | 2 | 2 |
| 29 | 18 | 22 | 18 | 12 | 14 | 24 | 31 | 28 | 16 | 3 | 1 | 8 |
| 30 | 18 | 22 | 17 | 12 | 14 | 24 | 31 | 28 | 15 | 3 | 1 | 9 |

## III (f. 108ʳ)

Maxime equaciones epiciculorum planetarum; et hec est prima tabula rote primi faciei

| In auge | | In opposito augis | | | Tabula verarum augium planetarum pro anno Christi 1348 | | |
|---|---|---|---|---|---|---|---|
| g | m | g | m | | s | g | m |
| 5 | 53 | 6 | 37 | Saturni | 8 | 11 | 40 |
| 10 | 34 | 11 | 35 | Iovis | 5 | 21 | 54 |
| 36 | 45 | 47 | 0 | Martis | 4 | 13 | 20 |
| 44 | 49 | 47 | 16 | Veneris | 2 | 29 | 42 |
| 19 | 1 | 23 | 16 | Mercurii | 6 | 28 | 56 |
| 4 | 56 | 7 | 34 | Lune | | | |

|  |  | g | m |
|---|---|---|---|
| Equaciones Mercurii: ad 2 signa et 4 (?) signa | | 22 | 2 |
| ad 5 signa | | 23 | 31 |
| ad 4 signa | | 23 | 53 |

[The auges are, as near as may be judged, those marked on the Merton equatorium]

## IV (f. 107r-v)

### Iupiter

| | 0 g | 0 m | 1 g | 1 m | 2 g | 2 m | 3 g | 3 m | 4 g | 4 m | 5 g | 5 m | 6 g | 6 m | 7 g | 7 m | 8 g | 8 m | 9 g | 9 m | 10 g | 10 m | 11 g | 11 m |
|---|---|---|---|---|---|---|---|---|---|---|---|---|---|---|---|---|---|---|---|---|---|---|---|---|
| 5  | 4  | 30 | 1  | 43 | 29 | 42 | 29 | 3  | 27 | 57 | 2  | 22 | 5  | 33 | 8  | 33 | 10 | 31 | 10 | 54 | 9  | 43 | 7  | 24 |
| 10 | 9  | 1  | 6  | 19 | 4  | 29 | 4  | 5  | 5  | 17 | 7  | 52 | 11 | 5  | 13 | 59 | 15 | 41 | 15 | 49 | 14 | 24 | 11 | 57 |
| 15 | 13 | 32 | 10 | 57 | 9  | 19 | 9  | 11 | 10 | 38 | 13 | 23 | 16 | 37 | 19 | 22 | 20 | 49 | 20 | 41 | 19 | 3  | 16 | 28 |
| 20 | 18 | 3  | 15 | 36 | 14 | 6  | 14 | 19 | 16 | 1  | 18 | 55 | 22 | 8  | 24 | 43 | 25 | 55 | 25 | 31 | 23 | 41 | 20 | 59 |
| 25 | 22 | 36 | 20 | 17 | 19 | 3  | 19 | 29 | 21 | 27 | 24 | 27 | 27 | 38 | 0  | 3  | 0  | 57 | 0  | 18 | 28 | 17 | 25 | 30 |
| 30 | 27 | 9  | 24 | 59 | 24 | 3  | 24 | 41 | 26 | 54 | 0  | 0  | 3  | 6  | 5  | 19 | 5  | 57 | 5  | 1  | 2  | 51 | 0  | 0  |

## V (f. 107r-v)

### Sol et Venus

| | 0 g | 0 m | 1 g | 1 m | 2 g | 2 m | 3 g | 3 m | 4 g | 4 m | 5 g | 5 m | 6 g | 6 m | 7 g | 7 m | 8 g | 8 m | 9 g | 9 m | 10 g | 10 m | 11 g | 11 m |
|---|---|---|---|---|---|---|---|---|---|---|---|---|---|---|---|---|---|---|---|---|---|---|---|---|
| 5  | 4  | 49 | 3  | 48 | 3  | 4  | 2  | 50 | 3  | 11 | 4  | 3  | 5  | 12 | 6  | 17 | 7  | 0  | 7  | 9  | 6  | 44 | 5  | 53 |
| 10 | 9  | 39 | 8  | 39 | 7  | 59 | 7  | 51 | 8  | 18 | 9  | 14 | 10 | 24 | 11 | 26 | 12 | 4  | 12 | 7  | 11 | 33 | 10 | 43 |
| 15 | 14 | 28 | 13 | 31 | 12 | 56 | 12 | 53 | 13 | 26 | 14 | 25 | 15 | 35 | 16 | 34 | 17 | 7  | 17 | 4  | 16 | 29 | 15 | 32 |
| 20 | 19 | 27 | 18 | 23 | 17 | 53 | 17 | 56 | 28 | 34 | 19 | 36 | 20 | 46 | 21 | 42 | 22 | 9  | 22 | 1  | 21 | 21 | 20 | 21 |
| 25 | 24 | 7  | 23 | 16 | 22 | 51 | 23 | 0  | 23 | 43 | 24 | 48 | 25 | 57 | 26 | 49 | 27 | 10 | 26 | 56 | 26 | 12 | 25 | 11 |
| 30 | 28 | 57 | 28 | 9  | 27 | 50 | 28 | 5  | 28 | 53 | 0  | 0  | 1  | 7  | 1  | 5  | 2  | 10 | 1  | 11 | 1  | 3  | 0  | 0  |

Amendments to table V (cf. Commentary for explanation):
Under 7ˢ 30°, for 1° 5′ read 1° 55′;
under 9ˢ 30°, for 1° 11′ read 1° 51′;
under 10ˢ 10°, for 11° 33′ read 11° 37′.

## VI (f. 107r-v)
### Orbis Lune

| | 0 | | 1 | | 2 | | 3 | | 4 | | 5 | | 6 | | 7 | | 8 | | 9 | | 10 | | 11 | |
|---|---|---|---|---|---|---|---|---|---|---|---|---|---|---|---|---|---|---|---|---|---|---|---|---|
| | g | m | g | m | g | m | g | m | g | m | g | m | g | m | g | m | g | m | g | m | g | m | g | m |
| 5 | 5 | 45 | 10 | 7 | 14 | 15 | 27 | 24 | 17 | 53 | 13 | 5 | 3 | 10 | 24 | 38 | 21 | 51 | 23 | 27 | 27 | 4 | 1 | 20 |
| 10 | 11 | 29 | 15 | 50 | 19 | 52 | 23 | 42 | 22 | 30 | 16 | 39 | 6 | 28 | 28 | 49 | 26 | 55 | 28 | 58 | 2 | 45 | 7 | 3 |
| 15 | 17 | 13 | 21 | 33 | 25 | 27 | 27 | 56 | 26 | 54 | 20 | 8 | 9 | 52 | 3 | 6 | 2 | 4 | 4 | 33 | 8 | 27 | 12 | 47 |
| 20 | 22 | 57 | 27 | 15 | 1 | 2 | 3 | 5 | 1 | 11 | 23 | 32 | 13 | 21 | 7 | 30 | 7 | 18 | 10 | 8 | 14 | 10 | 18 | 31 |
| 25 | 28 | 40 | 2 | 56 | 6 | 33 | 8 | 9 | 5 | 22 | 26 | 50 | 16 | 55 | 12 | 7 | 12 | 36 | 15 | 45 | 19 | 53 | 24 | 15 |
| 30 | 4 | 23 | 8 | 36 | 12 | 0 | 13 | 4 | 9 | 22 | 0 | 0 | 20 | 38 | 16 | 56 | 18 | 0 | 21 | 24 | 25 | 37 | 0 | 0 |

## VII (f. 107r-v)
### Verus [motus] Lune

| | 0 | | 1 | | 2 | | 3 | | 4 | | 5 | | 6 | | 7 | | 8 | | 9 | | 10 | | 11 | |
|---|---|---|---|---|---|---|---|---|---|---|---|---|---|---|---|---|---|---|---|---|---|---|---|---|
| | g | m | g | m | g | m | g | m | g | m | g | m | g | m | g | m | g | m | g | m | g | m | g | m |
| 5 | 4 | 36 | 2 | 22 | 0 | 42 | 0 | 4 | 0 | 46 | 2 | 45 | 5 | 28 | 8 | 2 | 9 | 37 | 9 | 52 | 8 | 50 | 6 | 56 |
| 10 | 9 | 13 | 7 | 2 | 5 | 31 | 5 | 5 | 6 | 1 | 8 | 10 | 10 | 56 | 13 | 23 | 14 | 46 | 14 | 46 | 13 | 34 | 11 | 34 |
| 15 | 13 | 49 | 11 | 43 | 10 | 21 | 10 | 9 | 11 | 18 | 13 | 37 | 16 | 23 | 18 | 42 | 19 | 51 | 19 | 39 | 18 | 17 | 16 | 11 |
| 20 | 18 | 26 | 16 | 26 | 15 | 14 | 15 | 14 | 26 | 37 | 19 | 4 | 21 | 50 | 23 | 59 | 24 | 55 | 24 | 29 | 22 | 58 | 20 | 47 |
| 25 | 23 | 4 | 21 | 10 | 20 | 8 | 20 | 23 | 21 | 58 | 24 | 32 | 27 | 15 | 29 | 14 | 29 | 56 | 29 | 18 | 27 | 38 | 25 | 24 |
| 30 | 27 | 43 | 0 | 55 | 24 | 5 | 25 | 33 | 27 | 21 | 0 | 0 | 2 | 39 | 4 | 27 | 4 | 55 | 4 | 5 | 2 | 17 | 0 | 0 |

Amendment to table VII: Under 1$^s$ 30°, for 0° 55′ read 25° 55′.

## VIII (f. 108ᵛ)

Medii motus planetarum in anno cum quarta parte unius diei fere.

| s | g | m | s | Medius motus |
|---|---|---|---|---|
| 12 | 0 | 0 | 0 | Solis |
| 4 | 12 | 34 | 48 | Lune |
| 3 | 1 | 53 | 23 | Argumentum Lune |
| 0 | 19 | 20 | 29 | Caput Draconis |
| 0 | 12 | 14 | 5 | Saturni |
| 1 | 0 | 21 | 42 | Jovis |
| 6 | 11 | 24 | 42 | Martis |
| 7 | 15 | 10 | 40 | Argumentum Veneris |
| 1 | 24 | 42 | 0 | Argumentum Mercurii |

## IX (f. 108ʳ)

Quantitas subtractionis omnium circulorum a toto circulo zodiaci.

| | g | m | s | |
|---|---|---|---|---|
| | 6 | 54 | 24 | Saturnus |
| | 4 | 17 | 0 | Iupiter |
| | 0 | 46 | 27 | Mars |
| | 0 | 14 | 21 | Sol |
| | 0 | 17 | 1 | Venus |
| | 0 | 21 | 16 | Caput Draconis |
| De circulo longitudinis | 10 | 54 | 0 | Aux |
| | 29 | 6 | 0 | Longitudo media |
| | 27 | 2 | 0 | Oppositum augis |
| De circulo horarum | 4 | 34 | 24 | Aux |
| | 35 | 17 | 31 | Longitudo media |
| | 82 | 41 | 12 | Oppositum augis |

Amendment: opposite 'Caput Draconis', read 6° 21′ 16″.

## X (f. 108ʳ)    XI (f. 108ʳ)

| | Caput Draconis | | | | Saturnus | | |
|---|---|---|---|---|---|---|---|
| | s | g | m | | s | g | m |
| 1 | 0 | 18 | 37 | 1 | 0 | 29 | 25 |
| 2 | 1 | 7 | 14 | 2 | 1 | 28 | 51 |
| 3 | 1 | 25 | 50 | 3 | 2 | 28 | 16 |
| 4 | 2 | 14 | 27 | 4 | 3 | 27 | 42 |
| 5 | 3 | 3 | 4 | 5 | 4 | 27 | 7 |
| 6 | 3 | 21 | 41 | 6 | 5 | 26 | 33 |
| 7 | 4 | 10 | 17 | 7 | 6 | 25 | 58 |
| 8 | 4 | 28 | 54 | 8 | 7 | 25 | 24 |
| 9 | 5 | 17 | 31 | 9 | 8 | 24 | 49 |
| 10 | 6 | 6 | 8 | 10 | 9 | 24 | 14 |
| 11 | 6 | 24 | 45 | 11 | 10 | 23 | 40 |
| 12 | 7 | 13 | 21 | 12 | 21 | 23 | 6 |
| 13 | 8 | 1 | 58 | | | | |
| 14 | 8 | 20 | 35 | | | | |
| 15 | 9 | 9 | 12 | | | | |
| 16 | 9 | 27 | 48 | | | | |
| 17 | 10 | 16 | 25 | | | | |
| 18 | 11 | 5 | 2 | | | | |
| 19 | 11 | 23 | 39 | | | | |

Amend the last Saturn entry to 12ˢ 23° 6′.

### XII (f. 109ʳ)   XIII (f. 109ʳ)   XIV (f. 109ʳ)

| | Iupiter | | | Mars | | | | | Argumentum Veneris | | | | |
|---|---|---|---|---|---|---|---|---|---|---|---|---|
| | s | g | m | s | g | s | g | m | s | g | s | g | m |
| 1 | 0 | 11 | 51 | 0 | 6 | 0 | 11 | 17 | 0 | 6 | 0 | 9 | 36 |
| 2 | 0 | 23 | 43 | 0 | 12 | 0 | 22 | 34 | 0 | 12 | 0 | 19 | 11 |
| 3 | 1 | 5 | 34 | 0 | 18 | 1 | 3 | 51 | 0 | 18 | 0 | 28 | 47 |
| 4 | 1 | 17 | 26 | 0 | 24 | 1 | 15 | 8 | 0 | 24 | 1 | 8 | 22 |
| 5 | 1 | 29 | 17 | 1 | 0 | 1 | 26 | 25 | 1 | 0 | 1 | 17 | 58 |
| 6 | 2 | 11 | 9 | 1 | 6 | 2 | 7 | 42 | 1 | 6 | 1 | 27 | 33 |
| 7 | 2 | 23 | 0 | 1 | 12 | 2 | 18 | 59 | 1 | 12 | 2 | 7 | 9 |
| 8 | 3 | 4 | 52 | 1 | 18 | 3 | 0 | 16 | 1 | 18 | 2 | 20 | 44 |
| 9 | 3 | 16 | 43 | 1 | 24 | 3 | 11 | 33 | 1 | 24 | 2 | 26 | 20 |
| 10 | 3 | 28 | 34 | 2 | 0 | 3 | 22 | 51 | 2 | 0 | 3 | 5 | 56 |
| 11 | 4 | 19 | 26 | 2 | 6 | 4 | 4 | 8 | 2 | 6 | 3 | 15 | 31 |
| 12 | 4 | 22 | 17 | 2 | 12 | 4 | 15 | 25 | 2 | 12 | 3 | 25 | 7 |
| 13 | 5 | 4 | 9 | 2 | 18 | 4 | 26 | 42 | 2 | 18 | 4 | 4 | 42 |
| 14 | 5 | 16 | 0 | 2 | 24 | 5 | 7 | 59 | 2 | 24 | 4 | 14 | 18 |
| 15 | 5 | 27 | 52 | 3 | 0 | 5 | 19 | 16 | 3 | 0 | 4 | 23 | 54 |
| 16 | 6 | 9 | 43 | 3 | 6 | 6 | 0 | 33 | 3 | 6 | 5 | 13 | 5 |
| 17 | 6 | 21 | 34 | 3 | 12 | 6 | 11 | 51 | 3 | 12 | 5 | 22 | 41 |
| 18 | 7 | 3 | 26 | 3 | 18 | 6 | 23 | 8 | 4 | 18 | 6 | 11 | 52 |
| 19 | 7 | 15 | 17 | 3 | 24 | 7 | 4 | 25 | 4 | 24 | 6 | 21 | 27 |
| 20 | 7 | 27 | 9 | 4 | 0 | 7 | 15 | 42 | 4 | 0 | 7 | 10 | 38 |
| 21 | 8 | 9 | 0 | 4 | 6 | 7 | 26 | 59 | 4 | 6 | 7 | 20 | 14 |
| 22 | 8 | 20 | 52 | 4 | 12 | 8 | 8 | 16 | 5 | 12 | 8 | 9 | 25 |
| 23 | 9 | 2 | 43 | 4 | 18 | 8 | 19 | 33 | 5 | 18 | 8 | 19 | 0 |
| 24 | 9 | 14 | 34 | 4 | 24 | 9 | 0 | 51 | 5 | 24 | 9 | 8 | 11 |
| 25 | 9 | 26 | 26 | 5 | 0 | 9 | 12 | 8 | 6 | 0 | 9 | 27 | 47 |
| 26 | 10 | 8 | 17 | 5 | 6 | 9 | 23 | 25 | 6 | 6 | 10 | 6 | 58 |
| 27 | 10 | 20 | 9 | 5 | 12 | 10 | 4 | 42 | 6 | 12 | 10 | 16 | 33 |
| 28 | 11 | 2 | 0 | 5 | 18 | 10 | 27 | 16 | 6 | 18 | 11 | 5 | 44 |
| 29 | 11 | 13 | 52 | 5 | 24 | 11 | 8 | 33 | 6 | 24 | 11 | 15 | 20 |
| 30 | 11 | 25 | 43 | 6 | 0 | 11 | 29 | 13 | 7 | 0 | 11 | 29 | 43 |

## XV (f. 109ʳ)

Tabula dividens circulum geuzahar

| g | m | s | g | s | g | s | g | s | g |
|---|---|---|---|---|---|---|---|---|---|
| 0 | 0 | 0 | 0 | 6 | 0 | 6 | 0 | 12 | 0 |
| 0 | 26 | 0 | 5 | 5 | 25 | 6 | 5 | 11 | 25 |
| 0 | 52 | 0 | 10 | 5 | 20 | 6 | 10 | 11 | 20 |
| 1 | 18 | 0 | 15 | 5 | 15 | 6 | 15 | 11 | 15 |
| 1 | 42 | 0 | 20 | 5 | 10 | 6 | 20 | 11 | 10 |
| 2 | 7 | 0 | 25 | 5 | 5 | 6 | 25 | 11 | 5 |
| 2 | 30 | 1 | 0 | 5 | 0 | 7 | 0 | 11 | 0 |
| 2 | 52 | 1 | 5 | 4 | 25 | 7 | 5 | 10 | 25 |
| 3 | 13 | 1 | 10 | 4 | 20 | 7 | 10 | 10 | 20 |
| 3 | 32 | 1 | 15 | 4 | 15 | 7 | 15 | 10 | 15 |
| 3 | 50 | 1 | 20 | 4 | 10 | 7 | 20 | 10 | 10 |
| 4 | 6 | 1 | 25 | 4 | 5 | 7 | 25 | 10 | 5 |
| 4 | 20 | 2 | 0 | 4 | 0 | 8 | 0 | 10 | 0 |
| 4 | 32 | 2 | 5 | 3 | 25 | 8 | 5 | 9 | 25 |
| 4 | 42 | 2 | 10 | 3 | 20 | 8 | 10 | 9 | 20 |
| 4 | 50 | 2 | 15 | 3 | 15 | 8 | 15 | 9 | 15 |
| 4 | 55 | 2 | 20 | 3 | 10 | 8 | 20 | 9 | 10 |
| 4 | 59 | 2 | 25 | 3 | 5 | 8 | 25 | 9 | 5 |
| 5 | 0 | 3 | 0 | 3 | 0 | 9 | 0 | 9 | 0 |

## XVI (f. 108ʳ)

Tabula mediis coniunctionibus Solis et Lune

| Lumaciones | h | m | [Illegible] | | | | | |
|---|---|---|---|---|---|---|---|---|
| | | | s | g | m | s | g | m |
| 1 | 12 | 44 | 0 | 29 | 6 | 0 | 25 | 49 |
| 2 | 1 | 28 | 1 | 28 | 13 | 1 | 21 | 38 |
| 3 | 14 | 12 | 2 | 27 | 19 | 2 | 17 | 27 |
| 4 | 2 | 56 | 3 | 26 | 26 | 3 | 13 | 16 |
| 5 | 25 | 40 | 4 | 25 | 32 | 4 | 9 | 5 |
| 6 | 4 | 24 | 5 | 24 | 38 | 5 | 4 | 54 |
| 7 | 27 | 8 | 6 | 23 | 45 | 6 | 0 | 43 |
| 8 | 5 | 52 | 7 | 22 | 51 | 6 | 26 | 32 |
| 9 | 18 | 36 | 8 | 21 | 58 | 7 | 22 | 21 |
| 10 | 7 | 21 | 9 | 21 | 4 | 8 | 18 | 10 |
| 11 | 20 | 5 | 10 | 20 | 10 | 9 | 13 | 59 |
| 12 | 8 | 49 | 11 | 19 | 17 | 10 | 9 | 48 |

## XVII (f. 108$^r$)

Tabula vere distancie Solis et Lune equate pro coniunctionibus et oppositionibus eorumdem

|     | aux | | | longitudo media | | | oppositum augis | | |
| --- | --- | --- | --- | --- | --- | --- | --- | --- | --- |
|     | s | g | m | s | g | m | s | g | m |
| 1   | 1 | 2 | 55 | 1 | 2 | 26 | 1 | 2 | 5 |
| 2   | 2 | 5 | 49 | 2 | 4 | 51 | 2 | 4 | 9 |
| 3   | 3 | 8 | 44 | 3 | 7 | 17 | 3 | 6 | 14 |
| 4   | 4 | 11 | 38 | 4 | 9 | 42 | 4 | 8 | 19 |
| 5   | 5 | 14 | 33 | 5 | 12 | 8 | 5 | 10 | 23 |
| 6   | 6 | 17 | 27 | 6 | 14 | 33 | 6 | 12 | 28 |
| 7   | 7 | 20 | 12 | 7 | 16 | 59 | 7 | 14 | 33 |
| 8   | 8 | 23 | 16 | 8 | 19 | 24 | 8 | 16 | 38 |
| 9   | 9 | 26 | 11 | 9 | 21 | 50 | 9 | 18 | 42 |
| 10  | 10 | 29 | 6 | 10 | 24 | 15 | 10 | 20 | 47 |
| 11  | | | | 11 | 26 | 42 | 11 | 22 | 52 |
| 12  | | | | 0 | 29 | 6 | 0 | 24 | 57 |
| 13  | | | | | | | 1 | 27 | 2 |

## XVIII (f. 109$^r$)

De eclipsibus Solis et Lune semidiametri. Tabula ostendens quantitatem minutorum Solis et Lune, minutorum casus et more et umbre, simul in utraque, in auge et in eius opposito, que eclipsi (?).

| m | s | Semidiam. | m | s | Quantitas minutorum |
| --- | --- | --- | --- | --- | --- |
| 25 | 43 | Solis | 30 | 32 | Solis ⎫ in auge |
| 24 | 50 | Lune | 53 | 24 | Lune ⎰ |
| 38 | 34 | Umbre | 52 | 34 | Sole ⎫ exeunte in opp. aug. |
| 26 | 49 | Solis | 34 | 29 | Lune ⎰ |
| 17 | 40 | Lune | 63 | 36 | Solis in opposito augis |
| 45 | 56 | Umbre | 63 | 46 | Lune in auge Sole exeunte (?) |
| Minuta more ⎰ | | | 21 | 58 | Luna in auge ⎫ Sole exeunte (?) |
| | | | 28 | 16 | In eius opp. ⎰ in auge |
| Diversitas aspectus in circulo altitudinis ⎰ | | | 53 | 34 | In auge epicicli |
| | | | 63 | 51 | In auge . . . [Illegible] |
| | | | | | augis ep . . . [Illegible] |

# APPENDICES

# APPENDIX 1

## Principal events in the life of Richard of Wallingford

| | |
|---|---|
| 1291 or 1292 | Born, at Wallingford, Berkshire |
| c. 1302–4 | Orphaned, and adopted by William of Kirkeby, Prior of Wallingford |
| c. 1308 | Sent to Oxford |
| 1314 | Bachelor's degree in Arts |
| c. 1314 | Leaves Oxford for St. Albans |
| 1316 (18 Dec.) | Ordained deacon |
| 1317 (28 May) | Ordained priest |
| c. 1317 | Returns to Oxford |
| c. 1318 | *Canones* and *Tractatus* |
| c. 1317–19 | *Exafrenon* |
| c. 1319–25 | *Quadripartitum* |
| 1326 | *Albion* and *Rectangulus* |
| c. 1327 | Bachelor's degree in Theology |
| 1327 (29 Oct.) | Elected abbot of St. Albans |
| 1327 (23 Nov.) | Leaves for Avignon, arriving 4 Jan. 1327/8 |
| 1328 (Apr.) | Returns to St. Albans, after papal confirmation of his appointment |
| 1327–36 | *Tractatus horologii, De sectore,* and (?)*Kalendarium.* The building of the clock is begun |
| 1331–2 | Climax of the struggle with the citizens |
| 1333 | Bishop of Lincoln's commissioners investigate charges against Richard |
| 1336 (23 May) | Death of Richard of Wallingford |

# APPENDIX 2

## *John of Whetehamstede's 'Invenire'*

As explained on p. 18, vol. ii, MS. Cotton Nero C.vi contains references to Richard of Wallingford under the general heading *Invenire*. This section of John of Whetehamstede's *Granarium* begins (f. 147$^r$): 'Quia felices et vere felices illorum credimus fuisse animas . . .'. It is, by and large, a compilation based on such authorities as Isidore and Josephus, devoted to the liberal arts and mechanics, and ends with a long discussion of agricultural techniques. Where it is a simple compilation, it merely informs us of the sort of knowledge an educated writer of the time, with next to no astronomical expertise, could be expected to have. There are, however, one or two pieces of information, apparently obtained by hearsay, which do not appear to have reached us through any other channel; and therefore I reproduce them here with some passages from the first category.

One by one, giving sources in most cases, the inventors of the well-known astronomical instruments are listed, beginning with the astrolabe ('Quoad instrumenta . . .', last two words of f. 148$^{vb}$). In referring to an ascription of the planispheric astrolabe to Ptolemy, John makes the classic medieval mistake of calling him 'Rex Agipti Ptholomeus Philadelphus'. He is aware, however, that Ptolemy's astrolabe, as described in the first book of the *Almagest*, was 'astrolabium spericum, sive speram solidam', and he thus avoids a mistake which some modern authors have made. (Ptolemy's *Planisphaerium* contains, perhaps, evidence that he knew the plane astrolabe. See O. Neugebauer, *Isis* xl (1949), 242.) A footnote shows that he was aware of the ascription to Hipparchus (lit. 'Abrachis'), and mentions a curious derivation of the word 'astrolabium' (ἀστρολάβος):

(149$^r$) Secundum Haly ['Ali ibn Riḍwān] super *Quadripartitum* Ptholomei, inventor primevus instrumenti istius fuit insignis astronomus Abrachis, temporibus regis Salomonis; vel, ut volunt alii, vir nomine *Lab* a quo *astot*, quod dicitur *linea*, est vocatum. Unde *astrolabium*, quasi *linea Lab*, dicitur per interpretationem.

The author, John of Whetehamstede, is more or less correct in ascribing the 'old quadrant' to Sacrobosco (although he was merely the author of an early Latin text, describing two sorts of quadrant, both clearly deriving from the Islamic world), and the use of a cursor to Campanus (he should have said a cursor different in style from that described by Sacrobosco and others). He mentions Petrus de Adamaro (Peter of St. Omer) as having modified the new quadrant (*quadrans novellus* rather than *novus*) of Profatius, adding that Peter studied at Paris. This of itself is not enough to allow us to equate any two of the four Peters of St. Omer who are known to the history of the late thirteenth and early fourteenth centuries, but it is interesting as a categorical early statement that the author of the modified quadrant studied at Paris. (See Sarton's *Introduction*, ii. 996 and 1041. On this edition of Profatius's tract on the quadrant, see E. Poulle, 'Le Quadrant nouveau mediéval', *Journal des Savants* (1964), 148–67, 182–214. The best authority on Peter of St. Omer is Olaf Pedersen, of the University of Aarhus. See for instance, 'Peder Nattergal og hans astronomiske regneinstrument', *Nordisk Astronomisk Tidsskrift*, nr. 2 (1963). Pedersen claims most of the work ascribed to the various Peters for 'Peter of Dacia'. Some evidence from a manuscript in the National Library, Florence, is relevant here. MS. II.iii.24 (= Magliabechiana XI, 117) has a treatise on eclipses 'secundum Petrum de Odemaro' (ff. 206$^{rb}$–f. 208$^{rb}$) followed by a 'prohemium in tractatum eclipsorii Petri Daci', that is, dealing with an eclipse instrument. The complementary nature of the texts suggests at least that the names were used interchangeably here.)

After the annulus (in the present sense, a disc for assisting in the computation of Easter) is conventionally ascribed to John of Northampton, there is this passage with possible reference to the *navicula* (f. 149$^{ra}$): 'Navem primitus adinvenit quidam monachus monasterii Glastoniensis qui Petrus de Mucheleyo fuit vulgariter cognominatus.' The *navicula* is a type of sun-dial, related to the *Analemma* of Ptolemy, a work translated by William of Moerbeke in the thirteenth century. The only Peter from Glastonbury whose name has passed into the history of astronomy is Peter Lightfoot (Petrus Lightfote, according to Leland), who is said to have built a clock at Glastonbury during the abbacy of Adam of Sodbury (1323–34), which clock was later said to have been moved to Wimborne. This was also said to be identifiable with the famous astronomical clock of Wells Cathedral, the story going that the move was made at the time of the Reformation. R. P. Howgrave-Graham, in *Peter Lightfoot and the Old Clock at Wells* (Glastonbury, 1922), has shown that both stories are out of the question. (Cf. John Leland, *Itinerary*, vol. iii (1710), p. 83.) There is certainly not sufficient evidence for identifying Peter of Muchelney with Peter Lightfoot. (Muchelney was the site of a Benedictine abbey of some importance, not far from Glastonbury, in Somerset.) From this fragment we can in fact conclude very little: one 'Peter of Muchelney' is said to have first invented the *navis*, an astronomical instrument, which was almost certainly the *navicula*. The only other hint we have as to the originator of the instrument is referred to by R. T. Gunther in *Early Science in Oxford*, ii. 379: 'T. Allen is stated to have been the owner in 1622 of a manuscript 8vo codex, now lost, which contained a treatise *De compositione navis, quadrantis et cylindre* by an otherwise unknown author, John Slape.' Gunther asks whether Slape was perhaps the author of the *navicula*.

The remaining passages of interest in the *Granarium* come consecutively, and will be quoted without interruption (149 [ra–b]):

Figuram in plano pariete que docet per umbras horas diei certitudinaliter agnoscere, adinvenit primitus, quoad horas inequales, Albategni secundum aliquos, Arzachel vero secundum alios; quoad horas vero equales, adinvenit [149rb] illam primitus monachus monasterii Albanensis, qui apud suos Robertus Stikford fuerat nuncupatus.

Albeonam utique, que in se unica omnium aliorum instrumentorum commoditates legitur continere, adinvenit primitus Ricardus de Walyngfordia, abbas olim monasterii Albanensis; vir siquidem in arte astronomie adeo sufficienter eruditis, quod a suo tempore usque in presens non surrexit sibi similis quoad Anglos. Quoad instrumenta alia, ut puta quoad equatorium et rectangulum, directorium et chilindrum, armillas, sapheam et turketum, quia certitudinaliter nichil repperii, ideo taceo ad presens de inventoribus, nec volo certitudinaliter procedere ubi hucusque certitudinem non inveni.

[Marginal insertion in the same hand] Dicunt tamen alii quod Arzachel adinvenit chelindrum et sapheam, Campanus equatorium, armillam Ptholomeus, rectangulum et horologium astronomicum Ricardus abbas Albanensis, de quo fit mentio in premissis. Ista tamen inscriptis autenticis non repperi, et propterea pro certo hic inserere non curam.

John of Whetehamstede's caution, as shown in the marginal note (if the note was his) was judicious, for he is here doing little more than recount the names of a few early writers on the instruments he mentions, but not their inventors. Robert Stikford, the St. Albans monk who is supposed to have put equal hours on a mural dial, is unknown to modern writers, as far as can be seen.

# APPENDIX 3

## *The entry on Richard of Wallingford in British Museum, MS. Cotton Nero D.vii, ff. 20ʳ–20ᵛ*

Ricardus, abbas vicesimus octavus, divina et humana scientia preditus, construxit horologium quod, ut credimus, omnia huius regni horologia antecessit. Hic multas tribulationes sustinuit pro ecclesie sue jure, sed Deo favente cuncta que incepit ad bonum finem ipse perduxit. Hic habendo respectum ad paupertatem rerum conventus, contulit eidem una vice C quarteria brasii melioris in granario suo reperti, et XL summas melioris tritici in suo manerio de Norton tunc temporis existentis. Reparavit insuper anno primo et secundo sue prelationis pene omnia tecta Monasterii, tam ex parte sua quam ex parte conventus. Et molendina de Parco, de Mora, de Codicot, et de Luyton reparavit et fecit, a fundamentis molendina de Stankefelde, et ad brasium in villa de Sancto Albano.

There is nothing in this passage not obtained from the *Gesta Abbatum*. It is amusing to notice which parts of the *Gesta* were thought worthy of being repeated, in particular the statement relating to Richard's gift of large quantities of malt and wheat to the refectory. (See *Gesta*, ii. 280, for the original passage.)

# APPENDIX 4

## Leland and Tanner on Richard of Wallingford

NOT all editions of Leland's *Commentarii* have this passage, which is taken from Anthony Hall's edition (Oxford, 1709). Thomas Tanner's rendering of the text is virtually identical, his variant readings being noted below, without presupposing the correctness of either version. For the circumstances under which these works were written, and further references, see pp. 19–21, vol. ii. (Differences in punctuation are generally ignored.)

RICHARDUS (WALLINGFORDUS sive) VALINGOFORDUS, ab oppido eiusdem nominis in ipsa Tamesis ripa posito, ubi natus fuit, sic dictus, cuiusdam Gulielmi, fabri ferrarii in sua arte peritissimi, filius erat; a quo ad ludum literarium[1] transmissus, non levia, dum adhuc puer esset, acris ingenii indicia dedit. Quare postea, curante id patre, Isiacum [Oxford, the town of Isis], vicinam academiam, petiit; ubi in collegio Maridunensi, pro aetate impendio profecit. Iamque exegerat in bonis literis bonam adolescentiae partem, cum coelestis quidam monasticae vitae amor in eius animum irrepserit. Ille igitur tempus non differendum ratus, continuo D. Albani monasterium famosissimum, quod ruinis Verolamii municipii olim nobilissimi adiacet, adivit, et in Benedicti verba iuravit. Quod sanctum hominis propositum adeo studium, quo erat mirifice erga literas adfectus, non retardavit, ut multum certe promoverit. Nam a curis solutus omnibus, per otium licuit liberrime vaca re libris; quibus cum abbas, non malus rerum aestimator videret illum omnino deditum, iussit, ut mature Isidis Vadum repeteret: id quod fecit, atque lubens.

Quid ego hic loquar, qua postea usus sit industria, ut rerum causas penitius scrutaretur? Quid eius labores et longas vigilias, quas arithmeticae, astronomiae, et caeteris id genus artibus impendebat, commemorem? Certe hoc constat, tam numerose doctum fuisse, ut, cum post paucos annos ad coenobium suum reverteretur, solus ex tanto monachorum numero visus est[2] dignus, qui titulo scientiae in mortui abbatis locum succederet. Electus itaque in monasterii praesidem dignitatem potius quam animum, quem habebat literarum appetentissimum, mutavit. Namque cum iam per amplas licebat fortunas, voluit illustri aliquo opere non modo ingenii, verum etiam eruditionis ac artis excellentis miraculum ostendere. Ergo talem horologii fabricam magno labore, maiore sumptu, arte vero maxima compegit, qualem non habet tota, mea opinione, Europa secundum; sive quis cursum solis ac lunae

---

[1] Tanner has *literarum*    [2] Tanner has *sit*

seu fixa sidera notet, sive iterum maris incrementa et decrementa, seu lineas una cum figuris ac demonstrationibus ad infinitum poene variis consideret: cumque opus, aeternitate dignissimum, ad umbilicum perduxisset, *Canones*, ut erat in mathesi omnium sui temporis facile primus, edito in hoc libro, scripsit, ne tam insignis machina errore monachorum vilesceret, aut incognito structurae ordine sileret. Albion, si recte memini, hoc genus organi dictum est ab auctore, de quo Joannes Stubius mathematicus multa refert. Inciderat ille in exemplar, sed mutilum, unde quod deerat in Albione supplevit. Pars Albionis instrumenti *Saphe* dicta, Norembergae impressa est. Scripsit praeterea quadripartitum *De sinibus demensuratis*,[1] et, ut ego accepi, alia opuscula cum *De rebus arithmeticis, tum astrologicis*: atque adeo coniectura est libellum *De Computo*, quem ego titulo Rogeri abbatis Albanensis nuper vidi in bibliotheca Vimundensis coenobii, quod a Nordovico, urbe celebri, . . . millibus passuum distat, huius Richardi quanquam corrupto nomine, opus fuisse.

Sed heu casum et sortem miserabilem! inter studia incidit Vualingofordus in pessimam corporis habitudinem, quae postea illum foeda inamabili leprae scabie perfudit. Unde, a suis tantum non derelictus, coactus est monasterii septa deferre, et intra muros Verolamii tumultuario opere domum sibi aedificare; ubi non multo post tempore, ingrassante morbi contagione, diem obiit. Extat adhuc domus in similes, si casus tulerit, servata usus. Vixit regnante Richardo secundo.

Much of the above document is discussed elsewhere, but the following notes may be added here:

*Joannes Stubius.* Foster lists five men by the name of John Stubbs in the sixteenth and seventeenth centuries. None of them is described as a mathematician. It would be profitless to speculate on misspelling (Stadius?).

*Pars Albionis instrumenti Saphe dicta, Norembergae impressa est.* The treatise on the saphea [Arzachelis] has no connection with *Albion* as such. The reference might in fact be to *Problemata XXIX: Saphaeae nobilis instrumenti astronomici, ab Joanne de Monteregio Mathematicorum omnium facile principe conscripta*, 4°, 13 fols., Nuremberg, 1534.

*Bibliotheca Vimundensis.* This is not a reference to Windlesham in Surrey (where there was a St. Albans chapel of ease), but to Wimundesham (Wymondham) in Norfolk, where there was a Benedictine abbey of the Blessed Virgin Mary, originally a priory and cell of St. Albans. Only three manuscripts from this abbey have been identified, and none answers the description. Leland's *Collectanea* has '*Wimundesham*, ex fundatione Gulielmi de Albeni, cuius filius duxit relictam Henrici primi. Computus Rogeri, abbatus S. Albani, viri in Mathesi peritissimi, cuius

---

[1] Tanner distinguishes between two works: *De sinibus, De mensuratis.*

opus est Horologium insigne quidem illud in coenobio Albanensi. [Incipit:] *Assiduis petitionibus . . .*' (vol. iv, p. 27). Hearne, in his *addenda*, claims that this should be *Computus Richardi* 'ut monuit Amicus noster in Notis ad hoc opus'. (See the last volume of Hearne's edition of the *Collectanea.*)

# APPENDIX 5

## *Bale on Richard of Wallingford*

Ricardus Vualingforde, ab oppido eiusdem nominis in ipsa Tamesis ripa ubi natus erat, sic dictus: curante patre fabro ferrario, Oxonium, Academiam vicinam adolescens petiit, atque in Martonense collegio literis profecit. Ita monasticae vitae amor (Lelandi verba sunt) in eius postea irrepsit animum, ut D. Albani monasterium adiret, et in Benedicti verba iuraret. Quid eius labores et vigilias, quas Arithmeticae, Astronomie, Geometriae, et caeteris id genus artibus impendebat, commemorem? Certe hoc constat, tam numerose doctum fuisse; ut tandem ad coenobium reversus, solus ex tanto monachorum numero visus sit dignus, in mortui abbatis locum succedere. Talem eruditionis, ingenii et artis excellentis miraculo horologii fabricam magno labore, maiore sumptu, arte vero maxima compegit qualem non habet tota, mea opinione, Europa secundam: sive quis cursum solis et lunae, magno labore, maiore sumptu, arte vero maxima compegit qualem non habet tota, mea opinione, Europa secundam: sive quis cursum solis et lunae, seu fixa sidera notet, sive iterum maris incrementa et decrementa, seu lineas una cum figuris ac demonstrationibus ad infinitum pene variis consideret. Huic operi canones, edito libro, cui nomen Albion, addidit ne monachorum errore tandem vilesceret, ut erat in Mathesi omnium sui temporis primus. Albion enim, omnia per unum (all by one) Brytannice sonat, arridetque coenobii vocabulo. Unde composuit

Canones in Albionem Lib. 1. Albion est geometricum instrumentum
De sinibus demonstrativis Lib. 1. Quia canones non perfecte
De chorda et arcu Lib. 1. Arcus dicitur pars circumferent.
De iudiciis astronomicis Lib. 1. Ad perfectam notitiam eorum
De rebus astronomicis Lib. 1.
De rebus arithmeticis Lib. 1.
Exafrenon Lib. 1.
Rectangulum Lib. 1.
De computo Lib. 1.

Atque alia plura. Claruit anno Salutis humanae per Servatorem natum 1326, sub Eduardo secundo. Incidit demum in leprae scabiem, qua morbi contagio ingrassante, extremum obiit diem, in suo coenobio honorifice sepultus.

John Bale, *Scriptorum illustrium maioris Britanniae* (Basle, 1557–9) p. 397.

# APPENDIX 6

## Twyne on Richard of Wallingford's writings

BRIAN TWYNE, in his *Antiquitatis Academiae Oxoniensis Apologia* (Oxford, 1608), was as much concerned with authorship as with authors. His most original contribution in regard to Richard of Wallingford was his argument against Richard's authorship of the Exafrenon (lib. 2, p. 217):

Illius enim superioris, Richardus Wallingford sive de Sancto Albano Mathematicus talis, *qualis a suo tempore usque in praesens apud Anglos non surrexit*, ut loquitur Ioannes de Granario in parte operis sui de historiis et historiographis prima, in verbo Invenire (Albeonam enim fecit quasi *all by one*, id est, omnium aliorum instrumentorum commoditates, ad vocabulum Albani quoque alludens, Mertonensis sub Edwardo primo alumnus, in opere suo quod *Exafrenon prognosticationis temporum* vocat, vereor tamen ne Exafrenon pro Exameron legatur quod de Lyncolniensis Examero saepe apud veteres usuvenit, quod sic incipit, *Ad perfectam notitiam iudiciorum artis astrologiae quae non regulatae ex effectibus planetarum oriuntur, etc*: quod initium non Exafrenon, sed libello de iudiciis Astronomicis, ascribit Baleus) disertam his verbis facit mentionem cap. 2. *Verumtamen* (inquit) *si habuerit tabulas calculatas pro ascensionibus villae tuae, cuiusmodi sunt tabulae Manduith calculatae ad latitudinem Oxon., certissime poteris scire ascensionem hoc modo, etc*: haec incertus ille author quicunque fuerit; non enim (ut iam quod sentio dicam) eum, Richardum Wallingford esse posse extimo, si vel solam temporis rationem respicias.

Autor enim Exafreni quo tempore floruerit, ex calculatione sui temporis intelligitur: *quaesivi* (inquit) *locum planetarum anno Arabum 646 completo qui est annum Christi, 1249, mense quarto, die eiusdem mensis 15. hac est decimo septimo Cal. Maij etc*. Arabici anni completi mensis ultimus est Dilhaga: ex quo colligitur eum circa annum domini 1249 floruisse: At Richardus Wallingford Rectangulum suum sic inchoat, *Rectangulum in remedium taediosi et difficilis operis armillarum eodem tempore quo composuimus Albion, hoc est, anno Christi 1326 concepimus etc*: autor igitur Exafreni Richardo Wallingford S. Albani Abbate annis fere octoginta antiquior est: ut illud omittam Mertonense collegium non ante annum domini, 1274 (ut vulgo creditur) fuisse fundatum.

This document is discussed where appropriate elsewhere. Twyne shows that his knowledge of the texts goes beyond their opening words, and thus he is not relying solely on Bale, Leland, and so on. 'Ioannes de Granario' refers to the work by John of Whetehamstede, *De Granario*. (See Appendix 2.) It is usual to write 'Maudith' rather than 'Manduith', but the manuscripts of his works disagree.

# APPENDIX 7

## *John Pits on the Saint Albans clock*

JUST as Bale's compilation was inferior to Leland's original, so the polemical retort by the Catholic John Pits, posthumously published in 1619 as *Relationum Historicarum de Rebus Anglicis*, was inferior to Bale. Pits's elaboration of the account of the clock is an interesting specimen of his rather florid style:

. . . ad ostentationem artis et ingenii, imo et ad opum forsan iactantiam, in suo Monasterio tale fabricavit horologium, quale, sive mirabilis ingenii portentosum inventum, sive operis elaborati stupendum artificium, sive sumptus regali splendore dignos respicias, simile aut secundum Europa vix aut ne vix quidem unquam habuisse existimatur. In eo videre erat solis et lunae cursus, stellarum ortus et occasus, siderum fixorum et errantium motus et coniunctiones, maris fluxus et refluxus, omnium denique caelestium corporum figuras, operationes, effectus, affectiones, et similia pene infinita, quae vel apposite suis exprimere nominibus.

*Non opis est nostrae, maioremque exigit artem.*

Praeterea consulens operis immortalitati, ut permaneret *monumentum aere perennius,* scripsit quosdam canones quasi suae scientiae conservatores, titulum libro praefixit *Albion,* tum ut eo vocabulo ad sui Monasterii nomen, S. scilicet Albanum alluderet; tum etiam . . . All by one.

# APPENDIX 8

## *Fabricius on Richard of Wallingford*

RICHARDUS *Walingford*, Anglus, ex oppido Walingford ad Tamesin patre fabro ferrario natus, Oxoniae in Collegio Mertonensi literis incubuit, postea monachus et tandem Abbas Benedictinus in coenobio S. Albani, Mathematicus insignis, obiit a. 1326. Horologium in monasterio suo stupendo artificio elaboravit. Scripsit *Canones in Albion* (quomodo scil. horologium Albanense tractandum esset) *de judiciis Astronomicis, de sinibus demonstrativis, de chorda et arcu, de diametris, de eclipsibus solis et lunae, de rectangulo, Exafrenon, de rebus Arithmeticis, de computo*. Lelandus c. 470. Balaeus Centur. V. 19. Pitseus c. 487.

This passage is from the 1746 (Hamburg) edition of J. A. Fabricius's *Bibliotheca Latina Mediae et Infimae Aetatis Latinitatis*, vol. vi (P–Z), p. 250.

# APPENDIX 9

## On the date (1327) of Richard's accession

H. T. RILEY's edition of the *Gesta Abbatum* has propagated an unfortunate mistake, not only in the year of Richard of Wallingford's death, but of his election, which he gives as 1326. (Since the years of each abbot's term of office are placed at the head of every page devoted to it, each mistake is printed nearly 200 times. A consequence of the error is to be seen in Appendix 12.) There is ample evidence that the year of Abbot Hugh's death was 1327. Riley at one point took pains to explain that a certain date within Hugh's term of office was in January (rather than February) 1327 (*Gesta*, ii. 159 n. 3). Elsewhere there is a statement concerning the rising of the townsmen, which Riley wrongly places in January 1326, rather than January 1326/7. The author of the *Gesta* tells us that the state of lawlessness began 'Circa annum Domini millesimum trecentesimum vicesimum sextum, postquam Dominus Edwardus Karnervan abjudicatus fuisset a regno . . .', and 'Sub quo tempore, villani de Sancto Albano, conjurati, contra Abbathiam erexerunt calcaneum suum, post Festum Epiphaniae Domini . . .' (ibid., pp. 155–6). The year is confirmed by the Patent Rolls. Hugh died on 7 September. It would only have been a matter of days before the monks obtained permission to elect, and they first had letters with licence to do so on 20 September 1327 (*Cal. Pat.*, 1327–30, p. 167; cf. pp. 184, 191, 272).

# APPENDIX 10

## *On the date (1336) of Richard of Wallingford's death*

As mentioned in our biographical introduction, Richard's condition is said to have become worse after the storm of 29 November 1334 (*Gesta* ii. 293). He died 'cum nondum complesset plene in praelatione sua novem annos: siquidem die Jovis in hebdomada Pentecostes, videlicet, decimo Kalendas Junii, circa Solis ortum', according to the same source. The tenth day *before* the Kalends of June is 23 May. If he had not yet completed nine years of his abbacy, the year was 1336. Riley, however, taking the year of his accession as 1326 (see Appendix 9), decided on 1335. On checking the date of Pentecost, it is found that in 1335 the week of Pentecost (Whitsuntide) extended from 4 June to 11 June, which is not consistent with the date given in *Gesta*.

The next abbacy is recorded only with respect to Richard's: the election was said to have begun ten days later. The *Calendar of Patent Rolls* 1334–8, p. 270, gives 29 May 1336 as the day on which two monks came to the king with news of Richard's death. They were duly given letters of licence to elect. The very nature of these records (that is, their scroll form) makes it unlikely that the record is out of sequence. The *Calendar of Papal Letters*, ii. 531, reveals that Michael of Mentmore's election was confirmed on 18 November 1336, and we know from the *Gesta* that this confirmation was given without demur. It can be shown that the week of Pentecost in 1336 extended from 19 May to 26 May. We may therefore reject the years 1335 and 1334 (both of which are frequently quoted as the year of Richard's death) in favour of 1336.

# APPENDIX 11

## *Richard of Wallingford's grave*

ACCORDING to the *Gesta*, he was buried by John, abbot of Waltham; but the place was not specified. Ashmole visited St. Albans on 19 July 1657, and his diaries (Bodleian library, MS. Ashmole 784, in which see ff. 41$^v$–42$^r$) have misled some into thinking that a long inscription he quotes was from the tomb of Richard of Wallingford. It is that of John Stoke, once prior of Wallingford. The earlier book *Ancient Funerall Monuments* (London, 1631) by John Weever makes no mention of Richard of Wallingford's tomb. Richard Gough, however, in his *Sepulchral Monuments of Great Britain* i. (1786), p. 305, recorded these words from the tomb: 'Richard gist ici Dieu de sa alme eit merci. Vous ke par ici passes Pater e Ave pur l'alme prierunt . . . jours de pardun averunt . . .'. The position of the tomb is in fact known from a fifteenth-century source, published in translation by Ridgway Lloyd: *An Account of the Altars, Monuments, and Tombs, existing A.D. 1428 in St. Albans Abbey; translated from the original Latin, with notes* (St. Albans, 1873), p. 7:

Now, in the Presbytery of the Church, near the lowest steps of the altar, there lie in order, under marble slabs, distinguished by their epitaphs, four Abbats of this Monastery; namely, Dom Hugh Eversdon and Dom Richard Walyngforde the clockmaker, in the middle, and Dom Michael Mentmore, and Dom Thomas de la Mare, at the sides . . .

This makes Richard's tomb the second from the left, facing the altar, which is that with the longest and narrowest stone. The inscription is now lost.

# APPENDIX 12

## *Commemoration of Richard of Wallingford*

ON 28 November 1327 Richard of Wallingford was given a safe conduct to visit the Papal court. Perhaps connected with this date was the 'Commemoration of the life and work of Richard of Wallingford, Abbot of Saint Albans, 1326–1335, Astronomer, Scientist, Mechanician, Mathematician', held on Saturday 27 November 1926. As we have seen in Appendices 9 and 10, H. T. Riley mistook the years of Richard's election and of his death, and the date on which his life and work were commemorated must have puzzled the recording angel on duty at the time. From an invitation card of 1926 (Museum of the History of Science, Oxford), we learn that the Dean of St. Albans and the Mayor presided jointly; that at 4 p.m. Richard's own prayers were used, perhaps not entirely appropriately, at evensong in the abbey, the Astronomer Royal laying a wreath on his grave; that at 5.30 p.m. there was a meeting in the Town Hall at which 'A collection of mediaeval scientific instruments (many of which were Wallingford's invention)' was on view, and that these included 'ABBOT WALLINGFORD'S ASTRONO-MICAL CLOCK, a unique scientific curio'. A collection was taken to cover expenses. The speakers were H. H. Turner, Savilian Professor at Oxford (Introductory Address), the Revd. W. A. Wigram ('Walling-ford as Abbot'), 'Professor Garrard, Merton College, Oxford', namely H. W. Garrod ('Wallingford and Merton College'), R. P. Howgrave-Graham ('Early Clocks'), and finally the man most probably responsible for the idea of such a commemoration, R. T. Gunther ('Wallingford's Astronomical Instruments'). So far as is known, no record of any of the addresses survives, although a group photograph and report are to be found, together with a picture of the clock, in the next issue of the local press. The clock illustrated has the appearance of a small, seventeenth-century, silver table clock.

# APPENDIX 13

## *On an astronomical manuscript probably owned by Richard of Wallingford*

THE manuscript in question is Trinity College, Dublin, MS. D.4.30 (444). With the exception of four leaves, it was copied in the second or third quarter of the thirteenth century. The four leaves ff. 91–4 were apparently written at the very end of the thirteenth or during the first part of the fourteenth century. On f. 2ʳ, in a fourteenth-century hand, are the words 'Liber Sancti Albani' and, written on a different occasion but at the same period, 'De studio Abbatis'. Also of this period, on f. 1ᵛ, is 'tabulas [*sic*] astronomie' and 'cum canonibus', with 'pretii XX', giving the price in pence. On the same page is evidence that the codex was given away by the abbot after the dissolution: 'Joannes Deeus 1553 28 Januarii, ex dono magistri Doctoris quondam Abbatis Sancti Albani'.

By reason of the fact that between the dates of composition and acquisition by the abbot's study there was only one abbot fully capable of appreciating what is a relatively difficult volume, we are inclined to suppose that Richard of Wallingford acquired the work for his own purposes, perhaps even whilst he was at Oxford. The only mark which might relate to earlier ownership is at the foot of f. 96ʳ, where the name of one Johannes de Fieffes occurs, possibly even as the name of the scribe. There is evidence that the volume was in use in 1289, for on f. 95ʳᵛ, pages which were formerly blank, there are pencilled records of mean motus and argument of the planets. Not all of this is now legible, but from the mean motus of Saturn, Jupiter, Caput Draconis, and the Sun, we may derive a date of 21 March 1289. There are other pencilled or inked marginal columns of figures on ff. 45ʳ, 62ᵛ, and 75ʳ. A very unusual indication of the quality of the scribe is to be found on ff. 79ᵛ–80ʳ, where alternative readings are given opposite some entries of a table, in the form 'alibi 10 per 11'. There is also the gloss 'In tabula erant 28, sed invenimus componendo tabulam quod debetur esse 27 dies, que enim 76ᵘˢ . . . etc.', continuing in the same vein. This is not a question of late alteration, for the figure 28 was never written in the table. The tables as a whole do, nevertheless, contain a number of uncorrected errors.

We know that before Richard of Wallingford was made abbot he used only the Toledan tables, rather than the Alfonsine. The canons we believe him to have written to the tables of John Maudith shows evidence of the use of the Cremona version of the canons to the Toledan tables, and such a copy is to be found here: (f. 4$^r$) 'Quoniam cuiusque actionis quantitatem temporis . . .' Later in the manuscript, following the Toledan tables, there is a set of tables for Toulouse, beginning with (f. 24$^r$) 'Tabulam ad inveniendum diem et horam introitus Solis in Arietem, Cancrum, Libram, atque Capricornum ad longitudinem civitatis Tolose que est 40 grad. min. 30 ab occidente.'

Radices for the tables generally are for the years 1008, 1032, . . . 1296, 1320, . . . Does this explain the choice of the year 1296 as a radix in *Exafrenon*, in connection with the table giving the Sun's entry into Aries? If so, we may perhaps suppose that *Exafrenon* was composed between 1310 (since the Maudith tables are mentioned) and 1320 (the next radix date in the series).

There is another weak link between *Exafrenon* and the Dublin manuscript. In the four added folios are three tables which are found in the former work:

(f. 91$^r$)   *Tabula de apparicione stellarum erraticarum superiorum;*
(f. 91$^v$)   *Visio Veneris et occultacio eius;*
          *Visio Mercurii et occultacio eius.*

(The remaining three folios contain 'tabule proieccionis radiorum', one table for each face of a sign, making thirty-six in all.) One noteworthy aspect of the tables is that at the top of the *occasus matutinus* column for Mercury there is a nonsensical 'a 249 in 249'. This, or a slight variant, is common to most *Exafrenon* manuscripts. Further notes on these tables will be found at the end of our commentary to *Exafrenon*, cap. 4.

A sexagesimal multiplication table at ff. 88$^r$–90$^v$ is exactly like that of St. Albans MS. Ashmole 1796, ff. 88$^v$–90$^r$.

# APPENDIX 14

## *Richard of Wallingford and Merton College*

THERE are at least two sources of the story that Merton College, which was one of the most illustrious centres of secular learning in the fourteenth century, numbered Richard of Wallingford among its fellows. One source is the so-called 'Old Catalogue' of fellows of Merton College, compiled in or around 1420, and used for subsequent compilations—which it is not necessary to invoke for this reason. The second source is a rubric on an early manuscript.

On the very first folio of the Old Catalogue, after a statement that the list begins with the reign of Edward I (1272–1307; the foundation charter of the College was issued in 1264), and the name of the first Warden, Peter of Abingdon, there is this entry: 'Ricardus Sancto Albano post abbas de Sancto Albano profundus astronomus fuit et fecit instrumenta astro[nomie] diversa et plures libros collectos(?) dedit.' A later marginal gloss explains: 'Cognominatus Wallyng Forde'. The Savile (1586) and Wilson (*c.* 1580) Catalogues recognize only 'Richard Wallingford', whom they still claim as a fellow. G. C. Brodrick, in *Memorials of Merton College* (Oxford, 1885), p. 171, half accepts this evidence, although he writes that 'there is some reason to believe that the name originally entered in the Old Catalogue was that of Robert of St. Albans, and that it was afterwards altered into *Richard* by some person anxious to identify this early fellow of Merton with the Abbot of St. Albans'. These are substantially Anthony Wood's findings, according to Tanner's *note b* to Leland's text. Brodrick does not give the catalogue entry, but acknowledges it implicitly in a twelve-line biography of Richard. In the original manuscript, however, all but the initial 'R . . . sancto Albano' has the appearance of a later addition, and the gloss is later still. But the most transparent error here concerns the date at which Richard is supposed to have been a fellow. Quite apart from the fact that his name comes first on the list—which was more chronological than honorific—Richard could not have been a fellow during the reign of Edward I, unless he reached this lofty position by the age of 15 or so. There are other men by the name of Robert of St. Albans, any of whom might have been the fellow whose name was originally altered. The evidence of the Old Catalogue will thus be treated as worthless. It is possible to trace to this source those claims for Richard of Wallingford's Merton fellowship made by Bale, Twyne, Pits, and possibly Leland too.

The other manuscript evidence is perhaps not even as early as the altered version of the College records. Remarking upon the fact that Leland was wrong in assigning Richard to Merton (see below), and stating what was presumably Wood's belief that Robert 'quondam de S. Albano' was a fellow *c.* 1299, Tanner adds that in MS. Norwic. More 820 (= Cambridge University, MS. Ee.III.61 (1017)), a note by Lewis of Caerleon says Richard of St. Albans 'socium esse domus scholarium de Merton'. Lewis flourished, however, at the end of the fifteenth century (died *c.* 1494), and it is more than likely that he took the idea from one of a series of manuscripts of the *Quadripartitum*, Richard's principal trigonometrical work. Of the ten or so manuscripts which survive of this, two (the older, now partly erased, being MS. Digby 178, f. 38$^r$ the other being MS. Trinity, Cambridge, O.9.6, f. 255$^v$) end as follows: 'Explicit quartus tractatus de corda recta et versa quem composuit Frater Ricardus Wallingforde, quondam Abbas Sancti Albani ac prius socius Collegi Walteri de Merton Oxonie, summus astronomus ac geometer eximius, cuius anime Deus propicietur excelsus.'

The statutes of Merton College forbade the granting of a fellowship to members of religious orders. The Benedictine Rule, as then applied at St. Albans, required its members not only to live in a Benedictine house when at a university, but also to take instruction only from members of the order. Inception under a secular master was repeatedly forbidden by the edicts of the Provincial Chapters (see Pantin, 63, 82, 212, for example). The first Benedictine to take a doctor's degree at Oxford was in something of a dilemma, which he resolved by taking the Chancellor himself as Master. (See H. C. M. Lyte, *A History of the University of Oxford* (London, 1886), p. 104.) Richard was at Oxford between the ages of about 16 and 22, when he would not have been a fellow, and between 25 and 34, when, being a Benedictine monk, he must have resided elsewhere, probably at Gloucester College. (Lyte, loc. cit., without any explanation, claims Richard for Gloucester College during both periods of residence.)

There remains the possibility that Richard came to Oxford as a young scholar at Merton. This might appear unlikely, since he was prepared for Oxford, and maintained there, by the Benedictine Prior of Wallingford (Prior, that is, of a cell of St. Albans), William of Kirkeby (*Gesta*, ii. 182). Leland, however, seems to suggest in the *Commentarii* that Richard went to Merton from his native town. He writes 'in collegio Maridunensi', which one might have construed as 'Caermarthen College' had there been any such place. Hall's index explains that 'Maridunense Collegium' and 'Merton College' are equivalent, but this was not necessarily Leland's intention. (Bale, in his *De Scriptorum Illustrium* . . . , changed Leland's phrase to 'in Martonense collegio', but he was relying on the Old Catalogue.) Could Leland have meant to place Richard in one of the several 'Maiden Halls' in being at the time? The use of 'collegium' makes this unlikely. His failure to mention the post of fellow is one reason for wondering whether he had access to documents not considered here.

# APPENDIX 15

## *Instruments mentioned in Merton College lists of the fourteenth, fifteenth, and sixteenth centuries. William Rede's Albion*

IN the catalogues and 'electiones' included by F. M. Powicke in his *The Medieval Books of Merton College* (Oxford, 1931), instruments are frequently mentioned. The following lists are abstracted from his transcriptions:

| MS. | Date | Item no. | Value | |
|---|---|---|---|---|
| P | early 14th cent. | 46 | 13s. 4d. | Item astrolabium inclusum in quodam casu de grosso corio |
| | (catal.) | 51 | 10s. | Astrolabium inclusum in quadam casula de corio grosso |
| D | c. 1410 | 7 | | Astrolabium magnum cum quinque tabulis |
| | (electio) | 16 | | Speram solidam cum capsula |
| | | 25 | | Quadrantem Campani |
| | | 37 | | Equatorium ligneum cum epiciclo |
| | | 50 | | Equatorium eneum cum epiciclo |
| | | 62 | | Astrolabium antiquum cum una tabula |
| | | 75 | | Speram communem eneam |
| E | 1418 (elec. of 10 Aug.) | 153 | | Quadrantem eneum |
| F | 1452 | 4 | | Astrolabium [magnum] cum 5 laminis |
| | (elec. of 13 Jan.) | 27 | | Triangulum Rede cum capsula in quo inscribatur nomen Rede |
| | | 48 | | Speram solidam cum capsula |
| | | 69 | | Triangulum cum cursore mobili |
| | | 93 | | Astrolabium magnum et medium(?) cum tractatu eiusdem |
| | | 114 | | Cartam marinam cum capsula et circino de laton |
| | | 136 | | Equatorium eneum cum epiciclo in dorso cum volvellis Solis et Lune |
| | | 158 | | Saphea cum allidada lignea |
| | | 180 | | Astrolabium antiquum cum duplici rethe |
| | | 202 | | Astrolabium cum una [tabula] |
| | | 225 | | Cartam marinam |
| — | 1519 | Langley 20 | | Spera cum laminis |
| | (elec.) | Norice 20 | | Spera parva |
| | | Hoper 19 | | Astrolabium antiquum |
| | | Seniori pertinentes | | Spera solida<br>Quadrans cum capsula |
| | | Tut | | Magna spera plat' cum laminis (more than one item?) |
| | | Serlis | | Spera fracta |
| Registrum | 1520 (13 Sept.) | | | . . . recepi a Magistro Moscroffe (Thos. Mosgrove) 14 instrumenta astronomie. |

The last reference is taken from H. E. Salter, *Registrum Annalium Collegii Mertonensis*, 1483–1521 (Oxford Historical Society, 1923), p. 499.

The will of Simon Bredon, a fellow of the college, is also printed by Powicke (op. cit., pp. 82–6; the will was made in 1368 and proved in 1372). This extract is of some interest: 'Item astrolabium maius lego aule de Merton et astrolabium minus lego Magistro Willelmo Reed.' William Rede in turn willed many of his possessions to the college, but he was a man of more substance, and his will is less specific, mentioning no instruments, by name or otherwise. On the other hand, he had made handsome gifts to the college (almost certainly in 1374) before his death, including money, books, and the following instruments: *albion, equatorium planetarum, quadrans, chilindrum, spera materialis, spera solida, tabula ymaginum celestium, carta maris, lapis calculatorius*, and *tabule dealbate pro tabulacione librorum*. (This list is taken from A. B. Emden's *A Biographical Register of the University of Oxford to A.D. 1500*, vol. iii (1959), p. 1558, and was taken in turn from Mert. Coll. Arch. P.1. Note that there is no mention of anything which could be Bredon's smaller astrolabe. Note also the ambiguity of *spera solida* which here—in contraposition to *spera materialis*—might have referred to an armillary sphere. The reference to a star map is intriguing: could it be the '*magna spera plat*'' of the 1519 election? The marine chart was perhaps a portolan, while the *lapis calculatorius* might have been some sort of exchequer board, or merely a slate. For the 'whitewashed tables for tabulating books' it is possible to make several suggestions, none wholly convincing. We recall that Rede's most important astronomical work was an edition of the Alfonsine tables, and the *tabule dealbate* were perhaps designed to ease the labour of ruling or copying in some way.)

Finally, in the fine copy of Chaucer's *Treatise on the Astrolabe* in MS. Bodley 619, at Book I cap. 21, there is a gloss referring to the presence of certain stars on the rete of a Merton College astrolabe. It is not improbable that knowledge of this note subsequently gave rise to the nonsensical tradition that the college possessed—and indeed still possesses—Chaucer's own astrolabe.

Merton College today possesses the following medieval instruments, illustrations of which—with descriptions not entirely unreliable—will be found in R. T. Gunther, *Early Science in Oxford*, vol. ii (Oxford, 1923), facing pp. 207, 209, 210, 211, 167, 170, 241:

(1) A large astrolabe, of over 14 inches diameter, without provision for separate plates, the mater inscribed for the latitude of Oxford. The mater is inscribed as an equatorium on the dorsum, and it seems probable that—since this was an unusual feature—the instrument would

have been catalogued in the Middle Ages as an equatorium. The epicycle of the equatorium is now lost. (See vol, ii, p. 256.) Date: 1350, or thereabouts.

(2) A heavily built astrolabe, with three plates surviving, and no rete. Diameter: $9\frac{3}{4}$ inches. Date: 14th cent. The astrolabe carries a pointer taken from another instrument of the type indicating lunar phase through a circular aperture.

(3) A Profatius New Quadrant. Radius: 12 inches. Date: 14th cent., perhaps first half.

(4) A circular instrument (diameter 6 inches). We may follow tradition in calling it a 'physician's quadrant', since on one side is inscribed a characteristic Old Quadrant together with a Zodiac Man of the sort to be found in very many calendrical manuscripts, especially of the late fourteenth century and after. The dorsum carries a number of solar scales. Note that Gunther's 'circle 9' (op. cit.) gives rough values for the solar declination at any time of year. On the dorsum are lunar and solar pointers. The main disc is pierced with holes to take a plumb line and sights, now lost, and with 72 peripheral holes at 5-degree intervals. It does not appear to have been previously noticed that the device is very rare indeed if the affinities are more than apparent between it and the prototype nocturnal of the now lost illustration of MS. Chartres 214 (173) (early 12th cent., reproduced with comments in Francis Maddison, 'Medieval Scientific Instruments and the Development of Navigational Instruments in the XVth and XVIth centuries', *Agrupamento de Estudos de Cartografia Antiga*, xxx (Junta de Investigaçõnes do Ultramar–Lisboa, Coimbra, 1969), 31).

We may make some tentative identifications here. Item 50 of MS. D and item 136 of MS. F were perhaps the 14-inch equatorium/astrolabe, and very probably that presented by Rede. It is possible that item 136 was two instruments, the solar and lunar volvelles being now represented by the alien volvelle on the back of the smaller astrolabe. This latter astrolabe might be the *'astrolabium magnum cum 5 tabulis'* of item 7, MS. D, and item 4, MS. F, and could well be the astrolabe bequeathed by Simon Bredon to the college. Either of the quadrants named in the elections of 1418 and 1518 could have been Rede's Profatius quadrant, although the brevity of the descriptions makes this no more than bare conjecture. One of the two marine charts in MS. F could have been Rede's, and the 'magna spera plat' ' of the 1519 election might well have been his 'tabula ymaginum celestium'.

In summary, not a single instrument surviving to the college can be identified, with anything approaching certainty, with one of the instruments named in the early records. It seems very probable, however, that the large equatorium/astrolabe, one of the finest of all surviving medieval instruments, was that given by Rede, while the other astrolabe was Bredon's. What should not escape notice is that no fewer than eighteen instruments of one sort and another—and perhaps several more—are today unaccounted for. And this total includes, sad to say, the most valuable instrument of all, Rede's albion.

Reference to Appendix 36, on John of Gmunden's will, reveals that there was much in common between the instruments in the possession of the Viennese astronomer and those used in Oxford. In particular, notice the use of wood, a substance which—if the contents of modern museums were to be taken as a guide—might be imagined to have been seldom used by the instrument-maker of the Middle Ages.

# APPENDIX 16

## '*Fiat triangulus rectangulus*'

THE short text of MS. Ashmole 1796, ff. 116ᵛ–118ʳ, which contains two figures, immediately precedes the *Albion* treatise in this important St. Albans codex, the contents of which are listed more generally elsewhere (vol. ii, pp. 310–15). The fact that the short text is of the nature of mere jottings on three instruments (for taking solar and stellar altitudes) makes it not improbable that Richard of Wallingford was responsible for it. It is also to be found in Cambridge University MS. Ee.3.61, a manuscript with a Lewis of Caerleon association, and including the *Rectangulus*. An aside, which is repeated, points out that it is immaterial whether one adopts the convention that the standard radius is of 60 units, or the convention which makes it 150 units. The latter, which was spread as a consequence of the wide use of the Toledan tables, and which had been Arzachel's choice, is not commonly found in English manuscripts before the latter part of the thirteenth century.

The text is of interest inasmuch as it discusses indirectly the accuracy attainable with two or three types of altitude-measuring device. Three such devices are discussed, but two are similar portable instruments, while the third is simply the gnomon, vertical or horizontal. The first portable instrument is a framework in the form of an isosceles right-angled triangle, with a plumb-line suspended from one of the acute vertices, the line passing over a scale of twelve equal parts on the opposite side. The plane of the triangle is made vertical, while the Sun or a star is observed through one of two pairs of perforated vanes. One pair is placed along the ungraduated short side, while the other pair are (each at the mid-point of a side) on a line parallel to the graduated side. One pair of sights permits the measurement of altitudes up to 45 degrees, the other pair from 45 to 90 degrees. This is all reminiscent of the *Tractatus* of our Section II.

The second portable instrument requires two scales, but only one pair of sighting vanes. It is simply a square frame with scales on two adjacent sides, and a plumb line from the vertex which is common to neither. The vanes are along either ungraduated side. It is pointed out that such an instrument may easily be tied into your book, and indeed one does often come across concavities cut away in the covers of medieval books to hold instruments, especially quadrants. The advantages claimed for these instruments (*machinamenta*, presumably meaning a device with moving parts, by contrast with the gnomon, which is said to be neither *machinacio* nor *ars*) and for the gnomon is that they may be accurately made, requiring, as they do, only a linear division of the scale(s).

The method of using the gnomon to determine solar altitude, and the converse method of working, which leads to a knowledge of shadow length, given the Sun's altitude, is typical of that to be found in any astrolabe treatise which explains the use of the shadow square. The gnomon is conventionally divided into twelve parts. The only tables which are assumed are of sines, or the equivalent. Tangents are effectively found from the division of sine by cosine, or, working in the opposite direction, by applying Pythagoras' theorem before effecting the division. It is pointed out that the accuracy of the gnomon derives from its size, and that Thebit used it in his investigations of the equinoxes (and of the obliquity).

Although surviving instruments testify to the abilities of medieval astronomers in dividing their instruments, the text makes the ambitions of at least one writer clear: if the side of the portable instruments be three or four feet, 'the twelfths may be divided into 60'. The resulting divisions, of a twentieth or fifteenth of an inch, are quite typical of surviving instruments, and by no means equal to the intervals on the best of them. Later, however, the possibility of dividing an inch into sixtieths is twice implied. Such a division is ambitious, but not implausible. The thread must, of course, have been of comparable thickness. It is further implied that a division of one fortieth of an inch is more reasonable.

There is discussion of a gnomon 63 feet in length, and unless this was thought to be the length of Thebit's, it is not clear why the figure was chosen. (It might have been the height of a pinnacle on the triforium or clerestory of a large church building, used as a gnomon.) At all events, the author points out how large a circular instrument would be required to equal the accuracy of a linear scale used in conjunction with a gnomon of this size.

The short text ends with some trifling notes on the multiplication of fractions, taking as an example the multiplication of $2\frac{1}{4}d$. by forty. Another lost art.

# APPENDIX 17

## *Abbots of St. Albans in the fourteenth and fifteenth centuries*

| | |
|---|---|
| 1291 | John of Berkhamstede |
| 1302 | John of Maryns |
| 1308 | Hugh of Eversdone |
| 1327 | Richard of Wallingford |
| 1336 | Michael of Mentmore |
| 1349 or 1350 | Thomas de la Mare |
| 1397 | John de la Moote |
| 1402 | William Heyworth |
| 1420 | John of Whetehamstede |
| 1440 | John of Stoke |
| 1451 | John of Whetehamstede (second period of office) |
| 1464 or 1465 | William Albon |
| 1476 | William Wallingford |
| (1484–92 Interregnum) | |
| 1492 | Thomas Ramryge |
| 1516 or later | Thomas Wolsey |

# APPENDIX 18

*Verses from the abbey windows, relating to the quadrivium*

BODLEIAN MS. Laud misc. 697 records some verses from the abbey windows. The following passages are from two of the library windows, apparently installed by John of Whetehamstede in about 1451 (from f. 27ᵛ):

[following Plato, representing ethics] . . .

[fourth-
window]  Ipseque Pictagoras do normas moribus aptas.

Crisippus  Dixi quis cubicus numerus sit quisque quadratus,
Nicomacus  Ac ego cur impar numerus mas femina sit par.
        Dixi quot Guido moduli sunt in monacordo.
        Quotque tenet citharo [docui] Vnchalus [Michalus?] ve
          viella.

[fifth
  window]  Euclides vocitor, magnus fueram geometer.
        Circi quadrator Archimenides ego dicor.
        Maximus astronomus reputatus eram Tholomeus.
        Magnus et Albumasar introductor vocitabar.

M. R. James, in the *Cambridge Antiquarian Proceedings and Communications*, viii (1895), 213–20, points out that in eleven of the twelve windows of the library were four figures of the men who, in the estimation of fifteenth-century scholars, stood at the head of the various branches of science and literature, each man being given one line. It is worth remarking upon the inclusion of Albumasar, the Arab astrologer, amongst the classical authorities on the liberal arts. The spelling of the name of Archimedes is the conventional medieval one. Guido the musician is Guido of Arezzo (990–1050), whilst the man who 'taught how many sounds the harp or viol has' may be a Greek referred to by Philoponus (see James's article).

# APPENDIX 19

## Some 'calculatours'

AN interesting passage of Middle English, dating from the latter part of the fourteenth century, and chronicling men wise in the art of calculation, is to be found in Trinity College, Cambridge, MS. O.5.26, at f. 112$^{vb}$. Richard of Wallingford is apparently omitted from the list, which includes Grosseteste, William Rede, Simon Bredon, and Walter Elvedon. It also, however, includes 'Richard, Bischop of Herford'. The twelfth-century astronomer of Hereford was Roger, and he was not a bishop. He was responsible for the Hereford tables (e.g. in British Museum, Arundel MS. 377, ff. 77(?)-87), and was therefore most probably the 'calculatour' intended. The names mentioned are all too well known to require comment, apart from that of Walter Elvedon, who was at Gonville Hall, Cambridge, *c.* 1330, and whose *Kalendarium* survives in C.C.C., Cambridge, MS. Parker 37, f. 27.

The passage in question is reproduced below in its entirety.

*Here is shewyde the tyme of the forseide coniunccion of Saturni and Iovis.*

A speche of the sawes of olde holy fadres and religiouse men, the lond untiled tiliynge, and in wildernesse perfiteliche lyvynge, for mooste noble and moost excellent scientis of philosophie and of astronomye to be geten and to the laste poynte to be enserchide, as Plato, Aristotle, Ptholome, Albumazar, Hermes, Dorotheus, Haly, Abenragel, Betheny, Thebith, Abbategni, Albohaly, Tiberiadis, Alkabucius, Alkindus, Thechel, Ramraf, Jafar, Jergis, and of othere many wise men in the arte of Calculacioun or of acountinge. And amonge all Calculatours moost trewe was Alphons the noble and excellent kyng of Castile, and Arzachel, and Mayster Robart Grostet, Bischop of Lyncolne, and Richard, Bischop of Herford, Mayster Willyam Reede, Bischop of Chechestre, and Mayster Symond Bredone, and Maister Walter Elfedene, whiche thre in oure tyme florischiden. And anothere olde man that was seyde Humenidis, which calculed perpetuel tables of the almanak to Cleopatre, the dou*gh*ter of Ptholomei; and so dide Profacius the Jewe, in the Mount Pessulan. The mynde of alle whiche amonge wise men is in blessinge and reverence and worschipe; amonge whiche the worschipful Mayster Joon Eschendene oweth namely and principaly to be nempned, whiche made his moost verey pronosticacioun for the rete conjunccioun of the thre moost conjunccions that beth seyde bifore of Messahallah; and also of the universal eclips of the Mone, which bifil in Marche, in the yere of oure lord a thousand thre hundred and five and fourty, and also for the yere of the same lord, a thousand thre hundred and fyve and sixty. And so dide Mayster John Lyneris, and also many othere, whiche at this tyme cometh not alle to my mynde at this tyme (sic); and also it nedeth not, for these suffiseth as at this tyme.

After the moost trewe calculacioun of alle the forseide calculatours or acounteres after the most certayne computacioun, ther schulle come twey grete princes of wonderful wisdome, and also of moost excellent kynde and nature, to a general counseyl from fer lond for to schewe to the worlde many merveyles; and dredeful thinges withynne a schort tyme after the counseyl of hem yeven and confermed. And thei schul come in the yere of oure lord a thousand foure hundred and fyve, in Januarie, the eleventhe day and the eleventhe houre, the thre and fourty mynute, the six and twenty secunde, and the fourty thridde. And the place of conjunccion of Saturni and Iovis schal be in Aquario, the thre and twenty degre, the tenthe mynute, the seven and fifty secunde.

Here endeth the chapitre of the verey tyme of the moost conjunccion that is of Saturni and Iovis, the which schal be in the yere of oure lord a thousand foure hundred and five. *Explicit.*

# APPENDIX 20

*Abbreviations used in connection with spherical trigonometry*

A SHORT explanation is offered of the following abbreviations:

Crd *XY*, Sin *XY*, Cos *XY*, VCrd *XY*, VSin *XY*, Δ*ABC*/*DEF*.

Referring to the first diagram,

(i) *corda AC*, or *corda recta AC*, abbreviated Crd *AC*, is the straight line *AC* ($= 2a = 2R \sin AC/2$);

(ii) *sinus rectus AB*, or *sinus AB*, or *sinus* α, abbreviated Sin *AB*, is the half chord *a* ($= R \sin α$);

(iii) *sinus complementi AB*, abbreviated Cos *AB*, is the line *c* ($= R \cos α$);

(iv) *corda versa AC*, abbreviated VCrd *AC*, is the line *b*

$$(= R - R \cos α = R - \text{Cos } α = R - \text{Cos } AB);$$

(v) *sinus versus AB*, abbreviated VSin *AB*, and often misleadingly called *corda versa* by Richard of Wallingford and others, following Albategni III. 26, is equal to VCrd *AC*, that is, it is the same quantity but regarded as a function of a different argument: VSin *AB* = $R - \text{Cos } AB$;

(vi) *corda dupli arcus AB*, or *sinus duplatus arcus AC*, that is, *AC*, not specifically abbreviated here, is equal to 2 Sin *AB*.

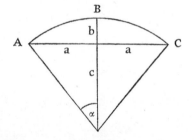 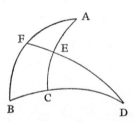

The theorem of Menelaus, which was seldom known by that name in the Middle Ages (when it was 'the theorem of the sector figure', or some equivalent name—see the commentary to *Quadripartitum* III), applies to plane and spherical triangles. In either case we denote by 'Δ*ABC*/*DEF*' a configuration such that *DEF* is a transversal (a straight line or great circle, as the case may be) to the triangle (formed of straight lines or great circles) *ABC*, *D* lying on *BC*, *E* on *CA*, and *F* on *AB*.

There are several equivalent ways of expressing the theorem itself. In modern notation it is usually expressed

$$\frac{\sin BD}{\sin DC} \cdot \frac{\sin CE}{\sin EA} \cdot \frac{\sin AF}{\sin FB} = -1.$$

Its original form may be written thus:

$$\frac{\text{Crd } 2CE}{\text{Crd } 2EA} = \frac{\text{Crd } 2CD}{\text{Crd } 2DB} \cdot \frac{\text{Crd } 2BF}{\text{Crd } 2FA}.$$

This, or the equivalent form wherein Crd $2CE$ is replaced by Sin $CE$, and so on, was the medieval arrangement of terms. No attention was paid to the order of letters within a pair, and of course the equation was expressed entirely verbally. Richard of Wallingford often leaves it in terms of arcs, that is, with the function tacitly understood.

# APPENDIX 21

## *Celestial coordinates*

THERE were three systems of coordinates in common use in the Middle Ages. Two of these, the systems of *right ascension/declination* and (ecliptic) *longitude/latitude*, are those in use today. The third system is easily confused with the first, owing to a shift of terminology. It is the system of *declination* (in the accepted sense, but often given the Latin name *latitudo* rather than *declinatio*) and *mediation* (as we shall call it), and is at least as ancient as the star catalogue of Hipparchus, in which it is extensively used. (See H. Vogt, 'Versuch einer Wiederherstellung von Hipparchs Fixsternverzeichnis', *Astr. Nachr.* cciv (1925), cols. 17–54.) The *mediation* of a star is the degree of the ecliptic which crosses the meridian at the same moment as the star. Although very occasionally known as the *mediatio celi*, it is one of the symptoms of medieval prolixity that writers were content to write out such a phrase as *gradus zodiaci cum quo stella mediat celum*, and to do so perhaps twenty times on a page. Since *declinatio* was interchangeable with *latitudo ab equinoxiali* (i.e. equatorial latitude), it was not uncommon to refer to the *mediation* of a star by the phrase *longitudo ab equinoxiali*. This was often abbreviated, especially in the middle of a text, to *longitudo*, providing an obvious source of confusion, particularly where a passage has been extracted from a longer work. Another source of confusion worth mentioning here is the use of the phrase *polus declinationis* to mean 'pole of the ecliptic', as opposed to *polus mundi*, the 'pole of the world' through which the circles of declination do in fact pass.

The different systems of coordinates, and the symbols often used for them in this book, are as follows (see the figure—the star is at *T*):

   I. Right ascension ($\alpha$) *EN*
      Declination ($\delta$) *NT*
  II. Ecliptic longitude ($\lambda$) *EK*
      Ecliptic latitude ($\beta$) *TK*
 III. Mediation ($\mu$) *EM*
      Declination ($\delta$) *NT*

The notation of the figure is a common medieval transliteration of Ptolemy's notation, as used, for example, in *Almagest* VIII. 5. *DEB* is the ecliptic, with pole at *H*. *CEA* is the equator (equinoctial), with pole at *Z*. *E* is the equinoctial point, or 'First point of Aries'.

Notice that whereas right ascensions are now given in units of time, in the Middle Ages they were generally given in degrees. Needless to say,

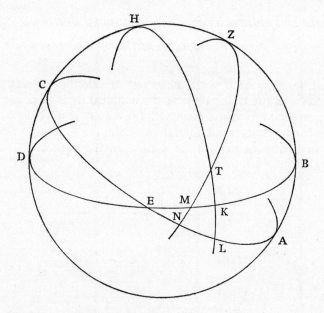

β and δ were reckoned north or south of the ecliptic and equator respectively, and were not given an algebraic sign.

Yet another coordinate by which star positions were given is explained in Appendix 26. For further remarks on right ascension, see Appendix 22; and for the conversion between one system and another see Appendix 24.

# APPENDIX 22

## *Right and oblique ascensions, ascensional difference, and difference in ascension*

PHENOMENA associated with the rising of celestial objects, and the determination of the time taken by the zodiacal signs in rising, were of great importance in early astronomy. Intimately connected with these problems are the concepts of the right and oblique ascension, which may be explained with the help of the accompanying figure.

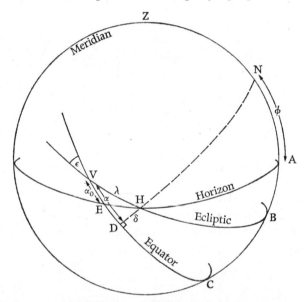

A point of the ecliptic $H$, of longitude $\lambda$, is rising. ($H$ is of course the point known in astronomy and astrology as the ascendent.) $N$ is the north pole, and $\alpha$ is the projection from $N$ on the equator of the arc $VH$, $V$ being the first point of Aries, the vernal point. The declination of $H$ is $\delta$. The relations between $\alpha$, $\delta$, and $\lambda$ involve the obliquity of the ecliptic ($\epsilon$), but have nothing to do with the latitude of the place, which we shall subsequently introduce as $\phi$. These relations, namely

$$\sin \delta = \sin \lambda \sin \epsilon,$$

and

$$\tan \alpha = \tan \lambda \cos \epsilon,$$

are equivalent to those obtained (in terms of chords) by Ptolemy in *Almagest* I. 14–16. The functions $\delta(\lambda)$ and $\alpha(\lambda)$ are tabulated in *Almagest* I. 15 and II. 8 respectively, the first for degree intervals, the second for ten degree intervals. The quantity is the *ascension on the direct circle* of the ascendent point *H*, or its *right ascension*, and of course this phrase has been extended in scope, so as to provide us with a coordinate for any point in the sky.

The equator meets the horizon in the easterly point *E*, which is, for a given observer, fixed—unlike the point *H*, which cuts the horizon at a point depending on $\lambda$. The arc *VE* is the ascension on the oblique circle, perhaps less misleadingly known as the *oblique ascension*. The difference $|\alpha - \alpha_0|$, where $\alpha_0$ is the oblique ascension, is the *ascensional difference*.

In *Almagest* II. 7 Ptolemy explains how to evaluate these quantities, with a view to tabulating them. It is stated in *Albion* IV. 17 that this passage was used in the construction of the table there given for ascensions at the latitude for Oxford. The underlying theory is dealt with in *Quadripartitum* IV. 7–8 (q.v.), and the resulting rules are included in the canons to the tables of John Maudith (see our section II, above). Albategni's chapters XI and XX cover the same ground, although more obscurely in the medieval Latin translation, as a result of the translator's confusion. Applying the theorem of Menelaus to $\Delta NCD/EHA$, an expression is obtained which is immediately reducible to

$$\sin|\alpha - \alpha_0| = \tan \delta \tan \phi.$$

Since the functions $\delta(\lambda)$ and $\alpha(\lambda)$ are known, $\alpha_0$ may now be calculated for any geographical latitude $\phi$.

If the quantities $\alpha$ and $\alpha_0$ for the ends of an *arbitrary* ecliptic arc are denoted by single and double primes, then an expression of the sort

$$(\alpha' - \alpha_0') - (\alpha'' - \alpha_0''),$$

may be rewritten            $$(\Delta\alpha - \Delta\alpha_0),$$

and it is this difference which we denote by $D(\lambda)$ in connection with *Quadripartitum* IV. 8 (commentary). The quantities $\Delta\alpha$ and $\Delta\alpha_0$ may be referred to as differences in ascension, but they must be carefully distinguished from ascensional differences, so-called. Each gives the rising time for an ecliptic arc between specified limits, $\Delta\alpha_0$ corresponding to a horizon at geographical latitude $\phi$, and $\Delta\alpha$ for an observer at the equator.

In Ptolemy's table of *Almagest* II. 8, the first column (after that which specifies the $10°$ limits of the ecliptic intervals) is for $\Delta\alpha$, while the second is for $\alpha$. In the remainder of the table there is a pair of columns for each of ten given geographical latitudes (selected zones, in fact, for which the longest day ranges from $12^h$ to $17^h$, at half-hour intervals). The first of each pair of columns is for $\Delta\alpha_0$ and the second for $\alpha_0$. Thus under 'sphaera recta' and opposite Taurus 10° ($\lambda = 40°$) we have $\Delta\alpha = 8°\ 4'$ and $\alpha = 30°\ 41''$, while the next entry, for Taurus 20° ($\lambda = 50°$) has $\Delta\alpha = 8°\ 31''$ and $\alpha = 39°\ 12''$. Although Ptolemy's table leads us to think of $\Delta\alpha$ as a function of the *end* of the ecliptic arc which rises (for clearly $8°\ 31' = 39°\ 12' - 30°\ 41'$), a more symmetrical curve is obtained (with respect to the coordinate axes) if we plot $\Delta\alpha$ against the *mean* longitude of the rising arc, and this is done in the figure on p. 70, vol. ii.

A better way of achieving symmetry, and one which allows us to confirm easily certain properties of the curves in question, is to take infinitesimal elements of the ecliptic, $\delta\lambda$, and to plot corresponding rising times (in angular measure). These we call $\delta\alpha_0$. As a suitable starting-point we take the equation

$$\cos \epsilon \ \cos \alpha_0 - \cot \lambda \sin \alpha_0 = \sin \epsilon \tan \phi,$$

obtained from the triangle between the horizon, the ecliptic, and the equator. In principle we might solve this for $\alpha_0$, differentiate, and thus derive an equation relating $\delta\alpha_0$ and $\lambda$, taking $\delta\lambda$ to be a constant interval. This would give an explicit equation representing the curves in which we are interested, but the equation turns out to be extremely cumbersome. Let $\delta\alpha_0$ be more conveniently written as $T$, and $\delta\lambda$ as $L$ (a constant). It is easy to show that for $\phi = 0$ (i.e. in the special case where $T = \Delta\alpha$), the curve has maxima at $\lambda = 90°$ and $270°$, with minima at $0°$ and $180°$. These properties are intuitively obvious. A property which is marginally less so is proved by differentiating the above equation. We obtain

$$(\cos \epsilon \sin \alpha_0 + \cot \lambda \cos \alpha_0)T - L \sin \alpha_0 \operatorname{cosec}^2\lambda = 0,$$

which shows that for $\lambda = 90°$ and $270°$, $T = L \sec \epsilon$ for all curves (since $\sin \alpha_0 \neq 0$). In short, all curves, regardless of the latitude, pass through the maxima of the curve for $\Delta\alpha$.

This result may also be proved in the case where rising times are taken for finite arcs of any length. Suppose that we consider the rising time ($T$) as a function of the mean longitude of the rising arc of the ecliptic, i.e. we take the oblique ascension to be $\alpha_0$ when $\lambda - L/2$ rises, and $\alpha_0 + T$ when $\lambda + L/2$ rises. We write $\cos \epsilon$ as $c$, and $\sin \epsilon \tan \phi$ as $a$, wherefore, setting $\lambda = 90°$ in the above equation for the fundamental triangle, we have

$$c \cos \alpha_0 - \tan(L/2)\sin \alpha_0 = a,$$

and

$$c \cos(\alpha_0 + T) + \tan(L/2)\sin(\alpha_0 + T) = a.$$

Subtracting, and noting that $\sin(\alpha_0 + T/2) \neq 0$, we find that

$$\sec \epsilon \tan L/2 = \tan T/2.$$

This equation for $T$ in terms of the finite length of the ecliptic arc and the obliquity of the ecliptic is again independent of the geographical longitude $\phi$, showing that all curves again pass through the two points (maxima) of the curve for *sphaera recta*.

# APPENDIX 23

## *Formulae connecting the sides and angles of a spherical triangle*

In stating here some of the fundamental formulae of spherical trigonometry, it is hoped to provide a standard for the comparison of historical examples, and to collect together formulae which are without widely accepted names.

A spherical triangle has six elements, namely the sides (arcs $a$, $b$, $c$) and the angles opposite them ($A$, $B$, $C$). The triangle is completely determined when any three of the elements are given. Three classes of formula may be distinguished, into which enter four, five, or six elements. The second and third classes (which include Napier's analogies, Delambre's analogies, Gauss's analogies, and related formulae) are not only historically irrelevant to the Middle Ages, but with the exception of one formula are not particularly useful. Apart from this exception (formula (V)) the formulae of the first class are therefore alone quoted:

(I) $\cos a = \cos b \cos c + \sin b \sin c \cos A$   (commonly referred to as the 'fundamental formula')

(Ia) $\sin \dfrac{A}{2} = \sqrt{\left\{\dfrac{\sin(s-b)\,.\sin(s-c)}{\sin b\,.\sin c}\right\}}$

(Ib) $\cos \dfrac{A}{2} = \sqrt{\left\{\dfrac{\sin s\,.\sin(s-a)}{\sin b\,.\sin c}\right\}}$   (derived directly from (I), $s$, as usual, being the semi-perimeter)

(Ic) $\tan \dfrac{A}{2} = \sqrt{\left\{\dfrac{\sin(s-b)\,.\sin(s-c)}{\sin s\,.\sin(s-a)}\right\}}$

(II) $\dfrac{\sin A}{\sin a} = \dfrac{\sin B}{\sin b} = \dfrac{\sin C}{\sin c}$   (commonly known as the 'sine rule', or the 'law of sines')

(III) $\cos a \cos C = \sin a \cot b - \sin C \cot B$   (the 'four parts formula')

(IV) $\cos A = -\cos B \cos C + \sin B \sin C \cos a$   (closely related to (I), and the source of formulae resembling (Ia, b, c))

(V) $\sin a \cos B = \cos b \sin c - \sin b \cos c \cos A$   (as mentioned above, this is the only formula quoted with five elements)

When one of the sides or angles is a right angle, the formulae are simpler, often considerably so. Important examples are:

When $A = 90°$, (I) becomes

$$(\text{I}') \quad \cos a = \cos b \cos c,$$

and (IV) becomes     $(\text{IV}') \quad \cos a = \cot B \cot C.$

When $C = 90°$, (III) becomes

$$(\text{III}') \quad \sin a = \frac{\tan b}{\tan B},$$

and when $B = 90°$, (III) becomes

$$(\text{III}'') \quad \cos C = \frac{\tan a}{\tan b}.$$

# APPENDIX 24

## *The transformation of celestial coordinates*

FOR reference purposes, we quote here equations relating the three principal systems of coordinates. In the triangle $HZT$, it will be seen that arc $HZ$ is the obliquity of the ecliptic, $\epsilon$, while $HT$ is the complement of $\beta$, $ZT$ is the complement of $\delta$, angle $ZHT$ is the complement of $\lambda$, and angle $HZT$ is $\alpha$ in excess of a right angle.

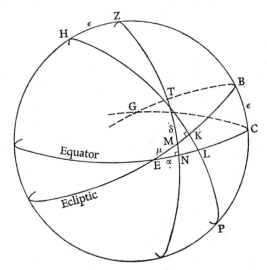

From formula (I) (see Appendix 23) directly, in $\triangle HZT$

$$\sin \beta = \sin \delta \cos \epsilon - \cos \delta \sin \epsilon \sin \alpha. \tag{1}$$

From formula (II) directly, in $\triangle HZT$,

$$\cos \beta \cos \lambda = \cos \delta \cos \alpha. \tag{2}$$

An alternative to (2), but more involved, can be obtained from formula (V), giving $\sin \lambda$ rather than $\cos \lambda$. In either case, (1) is to be applied first, to decide $\beta$, the value of which is then used in the remaining formulae.

Applying the same formulae in the alternative sense, we have

$$\sin \delta = \sin \beta \cos \epsilon + \cos \beta \sin \epsilon \sin \lambda, \tag{3}$$

and

$$\cos \delta \cos \alpha = \cos \beta \cos \lambda. \tag{4}$$

To relate $\alpha$ to $\mu$, apply (III″) to triangle $EMN$:

$$\tan \alpha = \cos \epsilon \tan \mu. \qquad (5)$$

(It will be seen from the accompanying graph, in which $\mu$ is plotted against $(\mu-\alpha)$, that $\mu$ and $\alpha$ never differ by more than $2\frac{1}{2}°$ or so.)

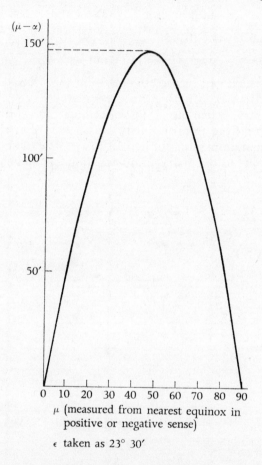

$\mu$ (measured from nearest equinox in positive or negative sense)

$\epsilon$ taken as 23° 30′

It is more pertinent in a critique of medieval astronomy (and in particular of star catalogues) to relate $\lambda$ to $\mu$, which is less easily done. Applying formula (III) to $\Delta HZT$ (sides $HT$ and $HZ$), and eliminating $\alpha$ by using (5), it will be seen that

$$\tan \mu = \tan \lambda - \tan \epsilon \tan \beta \sec \lambda. \qquad (6)$$

With (3), this allows the transformation of coordinates in one direction. There seem to be no *simple* converse equations. Applying (IV′) to

$\triangle EMN$, with angle $EMN$ denoted by $\phi$, and then applying (II) to $\triangle TMK$, we obtain

$$\cos \mu = \cot \epsilon \cot \phi,$$

and

$$\sin \beta = \sin \phi \sin \delta.$$

Eliminating $\phi$ between these two equations,

$$\sin^2\beta = \sin^2\delta/(1 + \tan^2\epsilon \cos^2\mu). \tag{7}$$

It is possible to rewrite (6) as

$$\sin(\lambda - \mu) = \tan \beta \tan \epsilon \cos \mu, \tag{8}$$

an equation which can be used to determine $(\lambda - \mu)$, and hence $\lambda$, once $\beta$ is calculated.

It is necessary, when using these formulae, to consider multiple solutions arising from the periodicity of the trigonometrical functions, quite apart from ambiguities of sign, as in (7). The following scheme is added as a guide to the above transformation formulae:

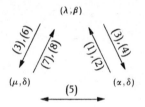

# APPENDIX 25

## *Thebit's theory of trepidation and the adjustment of John Maudith's star catalogue*

MEDIEVAL European astronomers usually refer to Thebit or to Arzachel, whom they believe based himself on Thebit, when they allude to the theory of a trepidational movement of the eighth sphere. On the basis of a Hebrew manuscript B. R. Goldstein has shown that Arzachel's theory was in point of fact more complex than Thebit's (see *Centaurus*, x (1964), 232–47). For a comprehensive discussion of Thebit's fundamental work see O. Neugebauer, *Proceedings of the American Philosophical Society*, cvi (1962), 290–9. The model differs in one small respect (not materially affecting positions assigned to the stars in medieval catalogues) from that which Goldstein and I (in a review in *Archives internationales d'histoire des sciences*, xx (1967), no. 78–9, 71–83) arrived at independently but wrongly. The difference concerns the constraints placed on the moving ecliptic, and Dr. Raymond Mercier points out to me that constraints placed in the manner Goldstein and I formerly supposed the (very obscure) text to intend were indeed the constraints of the periodic part of the Alfonsine model, according to Peurbach. According to Thebit, the vernal and autumnal points ($V$ and $A$) do not remain at fixed points of the celestial equator, nor at points moving steadily round the equator, but each rather moves at a uniform rate around its own fixed small circle of the sphere. Their radii we take as $r$. One of these small circles is centred on the so-called 'mean first point of Aries' ($V_0$), and the other on the opposite point $A_0$. The true ecliptic, according to Thebit, moved with $V$ and $A$, there being a mean ecliptic through $V_0$ and $A_0$ (obliquity $\epsilon_0$). The constraint on the moving ecliptic was such that it should pass through two points of the mean ecliptic at a distance of 90° from $V$ and $A$. The alternative model alluded to above is one in which the distances from $V_0$ and $A_0$ are 90°. The angle between the moving ecliptic and the equator will be denoted by $\epsilon$. In the accompanying figure, medieval terminology would have given to $\theta$ the name of *motus (accessionis et recessionis) octave spere*, with $E$ the *equatio capitis ab equatore*. A further quantity, not marked on the figure, was *equatio dimidii diametri circuli parvi* ($e$). This quantity seems to have been intended as the declination of $V$. $E$ is the quantity of principal interest, especially for work with changing stellar longitudes. (The stars are fixed with respect to $V$ and the *moving* ecliptic.)

It is not difficult to obtain general formulae determining $E$, $\epsilon$, and $e$ in terms of $\theta$, and to use them, for example, to show that Thebit's state-

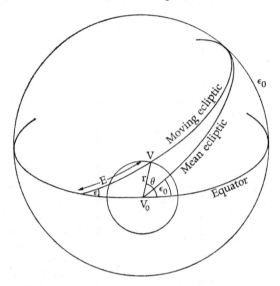

ments were not completely accurate. It may be shown, for example, according to Dr. Mercier, that

$$\sin E = \sin \theta \tan r \sqrt{\left\{\frac{1+\cos^2(\theta-\epsilon_0)\tan^2 r}{\sin^2(\theta-\epsilon_0)\tan^2 r+[\sin \epsilon_0+\tan^2 r \cos(\theta-\epsilon_0)\sin \theta]^2}\right\}}.$$

The expression I previously obtained (but which was misprinted in the review) for the alternative model is

$$\sin E = \sin \theta \sin r \sqrt{\left\{\frac{1+\tan^2 r \sin^2(\theta-\epsilon_0)}{\sin^2\epsilon_0+\tan^2 r \sin^2(\theta-\epsilon_0)}\right\}}.$$

It seems that Thebit and most of his successors took as exact the (good) approximations

$$e = r \sin \theta, \tag{1}$$

and

$$E = (r \operatorname{cosec} \epsilon_0)\sin \theta. \tag{2}$$

The parameters accepted by Thebit were $r = 4° \ 18' \ 43''$ and, as the multiplier of $\sin \theta$ in (2), $10° \ 45'$. He does not give equation (2) explicitly, nor does he ostensibly give the derivation of what is the multiplier in our equation. He leaves no doubt, however, that the maximum value of $E$ is to be $10\frac{3}{4}°$ (as he chooses to loosely state the figure.) In fact elsewhere he takes $\epsilon_0 = 23° \ 33'$, which would suggest a multiplier of

10° 47′ 31″. Strictly speaking, when $\theta$ is 90°, the value of $E$ on the basis of Thebit's construction is 10° 42′ 42″. The true maximum of $E$ is not, of course, attained when $\theta$ is precisely 90°.

It remains to fix the periodicity of access and recess ($\theta$), and the radix of the movement. If $t$ is the epoch in years of the Hegira (i.e. in Arab years), then Thebit's tables are closely summarized in the equation

$$\theta = (360°t/4,181\cdot5) + 1°\ 34′\ 02″. \tag{3}$$

(The daily movement was often quoted as $0″\ 52‴\ 28^{iv}\ 37^{v}\ 54^{vi}$, and 30 Arab years contain 10,631 days. If the year were taken as 354 days precisely, i.e. the ordinary year of the 30-year circle, a periodicity of 4,185·8 would be obtained. Bacon quoted the period as 4,181 years. Cf. p. 270 below.)

The star lists by the Oxford astronomer John Maudith, from which Richard of Wallingford appears to have derived his *Albion* list (see the commentary to *Albion* IV. 12), were expressly derived from Ptolemy's catalogue of *Almagest* VII and VIII by the addition of a constant 16° 40′ to the longitudes. It will be shown that this figure is a trepidation constant. (Essentially the same calculation was printed in my review in *Archives internationales d'histoire des sciences*, nos. 78–9 (1967), 71–83. Here I have made a change in the periodicity parameter at the suggestion of Dr. Mercier.) The beginning of the astronomer's year was usually taken as 1 March or 1 January, but by a curious chance the beginning of the Annunciation Year (25 March) in A.D. 1316 coincides with the beginning of A.H. 716. The heading to Maudith's tables runs as follows: 'Notandum quod anno Christi 1316, ad quem verificantur iste stelle, fuit annus Arabum 716, quo tempore in principio die mensis eorum anni 17 fuit motus octave spere 9 gra. 23 min. 10 sec.' Perhaps the proximity of the years' beginnings is why the epoch was chosen. For $t = 716$ we then have

$$\begin{aligned} E &= 10°\ 45′\ \sin(360° \times 716/4,181\cdot5 + 1°\ 34′\ 02″) \\ &= 10°\ 45′\ \sin(63°\ 12′\ 37″) \\ &= 9°\ 35′\ 46″. \end{aligned}$$

According to the quotation we should obtain 9° 23′ 10″. If we turn, however, to some of Thebit's tables in the Cambridge MS. Gg.VI.3 (f. 132ʳ), we find the radix wrongly stated as 1° 24′ 02″ (1° 34′ 04″ is often quoted elsewhere, while the original work has 02″ in the last place). Can it be that Maudith took this figure? Using it we find

$$E = 9°\ 34′\ 55″.$$

This is still larger than the quoted value, but referring to Maudith's astronomical tables elsewhere in the Cambridge manuscript (ff. 35ᵛ–40ʳ), we find 9° 31′ 42″ quoted as the motion of the eighth sphere for 1310 ('secundum Prefactium in almanac'). This is also the figure cited in the tables (f. 277ᵛ) in what I suggested (op. cit.) was his edition of the treatise on the new quadrant of Profatius. The figure 9° 23′ 10″ might therefore be miscopied from 9° 33′ 10″, or some other figure not very distant from it. At all events, it is clear that Maudith worked with virtually the original version of Thebit's theory. He might even have innocently made a straight interpolation in the table for ten-year intervals (which would give roughly 9° 32′, which could have been copied as 9° 23′) but this will be shown to be unlikely. What must first be proved is that the difference between these figures and 16° 40′ is close to what would have been calculated as applying between A.H. and the time of Ptolemy's catalogue (conventionally placed during the Middle Ages as A.D. 138), about 484 Christian years, or 499 Arab years.

Applying the formula we determine the quantity to be added to the *equatio capitis*. We now determine, that is to say, the amount of trepidation between the epochs of Ptolemy and Thebit:

$$10° \ 45′ \ \sin(360 \times 499/4,181{\cdot}5 - 1° \ 34′ \ 02″)$$

$$\simeq 10° \ 45′ \ \sin 41° \ 23′ \ 36″$$

$$\simeq 7° \ 06′ \ 29″.$$

Subtracting from 16° 40′ we get 9° 33′ 31″ which is so near to the figures mentioned above as to leave little doubt that 16° 40′ is correct, and that Maudith was not wrongly applying a simple theory of precession, but correctly applying a much more complicated theory of trepidation.

There is no indication at any point of the known writings of Richard of Wallingford that he adhered to Thebit's principles, but his constant of 17° for 1327 could have been loosely derived from these principles by a crude adaptation of John Maudith's figure for 1316, increased somewhat and rounded off to the nearest degree. See also p. 260, below.

Finally, I offer a simple method for finding the value of $E$ for any day, with Julian day number $N$. If the constants $0°{\cdot}0002429504$ and $471°{\cdot}8067$ are written as $p$ and $q$ respectively, then

$$E = 10°{\cdot}75 \sin(pN - q). \tag{4}$$

As an example of the application of (4), consider the 'Profatius' table referred to above. The figure quoted for 1300 is 9° 27′ 50″. For 1300 0 Jan., $N = 2195882$ and equation (4) yields 9° 27′ 49″. Taking the year to begin in March would mean adding 5″ to this figure, thus spoiling the excellent agreement. The figure in the manuscripts would be exact on the basis of (4) for a date of 10 Jan.

# APPENDIX 26

## *The astrolabe star coordinate* gradus longitudinis ex utraque parte *(marginal longitude)*

THE oldest European treatises on the astrolabe such as that by Llobet (Lupitus) of Barcelona (edited by N. Bubnov, *Gerberti Opera Mathematica* (Berlin, 1899), pp. 370–5, and by Millás, *Assaig d'història de les idees físiques i matemàtiques a la Catalunya medieval* (Barcelona, 1931), pp. 271–5) and that by Hermannus Contractus (edited by Bernard Pez, *Thesaurus Anecdotorum novissimus*, vol. iii, part 2 (Augsburg, 1723), pp. 95–106, and by Bubnov, op. cit., pp. 114–47, and included without bibliographical reference to its source (Pez) in R. T. Gunther's *Astrolabes of the World*, ii (1932), 409–22), recommend a method for placing the stars on the rete of an astrolabe making use of a coordinate not encountered in other contexts. Thus 'Typ III' and the first redaction of 'Typ XI' in Paul Kunitzsch's *Typen der Sternverzeichnissen* (Wiesbaden, 1966) both list this coordinate. The former type of catalogue is intimately related to the early treatises mentioned, as may be seen by reference to the sources cited by Kunitzsch, while the first redaction of 'Typ XI', despite the title ('*Tabula stellarum fixarum que ponuntur in astrolabio, certificata ad civitatem Parisius* . . . 1233') is essentially nothing more than a trivial rearrangement of the earlier list. More specifically, ten out of the twenty-seven stars in the earlier list have the same mediation as in the later, while the rest are all close, being neither systematically increased nor decreased (as would have been the case had precession been allowed for). Discounting a single serious case of miscopying, and four stars which are not identifiable with certainty, the average displacement of the mediations of the remaining twenty-two stars of 'Typ III' from 'Typ XI' (first redaction) is under 3′ arc. There is little doubt that miscopying, perhaps the miscopying of a common source, is responsible for all the divergences, both in the mediations and in the coordinate which we are about to consider.

The '1233' star list is found to have a column entitled *gradus longitudinis ex utraque parte* (a phrase here abbreviated as 'marginal longitude') which explains more fully than the 'altitude' of the earlier list the significance of the figures below it. This coordinate is not to be ranked with those—latitude, longitude, declination, mediation, and meridian altitude, for instance—introduced in Appendix 21. If the right ascension or the mediation of a star to be plotted on an astrolabe rete be specified, the star will lie on a known radius vector (*OP* in fig. a). The marginal

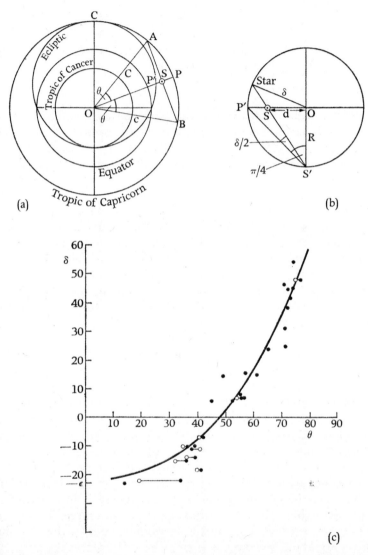

(a)

(b)

(c)

longitude $\theta$ was such that, constructing radii vectores *OA* and *OB* to each side of *OP* and at an angle $\theta$ to it, *A* and *B* being on the Tropic of Capricorn circle, the star (*S*) was correctly represented by the point of intersection of *AB* and *OP*. If the Capricorn circle is of radius *c*, the distance *OS* being *d*, then cos $\theta$ is equal to *d/c*. This ratio is readily expressed in terms of the obliquity of the ecliptic ($\epsilon$) and the declination ($\delta$). Turning to fig. b, which shows the stereographic projection of the star from the south pole $S'$ on the equatorial plane (seen edgeways-on in the diagram, that is, from a point in the plane at right angles to $OP$), it is clear from the construction line $S'P'$ that $d = R\tan(\frac{1}{4}\pi - \frac{1}{2}\delta)$, $R$ being the radius of the equatorial circle on the rete. As a special case of this relation, we see that $C$ (fig. a) is the projection of a point of declination $-\epsilon$, and that therefore $c = R\tan(\frac{1}{4}\pi + \frac{1}{2}\epsilon)$. Finally therefore,

$$\cos\theta = \cot(\tfrac{1}{4}\pi + \tfrac{1}{2}\epsilon) \cdot \tan(\tfrac{1}{4}\pi - \tfrac{1}{2}\delta).$$

It is this relation which is plotted as a continuous curve in fig. c, assuming for $\epsilon$ a value of $23° 30'$. (The constant cotangent factor is 0·65563.)

The derivative nature of the marginal longitudes and the seemingly inaccurate execution of the derivation in practice are surely not the only reasons for the eventual disappearance of the coordinate. Quite apart from their very limited use, there is good reason for thinking that the marginal longitudes of stars were found neither immediately from observations on stars nor by calculation from coordinates which had originally been so obtained. That they were in the first instance obtained in all probability from measurements made on a rete, or a drawing of such, the stars on which had previously been plotted in some other way, is suggested by the low level of accuracy attained in converting to them from declinations, assuming as a standard of comparison the declinations listed (from the second redaction of 'Typ XI') on the same page (p. 69) of Kunitzsch's book. Since the two lists of mediations there differ only trivially, and obviously again only as a result of miscopying, we may assume that the marginal longitudes ($\theta$) and the declinations ($\delta$) (there called *latitudines*) are meant to be consistent. The discrete points plotted on the same figure as the continuous ($\theta$, $\delta$) curve represent pairs of corresponding values of $\theta$ and $\delta$ taken from Kunitzsch's table. The more blatant errors may be corrected by taking the value of $\theta$ from 'Typ III', as may be seen from the figure. The errors are unsystematic, and suggest that the values of $\theta$ were not strictly calculated. The rete as traditionally constructed was usually somewhat larger than the Capricorn circle and by accepting the rim as equivalent to the Capricorn circle, errors would be introduced of precisely the same character as are observed in the figure. The full Capricorn circle might, admittedly, have been marked on the plate *before* it was fretted. Rival methods could be easily used, however, to position stars (by bending the pointers for instance) after the rete had been cut, whether or not a Capricorn circle was marked on it.

# APPENDIX 27

## *The treatise in MS. Digby 40 on the universal astrolabe: a description of 'Alī ibn Khalaf's astrolabe*

THE copy (item 1, ff. 1ʳ–8ʳ), which is apparently unique, dates from the late twelfth or early thirteenth century, and was written in England. It begins with the rubric: *In nomine domini pii et misericordis, incipit liber de composicione universalis astrolabii.* The text proper begins: 'Ptolomeus g(?) Mercurii incedens vestigiis in libro suo qui vocatur *Almagesti* de motu sic ait . . .' The prologue continues with a general discussion of Aristotelian tenets as to circular motions, and ends with a few rather vague distinctions between a special and a general instrument (which in the case of the astrolabe amounts to a distinction between an instrument useful at a restricted number of geographical latitudes and one applicable at all latitudes). The adjectives 'generale' and 'universale' are applied indiscriminately to the astrolabe here next described. There are six chapters in all, describing successively the mater, the construction of the almucantars and the azimuths (both in the form characteristic of the universal projection), the rete (spelt 'recte' and given in the Arabic as 'assanaka', presumably from 'assavaka' in the original, which in turn would derive from 'ash-shabaka' meaning 'the net'), the stars on the rete, and finally the zodiac. The text is liberally sprinkled with Arabic vocabulary, a fact which is explained by the final sentence: 'Explicit liber Ptolomei de composicione astrolabii universalis quem scilicet in civitate Londonie ex Arabico in Latinum transtulit era millesima centesima lxxxv.' This is therefore a copy of a translation made in London in A.D. 1147 (that is, the year 1185 of the era) from the Arabic. The use of this method of dating suggests a Spanish translator or at least a translator from those parts of Spain, Portugal, or southern France which had been ruled by the Visigoths. (The era used in these countries originates as commencing from the radix of some early Easter tables, 1 January 38 B.C.) It is worth asking whether the translator may not have been Abraham ben Ezra (b. Tudela 1092; d. Calahorra 1167). Between 1140 and 1148 he visited a number of towns in Italy, after which he travelled through France and to England (London and Winchester). In the same manuscript (MS. Digby 40, ff. 52–89ʳ) there is a work purporting to be by this very man. (Edited by J. M. Millás Vallicrosa, *El libro de los fundamentos de las Tablas astronómicas de R. Abraham ibn 'Ezra* (Madrid–Barcelona, 1947). There is an illustration of the script of Digby 40 in the second plate.)

Abraham ben Ezra was a prolific writer, and if one is to believe the claims that he composed in Latin as well as in Hebrew, a versatile one, for he certainly translated into Hebrew out of Arabic. There is a treatise on the astrolabe in two British Museum manuscripts which is in one of them closely associated with another work of his, and which has been claimed for him by its editor Millás (*Al-Andalus*, v (1940), 1–29). The opening words of the treatise are particularly interesting: 'Genera astrolabiorum duo sunt: dextrum et sinistrum. Numerus sinistrorum maior quam dextrorum est.' The distinction being made here, however, is not one between special and universal astrolabe, but between stereographic projection from the north and south poles. Stereographic projection from the south pole, leaving the stars of the northern hemisphere within the zodiac, is indeed the only convention recognized on medieval astrolabes. It is interesting to find that astrolabes following the alternative convention were not unknown in the twelfth century (or before, if the treatise in question was composed earlier). For some more general remarks on this subject, see *Horologium*, commentary to II. 4. 41.

The fact that the manuscripts (British Museum, MS. Cotton Vespas. A II 13, ff. 37$^{va}$–40$^{vb}$ (12th cent.) and MS. Arundel 377, ff. 63$^{ra}$–68$^{vb}$ (13th cent.)) were copied in England, that a comparison of the two Hebrew versions of Abraham ben Ezra's work on the astrolabe shows resemblances and parallels to the Latin text, that there are what Millás believes to be traces of Spanish style in the prose of the latter, and that (according to his biographer, J. L. Fleischer) Abraham was in England between 1158 and 1161, caused Millás to claim the work for him. R. Levy has since (*Speculum*, xvii (1942), 566–9) expressed doubt as to the validity of the claim. He sees no reason why one should not, even allowing the parallelism, argue that both Hebrew and Latin treatises were translations from an Arabic original. This is not an unreasonable position to take, and yet there is very little in common to the style of the disputed text and what we know to be a translation, namely the treatise on the astrolabe in MS. Digby 40; for the former is remarkably free from Arabic vocabulary, whereas the latter contains many such words. Although he moved freely around Europe, the evidence that Abraham was fluent in Latin is very weak. The two Hebrew texts on the astrolabe by him, differing chiefly in their length, were composed in 1146 and 1148. Is there any reason to suppose that the treatise of MS. Digby 40 could not have been translated by him in England between these dates? Although he was in England in 1158–61, is there any reason for thinking that he could not have come sooner? It is known that he was in Verona in 1146, Lucca in 1148, and Béziers in 1156. It seems very probable that the year 1147 found Abraham in Italy. It is not impossible that he was in London then, or that the explicit is in error as to the date, but the most we can say with anything approaching confidence is that the translator was a person with a similar background to his.

As explained in *Albion* II. 24. comm., the earliest type of universal astrolabe known is that described by ʿAlī ibn Khalaf in a treatise written early in the eleventh century and included in Castilian translation among the Alfonsine books two and a half centuries later. The Latin treatise of MS. Digby 40 is not only much shorter than that treatise, but bears no close resemblance to any part of it. The instrument described is nevertheless the same, and so far as is known there is no other medieval Latin treatise describing it. Since this copy is also apparently unique, the fact that there is no known (non-Hispanic) European example of the instrument is not surprising. The text is purely descriptive, providing no instructions for the use of the astrolabe, and the diagrams are incomplete; but the characteristic properties of the ʿAlī ibn Khalaf astrolabe are unmistakable (see *Albion*, II. 24. comm.). There is a rete, and this divided into a coordinate net on the one half and a star map on the other. The mode of graduation of the circles of the universal projection is the same.

There is some confusion in the final sentence, quoted earlier, where it is stated that Ptolemy is the author of the original. This information was obviously added quite gratuitously by the translator, and it was almost certainly prompted by the two occasions on which the word *Almagest* occurs in the text. The first occasion is in the opening words, quoted above. The second is in the sixth chapter, where we read 'siquis investigare voluerit, librum meum *Almagesti* perlegat . . .' The opening quotes Ptolemy, and is obviously therefore not his. The second quotation suggests not that Ptolemy is speaking, but rather an astronomer who has written a commentary on *Almagest*. At that time, almost any work of astronomy might have been regarded by its author as a commentary on the *Almagest*, but in a stricter sense there is in the required period (say 1020 to 1145) only one plausible choice. Towards the end of the period the Spanish Muslim astronomer Jābir ibn Aflaḥ wrote a 'Correction of the Almagest', Iṣlāḥ al-Majisṭī. Certainly he lived after Arzachel's modification of the universal astrolabe, but we know from the fact that later hybrid examples survive (see, again, *Albion* II. 24. comm.) that Arzachel's design was not universally accepted. There is thus a remote possibility that the text in question derives from a work in Arabic by Jābir, and that it was translated in London by a compatriot, in 1147, and perhaps within a decade or two of its completion.

# APPENDIX 28

## *Some possible applications of the albion saphea*

THE great versatility of those astrolabes which depend on the universal projection is a consequence of the several different interpretations which may be placed on each set of lines drawn on them. Imagine a sphere on the surface of which are the two usual sets of coordinate-lines, of 'latitude' and 'longitude'. Projecting stereographically from a point on the 'equator' upon a plane touching the sphere at the other extreme of the diameter, the point being the origin of the 'longitude coordinates', the result is the characteristic universal astrolabe projection. We may distinguish the two sets of coordinate curves by the names 'polar lines' and 'parallels'. The coordinates which we use them to represent will be given their usual symbols, for conciseness:

$\alpha$  right ascension
$\delta$  declination
$H$  hour angle, measured westwards from the meridian
$\lambda$  ecliptic latitude
$\beta$  ecliptic longitude
$\phi$  geographic latitude
$\epsilon$  obliquity
$a$  altitude
$A$  azimuth, measured westwards from the meridian

By such a phrase as 'the $\beta$-lines' we are to understand the curves along each of which $\beta$ is constant.

Richard of Wallingford recommended three (or more) sets of parallels and polar lines: a basic set, a set obtainable from this by rotating through the angle $\epsilon$, and a set obtainable by rotating through $\phi$. These may, again for convenience of reference, be designated by the terms 'set $O$', 'set $\epsilon$', and 'set $\phi$'. (In principle we may have any number of sets $\phi$, and it will be recalled that Richard recommended two. Arzachel had none, but required a construction operation using the brachiolus to replicate any such coordinate set. ʿAlī ibn Khalaf, and those who in the sixteenth century used a rete for a movable coordinate set, could use it for both set $\epsilon$ and set $\phi$, but of course a rete, besides being difficult to make really accurately, would have been difficult to add to the already crowded Albion.) In the last analysis, an astrolabe, apart from its use as an

instrument of observation, is a device for converting from one set of co-
—ordinates to another. The simplest application of the saphea involves
accepting the centre as the vernal point (or the autumnal point—the
disc necessarily represents both hemispheres simultaneously), with set $O$
as $\alpha$-lines and $\delta$-lines, and with set $\epsilon$ as $\lambda$-lines and $\beta$-lines. In summary,
the following table shows three such (two-fold) interpretations, the first
being that just explained:

| Use | Coordinate set | Centre of plate | Polar lines | Parallels |
|-----|---------------|----------------|-------------|-----------|
| 1 | set $O$ <br> set $\epsilon$ | $\big\}\lambda = 0°$ or $180°$ $\big\{$ | $\alpha$-lines <br> $\lambda$-lines | $\delta$-lines <br> $\beta$-lines |
| | (The stars are added to the plate in accordance with this interpretation.) | | | |
| 2 | set $O$ <br> set $\phi$ | $\big\}A = 90°$ or $270°$ $\big\{$ | $H$-lines <br> $A$-lines | $\delta$-lines <br> $a$-lines |
| 3 | set $O$ <br> set $\phi$† | ascendent ($\lambda$ pre- <br> sumed known) or <br> point of setting of <br> ecliptic $\big\{$ | $A$-lines <br> $\lambda$-lines ($\lambda$ measured <br> from ascendent) | $a$-lines <br> $\beta$-lines |

† Where $\phi$ is specially chosen to equal the zenith distance of the pole of the ecliptic.

There are two other useful interpretations of the lines of the saphea,
granted that it has a movable rete with coordinate net *or* a means (such
as a brachiolus or even a radial thread and bead) of reproducing any
point of such a net, starting from set $O$:

*Use 4.* The centre is interpreted as north (or south) pole. The two
polar-lines of the rete (real or imagined) of 'longitude' $\epsilon$ are selected.
(A rete with a one-degree mesh will not of course have such lines, which
must therefore be guessed at by interpolation.) Likewise we are to single
out the polar-lines of the plate of 'longitude' $\phi$. One of the two selected
lines of the rete will correspond to the ecliptic of the conventional planispheric astrolabe in equinoctial projection from one pole, while the
other will correspond to the other half of the ecliptic, projected from the
other pole. Likewise, the two arcs selected on the plate will together
comprise the *entire* horizon. The plate, in fact, with its full set of polar-
lines, may be considered a 'plate of horizons' in the traditional sense.
The outer limb of the plate corresponds to the equator. Thus, although
the scale of the astrolabe is greater than that of the conventional astrolabe, for a given plate size, by doubling the projection more of the
horizon is actually included. (But whereas normally the ecliptic may be
included as a single and complete circle, now it must be divided into
two arcs.)

A graduated rule (or thread and bead, by transferring to the central
scale) may be used to give the *declination* ($\delta$) and *hour angle* ($H$) of any of

the four points: ascendent, descendent, mid-heaven (*medium celi*), lower mid-heaven (*imum medium celi*). More generally, we may deduce δ given *H*, or conversely, for any point on the ecliptic.

*Use 5.* By analogy with 4, we take now a horizontal projection from the zenith (centre of the plate), the limb being the horizon. A suitable polar-line of the rete represents the ecliptic. (One must first discover the zenith distance of the pole of the ecliptic independently, by the application of uses 1–4.) A diametral rule suitably graduated (or thread and bead as before) then permits us to assign an azimuth to any point of the ecliptic of given altitude, or vice versa. In particular, we can find the altitude of mid-heaven and lower mid-heaven.

It is assumed here that sufficient has been said to allow those familiar with the workings of the conventional astrolabe to effect any of the coordinate transformations implicit in uses 1–5. Notice in particular, however, that whereas the daily rotation of an object (assuming a fixed position in the sky) makes its locus on the traditional astrolabe a concentric circle, this will now be so only with use 4. With uses 1 and 2 it will be the appropriate δ-line. Notice also that horizon effects are conveniently studied with a radial rule alone in uses 2 and 3, and that the problem of twilight could be easily taken into account in engraving the plate. Notice also that inaccuracies introduced through pivoting errors are, in the saphea of *Albion*, no longer significant, but that it would require great skill to perform uses 4 and 5 with nothing more than a thread and bead. Finally be it noted that our uses 1–5 do not exhaust the possibilities.

# APPENDIX 29

## *An outline of the Ptolemaic theory of planetary longitude, as applied in the Middle Ages*

W H A T follows is not a historical account of Ptolemy's thought but an illustration, by means of diagrams and rules for the use of astronomical tables, of the standard medieval methods (most of which originated with Ptolemy) of evaluating the apparent positions of Sun, Moon, and planets at any epoch. Although there are many good synopses of Ptolemaic planetary theory, such as those by Delambre and Dreyer, yet amongst those easily obtainable today, only that by Otto Neugebauer (*The Exact Sciences in Antiquity* (1957), pp. 191–207) is likely to assist anyone wishing to use the tables of the *Almagest*—and even then it was not Neugebauer's prime intention to provide a guide of this sort. The need for a guide to the tables in use in the Middle Ages is evident from the fact that reference is often made to Alfred Wegener's dissertation, *Die alfonsinischen Tafeln für den Gebrauch eines modernen Rechners* (Berlin, 1905). This is in many respects a useful work, giving rules for manipulating the Alfonsine tables, but without justifying the procedures advocated. He gave no formulae, whether exact or approximative, for the calculation of the tables, nor did he extract from them the fundamental constants on which they were based. What is more, in converting the sexagesimal notation of the originals into decimal, he made it very difficult for a historian to use his version of the tables to check the calculations of a medieval writer. The aim of the following note is therefore to supply what is not easily available elsewhere, and at the same time to collect together enough of the bones of medieval planetary longitude theory for readers to obtain a general impression of the skeleton. Exact trigonometrical expressions will be provided for certain important functions and this should make it easier to appraise traditional procedures, especially those used in *Albion* I.

### I. THE ESSENTIALS OF THE PTOLEMAIC THEORY

For the most part, the medieval astronomer was not concerned with the disposition of the planets in space: he was concerned with their positions in angle, but not in distance. In 'saving the appearances' he referred the planets to one of the traditional coordinate systems, most commonly ignoring movement in latitude (amounting to more than 5° in the case of the Moon), and citing only ecliptic longitude, with the equinox as origin. The ecliptic, at approximately $23\frac{1}{2}$° to the celestial equator, was divided into twelve signs of 30° each. Beginning from the vernal equinox, or 'First Point of Aries', that is to say, the ascending node of the Sun's apparent annual path (the ecliptic) on the celestial equator (defined, of course, by the diurnal motion), the order of the signs, with the more common conventional symbols for them, is as follows:

Aries ♈       (the symbol often also refers to the First Point of Aries, or vernal equinox, here denoted $V$)

Taurus ♉

Gemini ♊

Cancer ♋       (the symbol is often used for the summer solstice)

Leo ♌

Virgo ♍

Libra ♎       (the symbol is often used for the autumnal equinox)

Scorpio ♏

Sagittarius ♐

Capricornus ♑       (the symbol is often used for the winter solstice)

Aquarius ♒

Pisces ♓

The order runs from west to east (for an observer in the northern hemisphere, looking south), this being the direction of the Sun in its annual motion through the stars, and also the general direction of all planetary motions, overlooking their retrogradations, which are relatively short-lived.

The astronomical problem was that of accounting precisely for the irregularities in the motion of Sun, Moon, and planets, and especially for the retrogradations of all but the Sun and Moon. The highly involved solutions arrived at by Ptolemy and his precursors are summarized in figs. 1–7. In all cases the outer circle represents the zodiac, $T$ is the Earth, and the symbols within parentheses on the zodiac are the longitudes of the points to which they are attached. In general, Ptolemy placed each planet on the radius of an *epicycle* (which radius travels at constant angular velocity with respect to a certain line; see below), the centre of the epicycle being carried along an eccentric *deferent circle* (centre $C$), and the angular velocity of the centre of the epicycle being constant only about the *equant* point (denoted here by $E$). The explanatory scheme outlined in *Almagest* was not, however, a unified scheme, and specific differences between the explanations offered for the various planetary motions are indicated under the following headings.

*Sun*: In modern terms, the apparent movement of the Sun (against the background of stars) as seen from the Earth is the same (ignoring stellar parallax, proper motions, aberration, and so on) as the movement of the Earth referred to the Sun. The latter movement is along an ellipse of small eccentricity, but a simple eccentric circle provides a good approximation to it. In geocentric terms, this is shown in fig. 1. The Sun ($S$) moves with uniform angular velocity about $C$, the (eccentric) centre of the deferent. The line $TC$ defines the direction of the most

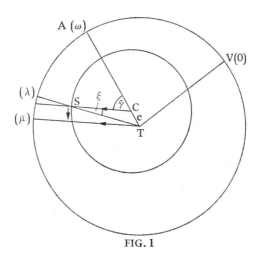

FIG. 1

distant part of the deferent circle from $T$, known as the *aux*. (On occasion this name refers to the point marked $A$, on the zodiac beyond the aux (apogee) of the deferent.) Line $TCA$ is accordingly referred to as the 'line of aux' (*linea augis*).

In calculating the longitude of $S$, that is to say the angle $VTS$ ($\lambda$), clearly the angle $ACS$ ($\bar{\gamma}$) is fundamental. The uniform motion of $S$ was measured, however, not with respect to the direction $TA$ (the longitude ($\omega$) of which was changing, although Ptolemy did not treat it as such), but with respect to the equinox. The line through $T$ parallel to $CS$ passes through a point of the zodiac known as the 'mean motus' (*medius motus*). The Latin phrase is doubly ambiguous, and can refer not only to this point, but to its longitude ($\bar{\mu}$) and to its motion or velocity. Here it is usually taken to have the character of a mean longitude.

The angle by which the apparent Sun differs from the mean Sun is known as the 'equation of centre' (*equacio centri*; symbol $\xi$). It is a simple matter (see below) to calculate $\xi$ as a function of $\bar{\gamma}$ ($= \bar{\mu} - \omega$), and to tabulate the function. There is such a table in *Almagest* III. 6, where the equation of centre is simply called the 'anomaly'. This is to be contrasted with Ptolemy's 'prosthaphairesis for anomaly', an anomaly called in the Middle Ages 'equation of the argument' (*equatio argumenti*). (The Greek word 'prosthaphairesis', commonly found in Renaissance astronomy, means simply 'that which is to be added or subtracted'. The Latin terms come from the Arabic.) Since in the case of the Sun there is only one equation or anomaly, it is customary to find it referred to simply as *equatio*, without further qualification. The mean motus of the Sun, how-ever, is often called *argumentum*, a word normally reserved elsewhere for an angle within an epicycle, and the angle $\xi$ might then be termed '*equatio argumenti*'. There can obviously be no confusion when there is no epicycle in the theory of the Sun which is used, but it seems that Ptolemy uses his equivalent words ('argument' and 'prosthaphairesis' for $\bar{\mu}$ and $\xi$) as a direct consequence of his considering the hypothesis of the simple eccentric together with the geometrically equivalent hypothesis of an epicycle travelling on a deferent centred at the Earth.

The maximum value of $\xi$ according to Ptolemy was $2° 23'$. The longitude of aux ($\omega$) was fixed at $65° 30'$. In the early fourteenth century, $\omega$ was correctly taken as being very close to $90°$, and the difference between this and Ptolemy's figure was more than could be explained in terms of precession (or trepidation) alone. (See Appendix 25.) These figures, together with the rather poor figure for the eccentricity ($e$) of $2° 30'$ (the distance $CT$, relative to a deferent radius of 60$^P$), were deduced without much difficulty from observations of the lengths of the seasons (i.e. the times spent by the Sun in the four quadrants of the zodiac, beginning at the vernal point). See *Albion* I. 1. The converse problem, namely that of calculating $\bar{\mu}$, $\bar{\xi}$, and hence $\lambda$, from a pre-determined set of parameters, is outlined in section 2 below.

*Moon*: The fact discovered by the Babylonians, perhaps before the third century B.C. (see Neugebauer, op. cit.), that the sidereal and anomalistic months differ appreciably, suggested a model with an eccentric but with a rotating line of aux. The appropriate parameters had been calculated (for example by Hipparchus) from eclipse records, that is to say, from data appropriate to the syzygies (conjunction and opposition), and Ptolemy found discrepancies between theory and observation near the quadratures. In terms of the epicyclic theory, there was seemingly a fluctuation in the size of the epicycle which was dependent on the angular separation of the Sun and Moon. In terms of the 'elongation' ($\bar{\eta}$) of the mean Moon from the mean Sun (i.e. $\bar{\eta} = \bar{\mu} - \bar{\mu}_\odot$), Ptolemy arranged matters so that the line of aux (on which lies $C$, the centre of the deferent) kept the same angular distance $\bar{\eta}$ from the mean Sun (see fig. 2). The distance of the epicycle from $T$ is thus a minimum at mean quadratures ($\bar{\eta} = 90°, 270°$), and a maximum at mean conjunction and opposition ($\bar{\eta} = 0°, 180°$).

Ptolemy found a correction to the anomaly to be necessary, this being most in evidence for values of $\bar{\eta}$ near the octants. He allowed for this by measuring the mean argument ($\bar{\alpha}$ in fig. 2), that is, the angle which the radius vector carrying the Moon ($M$) around the epicycle makes with some line through the centre of the epicycle from the line through $E'$, where $E'$ is the point diametrically opposite $C$ on the small circle of radius $e$, traced out by $C$. The mean argument is thus measured from

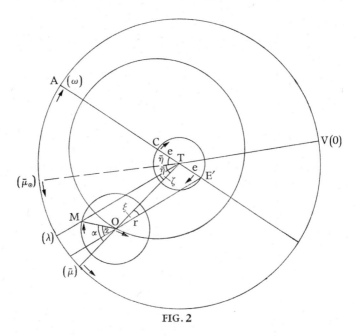

FIG. 2

the aux of the epicycle as seen from $E'$, as distinct from the true argument ($\alpha$), which is needed for computing the equation of the argument ($\zeta$), and which is measured from the (true) aux of the epicycle.

Notice that the Moon's movement is never retrograde, and that unlike all the remaining planets its revolution in the epicycle is contrary to the (direct) motion of the epicycle along the deferent.

Although the parameters of Ptolemy's lunar theory are collected together with other important parameters elsewhere, it is to be noticed that in *Almagest* (see V. 4 and V. 13 for example) he took the deferent radius as $49^P\ 41'$ ($= 60^P - e$), whereas in the *Canobic Inscription* (*Opera astronomica minora*, ed. Heiberg, vol. ii) he took the more usual standard radius of $60^P$, with consequent change in parameters:

| | | | |
|---|---|---|---|
| *Almagest* | $r = 5^P\ 15'$ | $e = 10^P\ 19'$ | $R = 49^P\ 41'$ |
| *Canobic Inscription* | $r = 6^P\ 20'$ | $e = 12^P\ 28'$ | $R = 60^P.$ |

(Parameters more strictly conforming with those in the *Almagest* but on
the convention $R = 60^p$ are:

$$r = 6^p\ 20'\ 24'' \qquad e = 12^p\ 27'\ 32''.)$$

Medieval astronomers were aware of the two different conventions, but
modern writers occasionally take the Moon's parameters of the *Almagest*
as though they corresponded to a radius of $60^p$. That they do not will
be confirmed in a later section.

*Venus, Mars, Jupiter, Saturn*: Before Ptolemy, the motion of each of
these planets had been explained as an epicyclic motion along an

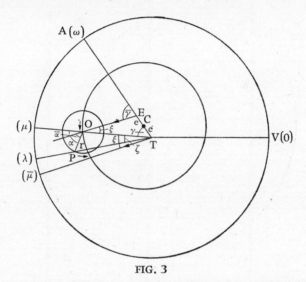

FIG. 3

eccentric deferent circle. We can today regard the double motion as a
compounding of the Earth's motion round the Sun with the planets'
round the Sun. The eccentricity is geometrically equivalent to taking
one or both of the orbits around the Sun as eccentric, and is in fact the
vector sum of the eccentricity-vectors of the Earth and planet. Ptolemy
introduced the equant $(E)$, a point of the now fixed line of aux, such that
$EC = CT\ (= e)$, and about which the motion of the centre of the epicycle was uniform. (In the case of the theory of the Moon, $T$ plays the
part of the equant, and in the theory of the Sun, $C$.)

In keeping with the well-known correspondence between the simple
heliocentric and geocentric (epicyclic) theories (see, for example,
Neugebauer, op. cit., pp. 122–5), the vector carrying the planet round
the epicycle remains parallel to the vector from $T$ to the mean Sun,
except in the cases of Mercury and Venus, for each of which $\bar{\mu} = \bar{\mu}_{\odot}$.
The sidereal period of the planet about the Sun corresponds, in the

cases of Mercury and Venus, to the period in the epicycle, and in the
cases of Mars, Jupiter, and Saturn, to the period of the epicycle centre
along the deferent.

*Mercury*: Assuming an epicyclic theory of the sort used for the other
planets, Ptolemy found that, like the lunar epicycle, there was a large
apparent variation in the size of Mercury's epicycle, the centre of which
was seemingly closest to the Earth for $\bar{\gamma} \simeq 120°$, $240°$. The explanation
offered by Ptolemy is illustrated in fig. 4. The centre of the epicycle

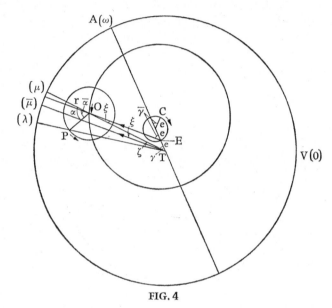

FIG. 4

moves in the usual (direct) sense round a deferent circle whose centre $(C)$
moves in the opposite sense on a fixed small circle of radius $e$ $(= ET)$,
passing through $E$, and with the centre on the line of aux. $E$ is the equant
point, such that the line from it to the centre of the epicycle is parallel
to the line of mean motus. The movement of $C$ is such that the two
angles marked $\bar{\gamma}$ are equal. The line from $E$ to the epicycle centre is
also that from which the mean argument $(\bar{\alpha})$ is measured.

It is easy to see that when $\bar{\gamma} = 120°$ or $240°$ exactly, $EC$ makes an
angle of 60° with the line of aux, and is therefore equal to $e$. The distance
of the centre of the epicycle from $E$ is then a minimum $(60^p - e)$, and it
is *near* these positions $(\bar{\gamma} \simeq 121°, 239°)$ that the minimum distances of the
epicycle from $T$ occur. The matter of geocentric distances is discussed
again below.

*Summary of the notation*: Barred quantities (as, for example, $\bar{\gamma}$, $\bar{\mu}$) are
those which increase at a uniform rate. The suffixes p, l, o, and m

(*propior, longior*, position at distance 60ᴾ, and position at minimum distance) are more fully explained in section 5 below. The suffix '$\odot$' is used of a solar quantity, to distinguish it from the corresponding planetary quantity only when confusion might otherwise be introduced (as in the theory of the Moon). In general, it is assumed that the context is sufficient to determine the planet to which a given term applies.

| | |
|---|---|
| $T$ | Earth |
| $C$ | centre of the deferent circle |
| $E$ | equant point (In the cases of the Sun and Moon, $C$ and $T$ act as equants, respectively.) |
| $E'$ | pole of the mean apogee (Moon only) |
| $A$ | aux of Sun, Moon, or planet |
| $O$ | centre of epicycle |
| $S$ | Sun |
| $M$ | Moon |
| $P$ | planet |
| $V$ | First Point of Aries, the vernal point |
| $r$ | epicycle radius, here taken (as is $e$ below) in ratio to the deferent radius of 60ᴾ |
| $e$ | see figs 1–4. $e$ ($= CT$) may be called the 'eccentricity' of the planet's deferent, but different writers use the term differently. With Mercury, $e$ is equal to the closest approach of $C$ to $T$. |
| $\lambda$ | true place of planet, longitude (along the ecliptic, i.e. ignoring changes in latitude) |
| $\bar{\mu}$ | mean motus, mean longitude of the centre of the epicycle (may be called 'mean argument' in the case of the Sun) |
| $\mu$ | true motus, true longitude of centre of the epicycle |
| $\bar{\alpha}$ | mean argument |
| $\alpha$ | true argument |
| $\omega$ | longitude of aux (*aux propria*) |
| $\bar{\gamma}$ | mean centre, i.e. the elongation of the centre of the epicycle from the line of aux (corresponds to $2\bar{\mu}$ (q.v.) in the case of the Moon) |
| $\bar{\eta}$ | elongation of mean Sun from mean Moon and line of aux ($2\bar{\eta}$, the elongation of the mean Moon from the line of aux, is known as the 'double elongation') |
| $\xi$ | equation of centre (The angle subtended at the centre of the epicycle by the line joining $T$ to the equant or quasi-equant, i.e. by $TC$ (Sun), $E'T$ (Moon), or $ET$ (Mercury, Mars, Venus, Jupiter, Saturn). For the Sun, $\xi$ may be called simply 'equation', or 'equation of argument'.) |
| $\zeta$ | equation of the argument, prosthaphairesis for anomaly |

$d$     distance of the centre of the epicycle ($O$) from the Earth ($T$)
$\theta$     angular diameter of the epicycle as seen from $T$
$\Pi$     proportional minutes
$\Delta$     *diversitas diametri* ($\Pi$ and $\Delta$ may denote several analogous but different functions, according to context. See section 5 below.)
$\delta$     proportional part (*pars proportionalis*; equal by definition to $\Delta . \Pi/60$)
$t$     epoch

## Ptolemy and the auges of the planets

There are at least two sets of aux positions in the *Almagest*, one set embodied in the text, the other—ostensibly more precise—heading the tables of equations (*Almagest* XI. 11). The two sets are as follows:

|        | Text | Epoch of the determining observations (A.D.) | Tables | Text reference |
|--------|------|----------------------------------------------|--------|----------------|
| Sun     | 65° 30'      | 141      | —        | III. 4 |
| Saturn  | 233°         | 137      | 224° 10' | XI. 5  |
| Jupiter | 161°         | 139      | 152° 9'  | XI. 1  |
| Mars    | 115° 30'     | 140      | 106° 40' | X. 7   |
| Venus   | 55°          | 138      | 46° 10'  | X. 2   |
| Mercury | 190° (app.)  | 135, 142 | 181° 10' | IX. 7  |

The differences between the parameters are (except for the Sun, where only one figure is given, and Jupiter, where a scribal error might have been intruded) 8° 50'. This difference corresponds to a precessional movement at the Ptolemaic rate of 1° per century over a period of a little over 883 years, that is to say, the period from the era of Nabonassar to A.D. 137. The aux positions above the tables refer to the former epoch (746 B.C., astronomical reckoning), which is Ptolemy's frequently used standard epoch. The aux of the Sun Ptolemy supposed to have a constant tropical longitude. The Moon's motion is of course totally independent of the motion of the eighth sphere.

## 2. ON DERIVING MEAN MOTIONS ($\bar{\mu}$, $\bar{\alpha}$), USING TABLES

The general method is to add the value of $\bar{\mu}$ or $\bar{\alpha}$ at some fundamental epoch (such values, or *radices*, being separately tabulated) to the increase in the corresponding quantity since that time. Such increments, for minutes, hours, days, months (thirty days in *Almagest*, for example, and calendar months in some of the Alfonsine tables), single years (*anni expansi*), and groups of years (*anni collecti*), are usually tabulated. *Anni collecti* may be grouped in 18s, as in *Almagest*, for example, or as in some versions of the Alfonsine tables they may be added to the fundamental epoch, giving in effect a whole series of radices, as for the last midnight of 1320, 1340, 1360, . . . Since the epochs are not always chosen as the end of a leap year, care must be taken with the motions in *anni expansi*, since as they were conventionally compiled in every fourth (leap) year —usually marked with the letter 'b' for 'bissextile'—the motion is appropriately greater than in a normal calendar year of 365 days.

Another system of reckoning time from the fundamental epoch which must be mentioned is that used in, for example, the European versions (including the early printed versions) of the Alfonsine tables. The time interval is there expressed in a purely sexagesimal system, with the day as unit, i.e. is expressed in terms of *secunda diei* ($60^{-2}$ days), *minuta diei* ($60^{-1}$ days), days, 60 days, $60^2$ days, etc. Tables are usually included to facilitate conversion from conventional reckoning to sexagesimal reckoning. The great advantage of the system is that only a single table of mean motion is required for each of $\bar{\mu}$ and $\bar{\alpha}$. Tables are therefore shorter. No Ptolemaic planetary tables are to be expected to tabulate $\bar{\mu}$ for the planets Mercury and Venus, or $\bar{\alpha}$ for the planets Mars, Jupiter, and Saturn, all of which quantities increase at the same rate as $\bar{\mu}_\odot$. In other words, the period of the centre of the epicycle of the first two, like the period of the planet in the epicycle for the other three, is the tropical year.

### 3. ON CALCULATING THE POSITION OF AUX FROM THE TABLES

As with $\bar{\mu}$ and $\bar{\alpha}$, those who believed the lines of aux of the planets to move with the fixed stars, i.e. to have a steady or trepidational motion with respect to the equinoxes, usually tabulated a *radix augis* for a fundamental epoch, to which radix the amount of precession or trepidation (this being known as *aux communis*) was added. The resulting longitude of aux (that is, $\omega$) was known as *aux propria*.

### 4. ON DETERMINING $\xi$, THE EQUATION OF THE CENTRE, FROM THE TABLES

The quantity is a relatively simple function of a single variable ($\bar{\gamma}$, or $2\bar{\eta}$ in the case of the Moon), given the eccentricity of the deferent circle. It is accordingly tabulated in an obvious way. The only important exception to this rule is in the tables of the *Almagest*, where Ptolemy separated $\xi$ into two components, separately tabulated but put in adjacent columns (column 3 and 4 in all tables). He did so, apparently, for no other reason than to reveal all the more clearly how the equation of the centre was made up. (For a modern trigonometrical expression for $\xi$ see section 7 below.) Theon combined the two columns into one in his version of the *Manual Tables*, and no later tables of any importance reverted to Ptolemy's distinction.

5. ON EVALUATING $\zeta$, THE EQUATION OF THE ARGUMENT,
FROM THE TABLES

The equation of the centre is clearly not only a function of the orientation of the epicycle radius, that is of $\alpha$, but also of the distance of the epicycle from the Earth $(T)$. It is therefore a function of $\alpha$ and $\bar{\gamma}$ (or $2\bar{\eta}$ in the case of the Moon). Since a full set of tables covering the double dependence would have been prohibitively long, Ptolemy and those who followed him adopted the method of tabulating the function $\zeta(\alpha)$ for some standard value of $\bar{\gamma}$, and correcting the values obtained by terms depending on $\bar{\gamma}$. When the epicycle is respectively at greatest distance from $T$ (*longitudo longior*), at least distance from $T$ (*longitudo propior*), and at a distance equal to the deferent radius, the value of $\alpha$ being constant $(a)$, the corresponding values of $\zeta$ will be denoted by $\zeta_1(a)$, $\zeta_p(a)$, and $\zeta_o(a)$. The corresponding distances $(d)$ of $T$ from the epicycle centre $(O)$ will be distinguished by the same suffixes.

There are now three somewhat different methods of using these quantities, the first applying to the Moon, the second to Venus, Mars, Jupiter, and Saturn, and the third to Mercury.

*The Moon*: Clearly $\zeta_1(a)$ is the minimum value of $\zeta(2\bar{\eta}, a)$, and $\zeta_p(a)$ is the maximum. It was usually assumed, following the method laid down by Ptolemy, that, for the general position of the epicycle,

$$\zeta(a) = \zeta_1(a) + \left(\frac{d_1 - d}{d_1 - d_p}\right).(\zeta_p - \zeta_1).$$

The equation will be written more generally here as

$$\zeta = \zeta_1 + \frac{\Pi.\Delta}{60},$$

where general functions are denoted, and where the ratio involving distances is expressed as $\Pi$ in sixtieths—these being termed *minuta proportionalia*. The function $\Pi$ $(= \Pi(2\bar{\eta}))$ is tabulated, following the procedure laid down by Ptolemy. The function $\Delta(\alpha)$, which is also tabulated, is known as *diversitas diametri*, for reasons not yet very obvious. The product $\Delta.\Pi/60$, the final correction to be applied to $\zeta_1$, and here occasionally denoted by $\delta$, was generally known as *pars proportionalis*.

The method of calculating $\Pi$ in terms of distance $(d)$ should be evident from fig. 5.

The assumption at the root of the above procedure for deriving $\zeta$ from $\zeta_1$ is best seen by rewriting the equations as

$$\frac{\zeta - \zeta_1}{\zeta_p - \zeta_1} = \frac{d_1 - d}{d_1 - d_p}.$$

This is not the same as that which was more or less explicitly asserted, without proof, in *Almagest* V. 7. Ptolemy expressed his proportion in terms not of $d$ but of the general angle ($\theta$) subtended at $T$ by the epicycle. For a comparison of the two, and a discussion of their validity, see section 8 below.

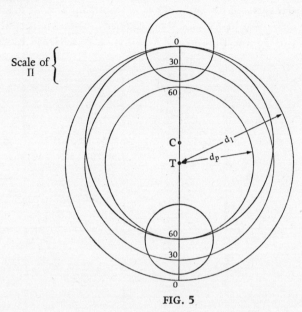

FIG. 5

*Venus, Mars, Jupiter, and Saturn*: Instead of adding an appropriate number of sixtieths of the whole *diversitas diametri* ($\zeta_p - \zeta_1$), the two ranges (one on each side of the line of aux) are each divided into two parts, the point of division being where $d$ is equal to the deferent radius, taken as of 60 units but here for didactic reasons written as $d_o$. Thus on the side of the *longitudo longior* we use the correction equation

$$\zeta = \zeta_o - \Delta_1 . \Pi_1/60,$$

where $$\Delta_1 = \zeta_o - \zeta_1 \quad \text{and} \quad \Pi_1/60 = \frac{d - d_o}{d_1 - d_o},$$

$\Delta_1 = \Delta_1(\alpha)$ and $\Pi = \Pi_1(\bar{\gamma})$ being tabulated as for the Moon. Notice that now $\delta$ is to be subtracted. On the side of *longitudo propior* we use the correction equation $$\zeta = \zeta_o + \Delta_p . \Pi_p/60,$$

where now $$\Delta_p = \zeta_p - \zeta_o \quad \text{and} \quad \Pi_p/60 = \frac{d_o - d}{d_o - d_p}.$$

Notice the change of sign in front of $\delta$. Again $\Delta_p$ and $p_p$ are tabulated, the latter as a continuation of the tables giving $\Pi_1$, which is in point of fact a different function. The method of evaluating $\Pi$ in terms of $d$ should be evident from fig. 6. The column of *minuta proportionalia* traditionally recorded the information '*long*' and '*prop*', according to position.

The justification for these procedures, or the near-equivalents where angles are used instead of distances in computing the values of $\Pi_p$, is not to be found, even in the *Almagest*, except to the extent that there in the case of the Moon the proportionality was at least more or less

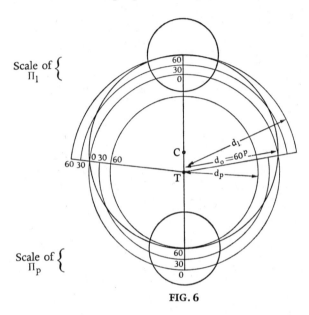

**FIG. 6**

explicit, and clearly meant to appear very plausible. Notice that the procedure used with the planets is not the precise mathematical equivalent of that used with the Moon, since the resulting figure for $\zeta$ is now theoretically exact at four points, rather than only two. A more careful approximation was necessary for the planets owing to the relatively large dimensions of their epicycles. The accuracy of both Ptolemy's and the later tacit hypotheses is discussed below in section 8.

*Mercury*: Although the Moon's deferent circle was not held to be stationary, this did not affect the calculation of $\zeta$ as a function of $2\bar\eta$, since $C$, $E$, and $T$ remain collinear and at a fixed separation. This is not so in the case of Mercury, the centre of whose epicycle describes not a circle with respect to the fixed aux-line, but an oval. This fact is not explicitly stated by Ptolemy, although he was probably not unaware of it. The compilers of the Alfonsine books drew an oval deferent in connection with the equatorium; Richard of Wallingford in *Albion* II. 9 took cognizance of a related oval form; and it was known to Peurbach

(d. 1461) in *Theoricae novae planetarum* (Nuremberg 1472, f. 21). The first thorough mathematical treatment was given by W. Hartner, 'The Mercury horoscope of Marcantonio Michiel of Venice', *Vistas in Astronomy*, ed. A. Beer, vol. i (1955), see especially pp. 109 seqq. Hartner expressed the oval (if the eccentricity had been large it would have been more to the point had we described the curve as pear-shaped, *pinnonada*) in terms of polar coordinates, but taking the *equant* as pole. He did *tabulate* the

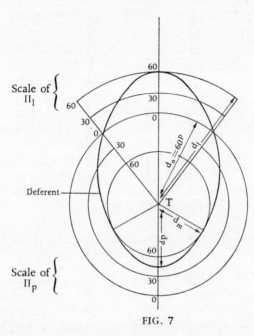

FIG. 7

geocentric distances (*d*; denoted by *s* in Hartner's paper) of the epicycle centre, however, and these are the distances appropriate to any discussion of the amendment of some fundamental set of $\zeta$-values by the method of proportional minutes. Because there are four turning-points in the function $d(\bar{\gamma})$, two of them where $d$ is less than $d_p$, the method of calculating $\delta$ ($= \Delta \cdot \Pi/60$) must be modified slightly (see fig. 7). As before, a division is made where $d = 60$, and the suffix 'o' applies to corresponding quantities. Towards *longitudo longior* there is no difference in procedure, and we write

$$\zeta = \zeta_o - \Delta_1 \cdot \Pi_1/60,$$

where $\qquad \Delta_1 = \zeta_o - \zeta_1 \quad \text{and} \quad \Pi_1/60 = \dfrac{d - d_o}{d_1 - d_o}.$

As before $\Delta_1 = \Delta_1(\alpha)$ and $\Pi_1 = \Pi_1(\bar{\gamma})$ are tabulated. On the side of *longitudo propior*, however, the earlier procedure must be modified, since the least value of $d$ is no longer at the opposite of aux, but at each of two points roughly 120° from aux. Denoting the value of the function $\zeta$ there by '$\zeta_m$', and the geocentric distance as $d_m$, the correction required is that given by the equation

$$\zeta = \zeta_o + \Delta_m . \Pi_p/60,$$

where now     $\Delta_p = \zeta_m - \zeta_o$   and   $\Pi_p/60 = \dfrac{d_o - d}{d_o - d_m}.$

As before, $\Delta_p = \Delta_p(\alpha)$ and $\Pi_p = \Pi_p(\bar\gamma)$ are tabulated. Notice that according to *Almagest* (bearing in mind that Ptolemy worked from angles) $\Pi_p = 39' 28''$, and (as explained in section 1 above) when $\Pi_p = 60'$ the tables make $\bar\gamma$ exactly 120° (or 240°).

(*Note.* To avoid confusion over the fact that, although there is only a *single* possible value of $\Delta$ for a given value of $\bar\gamma$, there are actually *two* $\Delta$ columns (for $\Delta_1$ and $\Delta_p$) running the whole length of the tables, it is perhaps worth pointing out that although $\Pi$ is a function of $\bar\gamma$, $\Delta$ is not, but is rather a function of $\alpha$; and $\alpha$ is not dependent on $\bar\gamma$ in any simple way. It is from the information recorded in the $\Pi$-column that we decide which of the two $\Delta$-columns ($\Delta_1$ or $\Delta_p$) to take.)

## 6. A SUMMARY OF THE PROCEDURE TO BE FOLLOWED IN USING THE TABLES TO DETERMINE THE TRUE PLACE OF A PLANET

The stages in the calculation are numbered, the numbers on the right referring to earlier stages in the calculation from which information is taken. The symbol 'T' indicates that information was taken from a table, or tables, not necessarily without a certain amount of calculation—as in the addition of a radix to such quantities as are taken from tables for groups of years, single years, months, days, etc. Thus the first line of the first summary is to be read: 'The mean motus is to be determined as a function of the epoch by reference to the tables.'

Throughout, angles which are negative in value, or which exceed 360°, are to have 360° added or subtracted respectively.

At certain points in the calculation alternative algebraic signs are indicated. A good set of tables should indicate whether addition or subtraction is called for, but copyists frequently make mistakes in this respect. Here the *upper* sign is to be taken when the argument corresponding to the equation which follows the sign ($\bar\gamma$ or $2\bar\eta$ corresponding to $\xi$, and $\alpha$ corresponding to $\zeta$) lies between 0° and 180°, and the lower sign when the argument lies between 180° and 360°.

Finally, it is important to note that the motion of the eighth sphere (or, as we should say, precession) is *not taken into account*. The reason is that conventions differ between the systems actually found. Ptolemy, for example, gives mean motus in tropical coordinates, while the Toledan tables are given in terms of sidereal.

*Sun*

(1) $\bar{\mu} = \bar{\mu}(t)$      T

(2) $\omega = \omega(t)$      T

(3) $\bar{\gamma} = \bar{\mu} - \omega$      1, 2

(4) $\xi = \xi(\bar{\gamma})$      T

(5) $\lambda = \bar{\mu} \mp \xi$      1, 4

*Moon*

(1) $\bar{\mu}_\odot = \bar{\mu}_\odot(t)$      T

(2) $\bar{\mu} = \bar{\mu}(t)$      T

(3) $\bar{\eta} = \bar{\mu} - \bar{\mu}_\odot$      1, 2

(4) $\xi = \xi(2\bar{\eta})$      T, 3

(5) $\bar{\alpha} = \bar{\alpha}(t)$      T

(6) $\alpha = \bar{\alpha} \pm \xi$      4, 5

(7) $\Pi = \Pi(2\bar{\eta})$      T, 3

(8) $\Delta = \Delta(\alpha)$      T, 6

(9) $\delta = \Delta . \Pi/60$      T, 8

(10) $\zeta_1 = \zeta_1(\alpha)$      T, 6

(11) $\zeta = \zeta_1 + \delta$      9, 10

(12) $\lambda = \bar{\mu} \mp \zeta$      2, 11

*The remaining planets*

(Mars, Jupiter, Saturn)      (1) $\bar{\mu} = \bar{\mu}(t)$      T

(Mercury, Venus)      (1) $\bar{\mu} = \bar{\mu}_\odot = \bar{\mu}_\odot(t)$      T

(Mercury, Mars, Jupiter, Saturn)      (2) $\omega = \omega(t)$      T

(Venus)      (2) $\omega = \omega_\odot = \omega_\odot(t)$      T

     (3) $\bar{\gamma} = \bar{\mu} - \omega$      1, 2

     (4) $\xi = \xi(\bar{\gamma})$      T

(Mercury, Venus)      (5) $\bar{\alpha} = \bar{\alpha}(t)$      T

(Mars, Jupiter, Saturn)      (5) $\bar{\alpha} = \bar{\mu}_\odot - \bar{\mu} = \bar{\mu}_\odot(t) - \bar{\mu}$      T, 1

     (6) $\alpha = \bar{\alpha} \pm \xi$      4, 5

     (7) $\Pi_p = \Pi_p(\bar{\gamma})$ or $\Pi_1 = \Pi_1(\bar{\gamma})$      T, 6

     (8) $\Delta_p = \Delta_p(\alpha)$ or $\Delta_1 = \Delta_1(\alpha)$      T, 6

     (9) $\delta = \Delta . \Pi/60$      7, 8

     (10) $\zeta_0 = \zeta_0(\alpha)$      T, 6

     (11) $\zeta = \zeta_0 + \delta_p$ or $\zeta = \zeta_0 - \delta_1$      9, 10

     (12 $\lambda = \bar{\mu} \mp \xi \pm \zeta$      1, 4, 11

## 7. THE CALCULATION OF PLANETARY TABLES

To some extent this subject was broached in sections 2–5. It is clearly a simple matter to draw up tables for $\bar{\mu}$ and $\bar{\alpha}$ once the mean periods of the planets along their deferents and epicycles have been determined. The tables for $\omega(t)$ were nothing beyond a list of predetermined aux positions (radices) for a given epoch together with (if the astronomer believed that the auges moved with the fixed stars, an idea which a few astronomers rejected) tables for the movement of the eighth sphere. This leaves the tables for $\xi(\bar{\gamma})$, $\zeta_1(\alpha)$, or $\zeta_0(\alpha)$, $\Pi(\alpha)$ and $\Pi_1(\alpha)$, $\Delta_p(\alpha)$ and $\Delta_1(\alpha)$. The following treatment is given in modern rather than historical terms. Except in small matters of approximation, as when a sine is used for the tangent of a small angle, the two approaches yield the same results.

(i) *The equation of centre $\xi$*

*Sun*: A single application of the sine rule in triangle $CTS$ gives the equation
$$\frac{\sin \xi}{e} = \frac{\sin(\bar{\gamma}-\xi)}{60}.$$
This is solved for $\tan \xi$:
$$\tan \xi = \frac{e \sin \bar{\gamma}}{(60+e \cos \bar{\gamma})}.$$
The maximum values of $\xi$ occur where $\bar{\gamma}$ is equal to $(90°+\alpha)$ and $(270°-\alpha)$, the angle $\alpha$ being the least positive value of arc $\sin(e/60)$. These (equal) maximum values of $\xi$ occur when $TS$ is at right angles to the line of aux, and consequently
$$\sin \xi_{max} = \frac{e}{60}.$$

*Moon*: Although the line of aux is a moving line, an expression for $\xi$ is now of greater complexity than that found in the case of the Sun, since $\xi$ is no longer the angle subtended at the deferent circle by the line $CT$. Using the latter angle as an auxiliary angle, applying the sine rule twice, and eliminating it between results, it is found that now
$$\tan \xi = \frac{e \sin 2\bar{\eta}}{\sqrt{(60^2-e^2 \sin^2 2\bar{\eta})}+2e \cos 2\bar{\eta}}.$$
The maximum value of $\xi$ can be found from this equation by differentiation, but it is more easily found by Euclidean methods. It may be shown that the maximum is attained when the centre of the epicycle is at either of the two points of the deferent where a circle through $T$ and $E'$ touches the deferent circle internally. It is then easily shown that
$$\sin \xi_{max} = \frac{60.e}{60^2-2e^2}.$$

The maximum occurs when $2\bar{\eta}$ is equal to $(90°+\alpha)$ and $(270°-\alpha)$, the angle $\alpha$ being the least positive value of

$$\arcsin\frac{2e}{60}\Big/\Big(\frac{1-e^2/60^2}{1-4e^4/60^4}\Big).$$

*Venus, Mars, Jupiter, Saturn*: Applying the sine rule first to triangle $ECO$ (introducing angle $EOC$ as an auxiliary) and then to triangle $COT$, and eliminating angle $EOC$, it is found after some simplification that

$$\tan\xi = \frac{2e\sin\bar{\gamma}}{\sqrt{(60^2-e^2\sin^2\bar{\gamma})}+e\cos\bar{\gamma}}.$$

Again it is most easily proved from considerations of elementary geometry that the maximum values of $\xi$, attained when $CO$ is perpendicular to the line of aux, are given by

$$\tan\Big(\frac{\xi_{max}}{2}\Big) = \frac{e}{60}.$$

The value of $\bar{\gamma}$ then exceeds $90°$ or falls short of $270°$ by the amount $\arctan(e/60)$.

*Mercury*: Of the several ways of deriving an expression for $\xi$ in this case, the following is perhaps the simplest. Joining $CE$ and noting that the angle between it and the (moving) line of aux is equal to $\bar{\gamma}/2$, apply the cosine formula to triangle $CEO$ and the sine formula to triangle $EOT$, on both occasions introducing the side $OE$ into the formulae. Eliminating $OE$ between the equations, after some simplification the general formula for $\xi$ is found:

$$\tan\xi = \frac{e\sin\bar{\gamma}}{eh+\sqrt{(60^2+e^2k)}},$$

where $h = 2\cos\bar{\gamma}+\cos 2\bar{\gamma}$

$\qquad = 2\cos^2\bar{\gamma}+2\cos\bar{\gamma}-1,$

and where $k = \frac{1}{2}(\cos 4\bar{\gamma}+2\cos 3\bar{\gamma}+\cos 2\bar{\gamma}-2\cos\bar{\gamma}-2)$

$\qquad = 4\cos^4\bar{\gamma}+4\cos^3\bar{\gamma}-3\cos^2\bar{\gamma}-4\cos\bar{\gamma}-1.$

I am not aware of a simple geometrical construction for $\xi_{max}$, and an analytical approach involves the solution of a quintic in $\cos\bar{\gamma}$. Hartner adopted a rapid numerical approach to the problem, although for another purpose, namely to calculate the values of $\xi$ via the lengths of the radii vectores $EO$ (op. cit.; see especially pp. 115–16. Hartner's '$\phi-v$' corresponds to our '$\xi$').

### (ii) *The equation of argument*

Although the problem of expressing $\zeta$ as a function of $\bar{\alpha}$ and $\bar{\gamma}$ is in general somewhat more difficult than that of expressing $\xi$ as a function of

$\bar{\gamma}$, in practice it is necessary only to express $\zeta$ as a function of $\bar{\alpha}$, $\bar{\gamma}$, and $\xi$. (This is because, in using the tables, $\xi$ is generally found before $\zeta$.) Doing so is a relatively simple matter, even in the case of Mercury. The method is always the same: to drop a perpendicular from $P$ (or $M$ in the case of the Moon) on $TO$, and to express $\tan \zeta$ as the ratio of this perpendicular to the projection of $TP$ (or $TM$) on $TO$. $TO$ itself is invariably found using the sine rule in an obvious way. In this section, the general result is written first, followed by certain special simplified cases required by the Ptolemaic procedure (that is, whatever is appropriate of $\zeta_p$, $\zeta_1$, $\zeta_o$, $\zeta_m$). For convenience, $\alpha$ ($=\bar{\alpha}\pm\xi$; see section 6 above) is used in the expressions for $\tan \zeta$, except where $\xi$ is zero, when $\bar{\alpha}$ is used. Needless to say, details of maximum values are given only in the special cases, for only they (or combinations of them) were traditionally tabulated. (The Sun, of course, does not have this equation.)

*Moon*

$$\tan \zeta = \frac{r \sin \alpha}{r \cos \alpha + e \, \mathrm{cosec} \, \xi \sin(2\bar{\eta}-\xi)};$$

$$\tan \zeta_p = \frac{r \sin \bar{\alpha}}{r \cos \bar{\alpha} - e + 60}; \qquad \max \zeta_p = \arcsin\left(\frac{r}{60-e}\right);$$

$$\tan \zeta_1 = \frac{r \sin \bar{\alpha}}{r \cos \bar{\alpha} + e + 60}; \qquad \max \zeta_1 = \arcsin\left(\frac{r}{60+e}\right).$$

*Saturn, Jupiter, Mars, Venus*

$$\tan \zeta = \frac{r \sin \alpha}{r \cos \alpha + 2e \, \mathrm{cosec} \, \xi \sin \bar{\gamma}};$$

$$\tan \zeta_p = \frac{r \sin \bar{\alpha}}{r \cos \bar{\alpha} - e + 60}; \qquad \max \zeta_p = \arcsin\left(\frac{r}{60-e}\right);$$

$$\tan \zeta_1 = \frac{r \sin \bar{\alpha}}{r \cos \bar{\alpha} + e + 60}; \qquad \max \zeta_1 = \arcsin\left(\frac{r}{60+e}\right);$$

$$\tan \zeta_o = \frac{r \sin \alpha}{r \cos \alpha + 60}; \qquad \max \zeta_o = \arcsin\left(\frac{r}{60}\right).$$

*Mercury*

$$\tan \zeta = \frac{r \sin \alpha}{r \cos \alpha + e \, \mathrm{cosec} \, \xi \sin \bar{\gamma}};$$

$$\tan \zeta_p = \frac{r \sin \bar{\alpha}}{r \cos \bar{\alpha} - e + 60}; \qquad \max \zeta_p = \arcsin\left(\frac{r}{60-e}\right);$$

$$\tan \zeta_1 = \frac{r \sin \bar{\alpha}}{r \cos \bar{\alpha} + 3e + 60}; \qquad \max \zeta_1 = \arcsin\left(\frac{r}{60+3e}\right);$$

$$\tan \zeta_0 = \frac{r \sin \alpha}{r \cos \alpha + 60}; \qquad \max \zeta_0 = \arc \sin\left(\frac{r}{60}\right);$$

$$\tan \zeta_m = \frac{r \sin \alpha}{r \cos \alpha + d_m}; \qquad \max \zeta_m = \arc \sin\left(\frac{r}{d_m}\right).$$

It is not impossible to express the value of $d_m$, the minimum value of $OT$, analytically in a relatively short space, but it was never expressed in a remotely correct mathematical way in antiquity or in the Middle Ages. It will probably be of greater use if we write down an expression for the value of $d$ ($d_{120}$, say) when $\bar{\gamma} = 120°$, close to the minimum value of $d$:

$$d_{120} = e \sqrt{\left(3 - 3 \cdot \frac{60}{e} + \frac{60^2}{e^2}\right)}.$$

Taking Ptolemy's value for the eccentricity ($e = 3^p$), this gives

$$d_{120} = 55^p \ 33' \ 39''.$$

In fact $d_m$ is less than 2 parts in 10,000 below this figure. (Notice, in calculating $d_{120}$, that $C$, $E$, and $O$ are collinear, and therefore immediately, in this case, $EO = 60 - e$.)

### (iii) *Diversitas diametri*

The functions $\Delta$, defined in section 5 above, are derived from the functions $\zeta_p$, $\zeta_1$, $\zeta_0$, and $\zeta_m$ of part (i) of this section. (Remember that all are functions of $\alpha$, when $\xi = 0$. Here we write the functions $\Delta$ in terms of $\alpha$.)

*Moon*

$$\tan \Delta = \frac{2re \sin \alpha}{60^2 + r^2 - e^2 + 120r \cos \alpha}.$$

*Mercury*

$$\tan \Delta_1 = \frac{3re \sin \alpha}{60^2 + r^2 + 180e + (120 + 3e)r \cos \alpha};$$

$$\tan \Delta_p = \frac{(60 - d_m)r \sin \alpha}{60d_m + r^2 + (60 + d_m)r \cos \alpha}.$$

*Venus, Mars, Jupiter, Saturn*

$$\tan \Delta_1 = \frac{re \sin \alpha}{60^2 + r^2 + 60e + (120 + e)r \cos \alpha};$$

$$\tan \Delta_p = \frac{re \sin \alpha}{60^2 + r^2 - 60e + (120 - e)r \cos \alpha}.$$

(iv) *Proportional minutes*

The functions $\Pi$ are here taken to be derived from expressions for the functions $d$. The values of $d$ were usually calculated in stages, with the evaluation of the length $EO$ (written as $f$ below) as an essential step. Expressing $\Pi$ in a single equation would occasionally lead to a somewhat cumbersome equation, hence the form of some of the following expressions for $\Pi$. (For the definitions of $\Pi$ see section 5 above.) There are obviously many alternative ways of expressing $\Pi$, some of them simpler than these, but obtained less directly. See also the next section.

*Moon*

$$\Pi = \frac{60}{2e} \cdot (60 + e - d)$$

where

$$d = e \cos 2\bar{\eta} + \sqrt{(60^2 - e^2 \sin^2 2\bar{\eta})}.$$

*Venus, Mars, Jupiter, Saturn*

$$\Pi_1 = \frac{60}{e}(d - 60),$$

and

$$\Pi_p = \frac{60}{e}(60 - d),$$

where

$$d = +\sqrt{(2 \cdot 60^2 + 2 \cdot e^2 - f^2)},$$

with

$$f = -e \cos \bar{\gamma} + \sqrt{(60^2 - e^2 \sin^2 \bar{\gamma})}.$$

*Mercury*

$$\Pi_1 = \frac{60}{3e}(d - 60),$$

and

$$\Pi_p = 60 \cdot \left(\frac{60 - d}{60 - d_m}\right)$$

where

$$d = +\sqrt{(e^2 + f^2 + 2ef \cos \bar{\gamma})},$$

with $f$ a root of the equation

$$f^2 - 2ef(\cos \bar{\gamma} + \cos 2\bar{\gamma}) - (60^2 - 2e^2(1 + \cos \bar{\gamma})) = 0,$$

namely    $f = e(\cos \bar{\gamma} + \cos 2\bar{\gamma}) + \sqrt{(60^2 - e^2(\sin \bar{\gamma} + \sin 2\bar{\gamma})^2)}.$

(On the evaluation of $d_m$, see the remarks at the end of part (ii) of this section.)

## 8. THE DISTANCE FUNCTION $(d)$ AND THE VALIDITY OF THE METHOD OF PROPORTIONAL MINUTES

For the application of the method see section 5. To investigate fully the accuracy of the method would require space out of all proportion to its importance, and only an outline of a possible approach will be given here. As already explained, the method of determining $\zeta$ as the sum of some standard function ($\zeta_0$, $\zeta_1$, etc.) and a product of appropriate $\Pi$- and $\Delta$-functions, is equivalent to upholding a whole set of proportionalities, of which a typical example is that for the Moon:

$$\frac{\zeta - \zeta_1}{\zeta_p - \zeta_1} = \frac{d_1 - d}{d_1 - d_p} \quad \text{or} \quad \frac{\theta - \theta_1}{\theta_p - \theta_1}$$

(where $\theta$ is the angle subtended at $T$ by the epicycle). Note that for each of the remaining planets there are two such relations, one for *longitudo propior* and one for *longitudo longior*. None of this applies to the Sun, which is not considered to have an epicyclic motion.

In each case, the acceptance of such an equation is equivalent to maintaining that the ($\zeta$, $d$) graph or ($\zeta$, $\theta$) graph is linear. It does of course also guarantee that the end-points ($\zeta_p$ and $\zeta_1$ in the above example) are accurately represented.

To investigate the maximum error involved in this linear representation, suppose the graph of the function $\zeta(d)$ to be drawn correctly through the end-points, which in the example taken are ($\zeta_1$, $d_1$) and ($\zeta_p$, $d_p$). The error in $\zeta$ is a maximum when the ordinate intercepted between the straight line through the end-points (for which we reserve suffixes '1' and '2') and the theoretical $\zeta$-curve is a maximum, that is to say, when the tangent to the curve is parallel to the line. Regarding $\zeta$ and $d$ as functions of $\bar{\gamma}$ (they are also functions of $\alpha$, of course, but this is taken to be held constant), the condition for maximum error in the curve is

$$\frac{\partial \zeta}{\partial \bar{\gamma}} \bigg/ \frac{\partial d}{\partial \bar{\gamma}} = \left( \frac{\zeta_1 - \zeta_2}{d_1 - d_2} \right).$$

Information as to the interpretation of the suffixes in the cases of the different planets can be extracted from section 5. The functions $\zeta(\bar{\gamma})$ will be found in section 7 (ii). The functions $d(\bar{\gamma})$ have not all been previously written down explicitly. They are:

*Moon*

$$d = e \cos 2\bar{\eta} + \sqrt{(60^2 - e^2 \sin^2 2\bar{\eta})};$$

*Venus, Mars, Jupiter, Saturn*

$$d = \sqrt{(60^2 + 2e^2 - e^2 \cos 2\bar{\gamma} + 2e \cos \bar{\gamma}\sqrt{(60^2 - e^2 \sin^2\bar{\gamma}))}};$$

*Mercury*

$$d = \sqrt{\{60^2 + e^2(c_4 + 3c_3 + 2c_2 + c_1 + 2) + 2e(2c_1 + c_2)\sqrt{(60^2 - e^2(s_1 + s_2)^2)}\}},$$

where                    $c_4 = \cos 4\bar{\gamma}$,      $s_2 = \sin 2\bar{\gamma}$, etc.

Now it should be evident that, even in the case of the Moon, numerical methods promise to be much less intractable than the method outlined. There is a compromise solution, however. Since in the case of most of the planets $\zeta$ varies by a few degrees at most, the left-hand side of such an equation as the first of this section (which is only an example) may be written to a high order of accuracy as $\tan(\zeta-\zeta_1)/\tan(\zeta_p-\zeta_1)$. It is a relatively straightforward matter to show that in all cases this ratio is equal to the right-hand side of the same equation (in $d$), multiplied by a quotient of two expressions, one of which is a function of $d$ (or $\bar{\gamma}$). This 'error factor' may then be considered, for the light it throws on the value of the method of proportional minutes. We shall illustrate the procedure with a single example—one of the planets Venus, Mars, Jupiter, or Saturn, towards *longitudo propior*.

Using the results of section 7 (ii), but occasionally retaining the symbols $d$, $d_o$, and $d_p$, for which other terms have been substituted there, it may be shown that

$$\frac{\tan(\zeta-\zeta_o)}{\tan(\zeta_p-\zeta_o)}\left(\simeq\frac{(\zeta-\zeta_o)}{(\zeta_p-\zeta_o)}\right)=\left(\frac{d_o-d}{d_o-d_p}\right)\cdot\frac{A}{B},$$

where 
$$A=r^2+r(120-e)\cos\alpha+60(60-e),$$

and 
$$B=r^2+r(60+d)\cos\alpha+60d.$$

(Notice that the first term on the right of the equation vanishes when $d=d_o$, as it should, for then $\zeta=\zeta_o$. Notice also that the right of the equation is unity when $d=d_p=60-e$, which shows that the approximation is valid at the other end of the range.) The *actual* error in the ratio on the left (which is almost exactly the actual error in $\zeta$ in radian measure divided by $\tan(\zeta_p-\zeta_o)$) is $(B-A)/A$ times the term on the right. The expression for this error is unwieldy, owing partly to the terms in $\alpha$. Since it should be clear that the error is a maximum when $\alpha$ is at or near $90°$ or $270°$, we can consider these cases. Both give as the (absolute) value of the error in $\zeta$ (radian measure)

$$\frac{60(d_o-d)(d-60+e)r}{(r^2+60d)(r^2+60(60-e))}\quad(\cos\alpha=0).$$

Taking as an example Ptolemy's parameters for the planet Mars ($e=6^p$, $r=39^p\ 30'$), the variation in the error may be investigated analytically. It may be shown that the error reaches a maximum (absolute) value when $d=56^p\cdot95$. Substituting in the formula for the error, we find it in this example to be approximately $3'\ 4''\cdot5$.

The size of this error gives some idea of the value of the method of proportional minutes, but its full significance is not yet completely obvious. It is to be remembered that the error represents the difference between the value of $\zeta$, correctly calculated *ab initio* on Ptolemaic planetary theory (and not from Ptolemy's tables), and the value of $\zeta$ correctly calculated on the basis of the method of proportional minutes but still without recourse to Ptolemy's tables (and with $\Pi$ calculated in terms of $d$ rather than $\theta$). Now calculating first the distance $EO$ ($f$, above) from a knowledge of the value of $d$ for maximum error ($d = 56^{\mathrm{p}}\cdot95$), and using it to obtain a value for $\cos\bar\gamma$ and hence $\bar\gamma$ ($\pm127°$ $32'$ is the angle calculated in this case), we find by the use of formulae in section 7 that (for $\alpha = 90°$)

$$\Delta_{\mathrm{p}} = 2° \; 49'\cdot6,$$

$$\Pi_{\mathrm{p}} = 30\cdot5,$$

and $$\zeta_0 = 33° \; 21'\cdot5.$$

We deduce that, to an accuracy of about half a minute of arc, the value of $\zeta$ as derived by the method is

$$33° \; 21'\cdot5 + \frac{30\cdot5}{60} \times 2° \; 49'\cdot5 = 34° \; 47'\cdot7.$$

Turning now to the exact formula for $\zeta$, we find that

$$\zeta = \mathrm{arc}\,\tan(39\cdot5/56\cdot95)$$

$$= 34° \; 44'\cdot7.$$

This is useful confirmation of our method of appraising the errors implicit in the method, for we have here shown that the error is indeed of the order of $3'$. (We should not have expected to arrive at exactly the same error as before, even had we worked to a greater number of significant figures, since that error was based on the tangent approximation.)

How do these theoretically accurate figures compare with those a medieval astronomer would have obtained from his tables for

$$\bar\gamma = 134° \; 32' \quad \text{and} \quad \alpha = 90°,$$

and how do they compare with what would have been found had $\Pi$ been defined in terms of $\theta$? We consider the latter possibility first, and assume the same values of $\bar\gamma$ and $\alpha$, not that we are in doing so assuming that they still give rise to a maximum error.

From the known distances of the epicycle, and its radius, we find that with the alternative definition of $\Pi_{\mathrm{p}}$

$$\Pi_{\mathrm{p}} = \left(\frac{\theta - \theta_0}{\theta_{\mathrm{p}} - \theta_0}\right).60 = 60\left(\frac{87° \; 49'\cdot8 - 82° \; 19'\cdot2}{94° \; 1'\cdot4 - 82° \; 19'\cdot2}\right) = 28\cdot25.$$

Using the same value of $\Delta_{\mathrm{p}}$ as before, we now calculate $\zeta$ to be

$$33° \; 21'\cdot5 + 1° \; 19'\cdot8 = 34° \; 41'\cdot3.$$

Strange to say, this is about as much below the correct figure as the earlier figure was above it.

Turning now to *Almagest* (Tables, Book XI), we find that by interpolation to $\bar{\gamma} = 127° 32'$ (the left-hand column is given only for groups of 6° (0°–90° and 270°–360°), and for groups of 3° (90°–270°)),

$$\Pi_p = 28° \; 08'\!\cdot\!4 = 28°\!\cdot\!14 \quad \text{(column 8)}.$$

The remaining functions are dependent on $\alpha$ only:

$$\Delta_p = 2° \; 45' \quad \text{(column 7)},$$

$$\zeta_0 = 33° \; 22' \quad \text{(column 6)}.$$

We then calculate $\zeta$ by the method of proportional minutes:

$$33° \; 22' + 1° \; 17'\!\cdot\!4 = 34° \; 39'\!\cdot\!4.$$

The error, of about 5' arc, stems largely from the figure quoted for $\Delta_p$, as may be seen. The fact remains that the method of proportional minutes, in the example taken, even used in conjunction with Ptolemy's slightly inaccurate tables, introduces an error of only a few minutes of arc into the final planetary position.

As explained earlier, Ptolemy defined $\Pi$ in terms of $\theta$, whereas it is elsewhere often treated as though defined in terms of $d$. Tests for the method used in compiling tables of equations are given in the following section.

9. ON THE ANALYSIS OF PLANETARY TABLES, AND, IN PAR-
TICULAR, SIMPLE METHODS OF ABSTRACTING $e$ AND $r$ FROM
TABLES OF PLANETARY EQUATIONS

(We discuss only those tables mentioned earlier. Tables of latitude, for example, are ignored.)

The extraction of mean motions from tables of mean motus should be self-evident. Generally speaking, the larger the unit of time the better. Notice, however, that since tables are usually compiled by repeated addition, mistakes might occur which vitiate the last remark. Notice also the alternation of ordinary and bissextile years, and indeed the use of different calendrical conventions. (For these, see, for instance, Robert Schram's *Kalendariographische und chronologische Tafeln* (Leipzig, 1908); or the much more condensed Göschen booklet (no. 1085) *Zeitrechnung* by D. H. Leitzmann and D. K. Aland (Berlin, 1956).)

Tables of planetary equations, which allow us to deduce values for $r$ and $e$, being more difficult to compile, are useful and interesting to the extent that they tend to change little, and then only at the hands of the more able astronomers. We proceed to give a number of ways in which $r$ and $e$ can be deduced from tables of equations or fragments of such tables. The methods are numbered, the prefixed letters indicating that the method applies to the Sun (S), Moon (L), the planets excluding

Mercury (P), or Mercury (M). At the outset it is worth pointing out that those methods which require a maximum value of some function are generally less reliable than the rest, since we have no guarantee that the actual maximum will occur at any argument to be found in the table of the function, that is, at an integral number of degrees. Notice also that since many tables are drawn up for $3°$ or $6°$ divisions of the argument, there is little point in formulating a method which depends on such a simplification as arises by taking an argument of $45°$; but arguments which are multiples of $30°$ will almost invariably be found.

All the methods which follow are applications of formulae in earlier sections. It is assumed throughout that the radius of the deferent is of 60 units, and the values of $e$ and $r$ which result will therefore not agree with those cited by the author of the tables compiled to another standard, unless suitably adapted.

In the second column below is a list of the data required from the tables, together with abbreviations introduced to simplify the formulae which follow, and to assist in tabulation. (The same abbreviation is used in different rules for comparable, if not identical functions.) Once again it must be repeated that the solutions offered are not always strictly equivalent to results which might be obtained from a textual study.

S.1 $\qquad$ $\xi(90°) = x_1$ $\qquad$ $e = 60 \tan x_1$

S.2 $\qquad$ $\begin{aligned}\xi(\bar{\gamma}) &= x_2 \\ \xi(180° - \bar{\gamma}) &= x_3\end{aligned}$ $\qquad$ $e = \dfrac{120 \operatorname{cosec} \bar{\gamma}}{\cot x_2 + \cot x_3}$

S.3 $\qquad$ $\max \xi = x_4$ $\qquad$ $e = 60 \sin x_4$

L.1 $\qquad$ $\xi(90°) = x_1$ $\qquad$ $e = 60 \sin x_1$

L.2 $\qquad$ $\max \xi = x_4$ $\qquad$ $\dfrac{60e}{(60^2 - 2e^2)} = \sin x_4$

L.3 $\qquad$ $\begin{aligned}\zeta_1(90°) &= z_1 \\ \Delta(90°) &= y_1 \\ \cot z_1 &= b \\ \cot(z_1 + y_1) &= a\end{aligned}$ $\qquad$ $e = 60 \cdot \left(\dfrac{b-a}{b+a}\right)$

L.4 $\qquad$ (as L.3) $\qquad$ $r = \dfrac{120}{(a+b)}$

L.5 $\qquad$ $\begin{aligned}\zeta_1(\alpha) &= z_2 \\ \zeta_1(180° - \alpha) &= z_3 \\ \Delta(\alpha) &= y_2 \\ \Delta(180° - \alpha) &= y_3\end{aligned}$ $\qquad$ $e = 60\left(\dfrac{b-a}{b+a}\right)$

$$\tfrac{1}{2}\sin\alpha(\cot(z_2+y_2)+\cot(z_3+y_3)) = a$$
$$\tfrac{1}{2}\sin\alpha(\cot z_2+\cot z_3) = b.$$

(Note: The same values of $a$ and $b$ should be obtained as in L.3, and the two methods can be combined or compared.)

| | | |
|---|---|---|
| L.6 | $z_1, z_2, z_3$ as before | Substitute for $\alpha$ (e.g. for values $30°$ and $60°$) in the full equation for $\zeta_1$, giving a linear equation in $e$ and $r$. |
| L.7 | $\Delta(90°) = y_1$ | $e^2+2re\cot y_1-60^2-r^2 = 0$ <br><br> (A quadratic equation for $e$ or $r$, the other being known.) |
| L.8 | $\Pi(90°) = p$ | $p = \dfrac{60}{2e}(60+e-\sqrt{(60^2-e^2)})$ <br><br> if $\Pi$ was calculated in terms of $d$ (and not $\theta$). May be used as a test, especially as providing negative evidence, of this fact; or as a quadratic equation for $e$. |
| L.9 | $\Pi(90°) = p$ <br> $\xi(90°) = x_1$ | $e = \dfrac{60^2}{2p+60(\operatorname{cosec} x_1-1)}.$ <br> (With the caveat of L.8.) |
| P.1 | $\xi(90°) = x_1$ | $e = 30\sin x_1$ |
| P.2 | $\max\xi = x_4$ | $e = 60\tan\tfrac{1}{2}.x_4$ |
| M.1 | $\xi(90°) = x_1$ <br> $1+\cot x_1 = a$ | $e = \dfrac{60}{\sqrt{(1+a^2)}}$ |
| M.2, P.3 | $\zeta_0(90°) = z_1$ | $r = 60\tan z_1$ |
| M.3, P.4 | $\zeta_0(\alpha) = z_2$ <br> $\zeta_0(180°-\alpha) = z_3$ | $r = \dfrac{120\operatorname{cosec}\alpha}{\cot z_2+\cot z_3}$ |
| M.4, P.5 | $\max\zeta_0 = z_4$ | $r = 60\sin z_4$ |
| M.5 | $\Delta_1(90°) = y_1$ | $60^2+e(180-3r\cot y_1)+r^2 = 0$ <br><br> (An equation for $e$ or $r$, the other being known.) |
| P.6 | $\Pi(90°) = p$ | $p = \dfrac{60}{e}(\sqrt{(60^2+3e^2)}-60)$ <br><br> (See remarks under L.8.) |

M.6                     $\Pi(90°) = p$            $p = \dfrac{60}{3e}(d-60)$, where
$$d^2 = e^2 + 60^2 - 2e\sqrt{(60^2 - e^2)}.$$
(See remarks under L.8.)

P.7                     $\xi(90°) = x_1$          $e = \dfrac{60^2}{120 \operatorname{cosec} x_1 - p}$
$\Pi(90°) = p$
(See L.8.)

M.7                     $\xi(90°) = x_1$          $e = \dfrac{60^2}{60 \operatorname{cosec} x_1 - 3p}$
$\Pi(90°) = p$

L.10, P.8, M.8     $\max \zeta_p = z_5$         $e = 60 - r \operatorname{cosec} z_5$

(An equation for $e$ or $r$, the other being known.)

L.11, P.9             $\max \zeta_1 = z_6$         $e = r \sin z_6 - 60$
(As above.)

M.9                     $\max \zeta_1 = z_6$         $3e = 60 - r \sin z_6$
(As above.)

## 10. ACTUAL TABLES OF EQUATIONS

When the methods of the last section are applied to the more important tables of equations in the history of Ptolemaic astronomy, the results are surprisingly uniform. This is not the place for a detailed survey, but there are a number of general results worth indicating. I shall denote five important sources by the following sigla: A—*Almagest*; M—*Manual Tables*; B—Al-Battānī's tables as found in the Arabic (ed. Nallino); T—Toledan tables, chiefly using MS. Laud (misc.) 644; O—Alfonsine tables as found in the Oxford versions (MS. Ashmole 1796, etc.). These are essentially the same as are found in versions of the Alfonsine tables in their continental form. To the five sets of tables of equations I could have added an interesting textual source, namely Ptolemy's *Hypotheses planetarum*. This differs from *Almagest* only in the parameters for Mercury, having $e = 2^P 30'$ and $r = 22^P 15'$.

On applying the methods of the last section, two surprising results emerge. First, a superficial application of only the simpler expressions (S.1, L.1, M.1, P.1) yields a remarkably consistent picture, if we approximate our results to the nearest minute. In fact, overlooking the different conventions concerning the definition of the Moon's parameters, and following here the convention of the Canobic inscription (see p. 172), we find that for every parameter there is at least one value shared by all tables except in the values of $e$ for the Sun and Venus. Since the value of $e$ for Venus was always taken to be half the solar eccentricity, this amounts to only one real difference; and we are reminded that the Ptolemaic figure for the solar eccentricity was rather poor.

The other fact appears only when the more involved procedures are applied, and is namely that there are inconsistencies (other than trivial ones which have arisen from miscopying or rounding off) in tables B, T, and O for the Moon and Venus, and in O for Jupiter. In the case of the Moon, for example, the table of proportional parts is incompatible, even in A and M, with the table of equation of the centre. In B, T, and O, the equations of the argument yield different results from the Ptolemaic value followed for the equation of the centre ($12^p\ 23'$ and $13^p\ 39'$ for $e$, as against Ptolemy's $12^p\ 28'$). For Venus, all follow Ptolemy, more or less, in accepting figures for the equation of the argument on the basis of an eccentricity of $1^p\ 15'$; but B and T calculate the equation of the centre with an eccentricity of $1^p\ 02'$ (to the nearest minute), while O takes $1^p\ 08'$. For Jupiter, O follows Ptolemy's $2^p\ 45'$ for the calculation of the equation of the argument, but takes a new value of $3^p\ 07'$ for calculating the equation of the centre.

These conclusions do more than inform us of the thoughtlessness of medieval compilers. They warn us against calculating planetary longitudes by inserting medieval parameters in a mathematically consistent way into the Ptolemaic models pure and simple. (The resulting errors would be at the most $14'$ (tables O) or $25'$ (B and T) in the longitude of Venus, and $42'$ in that of Jupiter (tables O only).) This is not to say that procedures cannot be devised for taking the twin eccentricities into account, of course, but there are inevitably disadvantages in all modern computing procedures which do not involve scanning the tables actually used, with all their blemishes; and even then, a modern calculator must build into his procedures the same techniques for rounding off as his medieval forebears are likely to have used. There are further difficulties, which are discussed in the next section.

The following tables of parameters are included for reference, with qualifications of the sort made above. They do not do full justice to the complete range of tables discussed. In approximating to the nearest minute, we do of course obscure the fact that tables might have been recalculated from slightly different parameters, and we mask the fact that some tables of equations (the solar and lunar tables of B, T, and O) are calculated to seconds, while others are not. One should be sceptical, nevertheless, about textual parameters quoted to seconds—such as those quoted from Albategni by Richard of Wallingford (*Albion*, I, 4). There is little evidence that tables of equations were recalculated on the basis of such pretentiously accurate parameters; nor, for that matter, is there any evidence of recalculation with the eccentricity of $6^p\ 30'$, often

quoted for Mars in the Middle Ages. (See p. 142, vol, ii; also Campanus
of Novara, and other writers in the tradition of *theoricae planetarum*.)
Even had the Albategni radii been used to calculate a new table of
equations, however, the greatest effect they would have had on predicted planetary longitudes would never have been large. (For Saturn,
the maximum effect would have been 42″; for Jupiter 15″; for Mars
4′ 30″; for Venus 15″; and for Mercury 2″.)

| Table of derived eccentricities | | | | Table of derived epicycle radii | | |
|---|---|---|---|---|---|---|
| Planet | Tables | p | ′ | Planet (all tables) | p | ′ |
| Sun | A M | 2 | 30 | Moon | 6 | 20 |
| | B T | 2 | 05 | Mercury | 22 | 30 |
| | O | 2 | 16 | Venus | 43 | 10 |
| Moon | all | 12 | 28 | Mars | 39 | 30 |
| | B T | 12 | 23 | Jupiter | 11 | 30 |
| | O | 13 | 39 | Saturn | 6 | 30 |
| Mercury | all | 3 | 00 | | | |
| Venus | A M O | 1 | 15 | | | |
| | B T | 1 | 02 | | | |
| | O | 1 | 08 | | | |
| Mars | all | 6 | 00 | | | |
| Jupiter | all | 2 | 45 | | | |
| | O | 3 | 07 | | | |
| Saturn | all | 3 | 25 | | | |

## II. ADJUSTMENTS FOR THE MOTION OF THE EIGHTH SPHERE

In deriving a planetary longitude from the tables in the manner
explained earlier in this Appendix, it is important to know whether the
end result is a sidereal or a tropical longitude. The difference, namely
the so-called 'motion of the eighth sphere', has been discussed in connection with Thebit's theory of trepidation in Appendix 25. In order to
derive a tropical longitude from the Toledan tables, it is necessary only
to add a quantity such as is calculated in the example at the end of that
Appendix. The mean motions derived from the tables of Ptolemy, on the
other hand, are tropical. They already incorporate a steady secular
motion of the eighth sphere (of one degree per century), and no final
adjustment is necessary. The Alfonsine mean motions include a secular
term of 26″·45, but it is necessary to make a final adjustment by adding
a periodic term somewhat analogous to that in Thebit's theory (see
Appendix 25).

The resolution of the movement of the eighth sphere into two parts
in the Alfonsine tables is done on very curious lines. The secular term

seems to have been chosen purely and simply to equalize the mean motion of the actual Sun and that of the Sun in the Julian calendar. It gives the Julian calendar, therefore, a very peculiar status. The figure of 49,000 years often quoted in connection with the Alfonsine theory is the time taken by the secular movement to cover 360°. We may derive it as follows. The length of the year as quoted in Alfonsine astronomy is $365\ 5^h\ 49^m\ 15^s\ 58^t\ 49^{iv}\ 46^v\ 26^{vi}$. This falls short of a Julian year of 365·25 days by an amount $0^d \cdot 0074539295$, and an Alfonsine calendar thus diverges from a Julian calendar by $1^d$ per 134·157426 tropical years, on this reckoning. Multiplying this figure by the length of the year in days gives the period, in tropical years, for the difference to amount to a complete revolution of the ecliptic, and the result turns out to be different from 49,000 years only in the fifth decimal place. Apparently the length of the tropical year was found approximately, and then rounded off to give this remarkably harmonious periodicity. (The number 49 was endowed with special powers in the eyes of medieval man. Is it a mere coincidence that the chapter in which the tables of the eighth sphere are explained is numbered 49?) It looks almost as though we have found two respects in which the Alfonsine astronomers gave God advice on the ordering of the universe.

The non-secular term in Alfonsine astronomy for the motion of the eighth sphere is of the form $9° \sin A$, where $A$ has a periodicity of 7,000 Julian years and was last zero at a radix date of A.D. 16 May 15·0 (a date supplied by Dr. Mercier). The value of the periodic term on Julian Day $N$ may thus be expressed $9° \sin(0 \cdot 0001408038(N-1727038))$.

If the mean motus extracted from the tables is tropical, then the values accepted for the aux positions must likewise be tropical, and if the one is to be sidereal then also the other. If a mean motus and an aux position are not compatible in this sense, then there will be errors in the equations, which are calculated in terms of the difference. In the printed Alfonsine tables the movement of the eighth sphere from the time of the radix chosen (*radix incarnationis*) must be added to the radix to give the aux of date (*aux propria*). In the Oxford tables there are tables in which the *aux propria* for each planet is given at intervals of 20 years. There is no allowance for any independent movement of the auges. Once the *aux propria* has been found, it must be subtracted from a tropical mean motion in longitude if it is to give the correct equations. The tabulated mean motions, however, do not include the periodic term—nor indeed could they be easily made to do so. This periodic term seems to have been quite wrongly ignored by those who compiled and used the tables, and this is another point at which a modern calculator, trying to reproduce medieval results, should avoid complete consistency.

Ptolemy's coordinates are tropical, both for auges and mean motions, but since there is no complication due to a periodic term, consistency is easily achieved. The auges are taken to move with the eighth sphere (1° per century), and the same movement is incorporated into the mean motions.

The Toledan tables use sidereal coordinates, and therefore the equations will be found from a mean centre calculated simply as the difference between aux position at the radix date and mean motus on the day in question; for this mean motus makes no allowance for the movement of the eighth sphere.

It is often said that the Toledan tables were replaced by the Alfonsine when it became clear that the inaccuracies of the former had become intolerable. When the two are compared, however, for fourteenth- and fifteenth-century dates, the results they give are usually very close if, in conjunction with the Toledan tables, a figure is accepted for the movement of the eighth sphere which is based on some other method than that laid down within the tables themselves. The positions of the equinoxes begin to diverge appreciably from those derived from the Toledan model especially after 1300, whereas the Alfonsine model follows reality more closely. (In 1400 the models were in error by about $1\frac{3}{4}$ degrees and $\frac{1}{2}$ degree respectively.) Despite this fact, there is ample evidence in the form of surviving manuscript copies of the Toledan tables from the fourteenth and fifteenth centuries that the earlier tables did not disappear from the scene at all quickly. One would like to know more about the practice of medieval calculators in regard to the adjustment for precession, for it seems not at all unlikely that an astronomer working out planetary longitudes would have operated over an appreciable period of time with a 'private' value for the movement of the eighth sphere, and that this value might have been more accurate than the value derived from the 'official' model. Many tables of star coordinates, for example, incorporate better values for the motion of the eighth sphere than would be expected on the basis of either Toledan or Alfonsine parameters, and it is not unlikely that at least some astronomers used similar values in planetary work.

## 12. ADJUSTMENTS FOR GEOGRAPHICAL POSITION

The tables for planetary longitude discussed here are appropriate to a definite geographical longitude, whether of Alexandria, Toledo, Cremona, Oxford, or elsewhere. Tables of mean motions and equations will be the same for all places, but the radices of mean motus and (in principle but hardly in practice) of aux position will differ as between different towns, for the simple reason that they are specified for a particular local time (midnight or noon) rather than on a scale of universal time. Conversion from one local time to another is done at the rate of

$1^h$ per $15°$ of longitude difference. Radices are then altered by adding or subtracting the movement appropriate to the time interval found. Thus radices for the *era regis Alfonsi* found in the printed versions of the Alfonsine tables correspond to the time of noon on the coronation date of 1 June 1252 (Julian Day number 2178503), while the Oxford Alfonsine tables give radices for (for example) 1320 complete, i.e. for the noon of 31 December 1320 (J.D. 2203553). Taking the motion in argument of Mercury, the printed tables give a radix of $213° 48' 39''$ (to the nearest second). The movement corresponding to the interval between the radices brings this up to $279° 48' 35''$, while the Oxford radix is $279° 46' 34''$. The difference of $2' 01''$ is covered (at a rate of $7' 46''$ per hour, approximately) in about $15^{m}\cdot6$. As William Rede says in his preface to the canons, he takes Oxford to be $4°$ removed from Toledo in longitude, 'quibus correspondent 16 min. unius hore'. Not all the Oxford adjustments were made so carefully, however.

# APPENDIX 30

*Series expansions for the equation of days and the modern
equation of time, assuming a simple eccentric solar orbit*

THE equation of time $E$ may be written $(\bar{\mu}-\alpha)$, with our usual notation
(see *Albion* II. 10. comm., etc.). The table of iomyn (*Albion* IV. 8) com-
bines this equation $E$ with the argument of the table, namely with $\lambda$, but
here we shall restrict our attention to $E$ alone. We may write

$$E = (\lambda-\alpha)-(\lambda-\bar{\mu}),$$

the first bracket containing the so-called 'reduction to the equator' and
the second the equation of the centre, $\xi$. The reduction to the equator
may be expressed as a trigonometrical series in $\lambda$. From an application
of the four-parts formula to the spherical triangle vernal point–Sun–
equatorial Sun, it is found that

$$\tan \alpha = \cos \epsilon \tan \lambda,$$

where $\epsilon$ is the obliquity of the ecliptic. Expressing the tangent functions
in terms of exponentials,

$$\frac{\exp(2i\alpha)-1}{i(\exp(2i\alpha)+1)} = \frac{\exp(2i\lambda)-1}{i(\exp(2i\lambda)+1)} . \cos \epsilon.$$

Writing $\cos \epsilon$ in the form $\left(\frac{1-t^2}{1+t^2}\right)$, where $t = \tan\frac{\epsilon}{2}$, and solving the last
equation for $\alpha$ (by solving for $\exp(2i\alpha)$ and taking logarithms),

$$2i\alpha = 2i\lambda+\log(1+t^2\exp(-2i\lambda))-\log(1+t^2\exp(2i\lambda)).$$

Using the logarithmic expansion, and observing that the series con-
verges, since $t$ is less than unity (it is of the order of 0·04),

$$\alpha = \lambda-t^2 \sin 2\lambda+\frac{t^4}{2} \sin 4\lambda-\frac{t^6}{3} \sin 6\lambda+...$$

(angles $\alpha$ and $\lambda$ in radian measure). It is easily verified that the maximum
value of the last term is of the order of only 5 seconds of arc (little more
than a third of a second of time).

We now proceed to express $E$ in terms first of $\bar{\gamma}$, then of $\bar{\mu}$, and finally
of $\lambda$. Turning to the equation of the centre, this is given by the equation
(see Appendix 29)

$$\tan \xi = \frac{e \sin \bar{\gamma}}{60+e \cos \bar{\gamma}}.$$

Expressed in exponentials in much the same way as before, and after some reduction, this is found to be equivalent to

$$\exp(2i\xi).\left(1+\frac{e}{60}\exp(-i\bar{\gamma})\right)=\left(1+\frac{e}{60}\exp(i\bar{\gamma})\right).$$

Taking logarithms, expanding in a (converging) series, and collecting terms into trigonometrical form,

$$\xi=\frac{e}{60}\sin\bar{\gamma}-\frac{1}{2}.\frac{e^2}{60^2}\sin 2\bar{\gamma}+\dots$$

(This could have been written down immediately to the required approximation, by expanding the right-hand side of the equation for $\tan\xi$, had we simultaneously justified the approximation $\tan\xi=\xi$.)

Combining the series for $\xi$ with that for $(\lambda-\alpha)$ is done preferably after $\lambda$ in the latter has been replaced by $(\bar{\mu}-\xi)$, $\xi$ being taken from the last equation. As stated earlier, $t^2$ is about 0·04, and $\frac{e}{60}$ is of the same order of magnitude. Keeping terms only of the second degree in these quantities, the expansion for $(\lambda-\alpha)$ becomes

$$t^2\sin 2\bar{\mu}-\frac{2t^2.e}{60}\sin\bar{\gamma}\cos 2\bar{\mu}-\frac{t^4}{2}\sin 4\bar{\mu}-\dots$$

Combining, finally, with the last expression for $\xi$, and simultaneously writing $\bar{\mu}$ in the form $(\omega+\bar{\gamma})$, where $\omega$ is, as before, the longitude of aux, we find that

$$E=\frac{e}{60}\sin\bar{\gamma}-\frac{e^2}{2.60^2}\sin 2\bar{\gamma}+t^2\sin 2(\omega+\bar{\gamma})-$$
$$-\frac{2t^2e}{60}\sin\bar{\gamma}\cos 2(\omega+\bar{\gamma})-\frac{t^4}{2}\sin 4(\omega+\bar{\gamma})+\dots$$

This may be written as a sum of sines and cosines of multiples of $\bar{\gamma}$, with constant coefficients:

$$E=a'\sin\bar{\gamma}+b'\cos\bar{\gamma}+c'\sin 2\bar{\gamma}+d'\cos 2\bar{\gamma}+f'\sin 3\bar{\gamma}+$$
$$+g'\cos 3\bar{\gamma}+h'\sin 4\bar{\gamma}+l'\cos 4\bar{\gamma}.$$

The coefficients are then (giving angles in radians):

$$a'=\frac{e}{60}\left(1-t^2\cos 2\omega\right),$$

$$b'=\frac{t^2e}{60}.\sin 2\omega,$$

$$c'=\frac{-e^2}{2.60^2}+t^2\cos 2\omega,$$

$$d' = t^2 \sin 2\omega,$$

$$f' = \frac{-t^2 e}{60} \cos 2\omega,$$

$$g' = \frac{-t^2 e}{60} \sin 2\omega,$$

$$h' = \frac{-t^4}{2} \cos 4\omega,$$

$$l' = \frac{-t^4}{2} \sin 4\omega.$$

It is, however, more convenient, and more in keeping with current astronomical practice, to express $E$ as a function of $\bar{\mu}$, since this angle is measured from the equinox. (When this practice began I cannot say, but it is found in al-Khwārizmī (Suter ed., Tables 67–8) and is of course equivalent to the practice of tabulating $E$ as a function of the time of year (ignoring precession and the movement of apogee).) Commencing from the penultimate equation for $E$, replacing $\bar{\gamma}$ by $(\bar{\mu}-\omega)$, and suitably expanding and collecting terms, it will be found that

$$E = a'' \sin \bar{\mu} + b'' \cos \bar{\mu} + c'' \sin 2\bar{\mu} + d'' \cos 2\bar{\mu} +$$
$$+ f'' \sin 3\bar{\mu} + g'' \cos 3\bar{\mu} + h'' \sin 4\bar{\mu},$$

where the coefficients are as follows:

$$a'' = \frac{e}{60} (1+t^2)\cos \omega,$$

$$b'' = \frac{-e}{60} (1-t^2)\sin \omega,$$

$$c'' = t^2 - \frac{e^2}{2 \cdot 60^2} \cdot \cos 2\omega,$$

$$d'' = \frac{e^2}{2 \cdot 60^2} \cdot \sin 2\omega,$$

$$f'' = \frac{-t^2 e}{60} \cdot \cos \omega,$$

$$g'' = \frac{t^2 e}{60} \cdot \sin \omega,$$

and

$$h'' = \frac{-t^4}{4}.$$

Finally, we express $E$ in terms of $\lambda$. Although it is not usually necessary to express $\xi$ in terms of $\lambda$, this is easily done directly by an application of the same rule to the triangle Earth–equant–Sun, giving

$$\sin \xi = \frac{e}{60} . \sin(\lambda - \omega).$$

Since $\dfrac{e}{60}$ is comparable in magnitude with $t^2$, we may ignore its cube, and write

$$\xi = \left(\frac{e}{60} \cos \omega\right)\sin \lambda - \left(\frac{e}{60} \sin \omega\right)\cos \lambda.$$

Combining this with the expression for $\alpha$ in terms of $\lambda$, we find the equation of time as a function of $\lambda$, to the required order of accuracy:

$$E = a \sin \lambda + b \cos \lambda + c \sin 2\lambda + h \sin 4\lambda,$$

where
$$a = \frac{e}{60} \cos \omega,$$

$$b = \frac{-e}{60} \sin \omega,$$

$$c = t^2,$$

and
$$h = \frac{-t^4}{2}.$$

With obvious reservations, any one of the three approximate expressions for $E$ might in principle be required in the analysis of a tabulated equation of time, assuming that the algebraic sign of $E$ be held open to change, and a constant be added to take into account the different conventions alluded to by Ibn Yūnus (see *Albion* IV. 8. comm. above). An examination of the three sets of coefficients (see also below) will show that a choice between two of the three formulae can scarcely be made on the grounds of a table alone. Textual evidence is required to ascertain which independent variable is in use. In Albategni's case, we know that $\lambda$ was the independent variable; but that there is another approach to the problem may be shown by considering his table as an example, that is to say, by taking our *Tabula iomyn* of *Albion* IV. 8, from each entry of which the tabular argument is supposed, for convenience, to have been subtracted, giving a true equation of time (see the commentary on *Albion* II. 10). We may form sets of simultaneous equations for the coefficients, assuming no terms beyond those of the three series found above, and disregarding all likely theoretical forms for those coefficients. As it happens, the first three coefficients (apart from an added constant which we shall call $K$) are very much more significant than the rest. Without making any rigorous numerical analysis, for example, it is easily deduced from the values of $E$ assumed when the argument of the table (which from the text we happen to know to be $\lambda$) is $0°$, $90°$, $180°$, and $45°$, that the first three coefficients and the constant are approximately $A = 15'(\cdot 5)$, $B = -117'$, $C = 148'$, and $K = 246'$. (where $A$ stands for $a$, $a'$, or $a''$, according to the series anticipated, and similarly for $B$, $C$, etc.) The additional assumption that $H$ is the only other significant coefficient does not alter these values, and from the tabular entry for argument $30°$ we find $H = -2'\cdot 7$, while from that for argument $60°$ we find $-4'\cdot 4$. With the advantage of foresight we might settle on a round figure of $3'$. Adopting these coefficients, ignoring other terms, and calculating intermediate values, the stated function fits the table tolerably closely. (As an example, when $\lambda = 120°$, $E$ is predicted as $3°\ 7'$, whereas Albategni has $3°\ 5'$. It must be emphasized here that we are not aiming at understanding his techniques of computation, for which his treatise is preferable and sufficient.)

Turning to the theoretical expressions for the coefficients, we may next utilize Albategni's figures for $\epsilon$ ($23°\ 35'$), $e$ ($1°\ 59'$), and $\omega$ ($82°\ 14'$), to derive ideal values for the coefficients. They are:

|  | $A$ | $B$ | $C$ | $D$ | $F$ | $G$ | $H$ | $L$ |
|---|---|---|---|---|---|---|---|---|
| $E(\bar{\gamma})$ | $118'\cdot 41$ | $1'\cdot 33$ | $-146'\cdot 22$ | $4'\cdot 01$ | $4'\cdot 77$ | $-1'\cdot 33$ | $-2'\cdot 80$ | $-1'\cdot 68$ |
| $E(\bar{\mu})$ | $16'\cdot 03$ | $-107'\cdot 69$ | $151'\cdot 63$ | $0'\cdot 50$ | $-0'\cdot 67$ | $4'\cdot 91$ | $-1'\cdot 63$ | |
| $E(\lambda)$ | $15'\cdot 36$ | $-112'\cdot 59$ | $149'\cdot 81$ | $0$ | $0$ | $0$ | $-3'\cdot 26$ | |

(These figures may be converted to seconds of time by multiplying by 4.) As will be seen, $A$ and $C$ for the last two series are close to the values derived from the table, but the values they suggest for $B$ differ more noticeably, and, of the two, the formula we know to be correct (for $E(\lambda)$) provides the better fit. It will also be appreciated that, had the middle formula been the correct one, we should not have been justified in claiming accurate results while ignoring $D$, $F$, and $G$.

Finally, the coefficients deduced from the table allow us in principle to derive values for the assumed obliquity, aux position, and even solar eccentricity. These cannot be more than approximate. Thus the obliquity is found from $C$ to be about $23°\ 27'$, and the eccentricity about $2°\ 12'$, neither, perhaps, to be taken too seriously. (There is doubtless no significance in the proximity of $2°\ 12'$ and $2°\ 14'$, al-Bīrūnī's maximum solar equation, which corresponds to an eccentricity of about $2°\ 20'$.) It is of interest to see, however, that from the quotient of $A$ and $B$ a longitude of apogee of $82°\ 27'$ is derivable. This is very close indeed to the figure quoted by Albategni ($82°\ 14'$), suggesting that he did not simply borrow the table from an earlier source unless he borrowed both. At the same time it reminds us that the occurrence of $\omega$ in the expressions for the coefficients meant that an astronomer like Richard of Wallingford who accepted the table at a much later date was involved in a venial form of inconsistency.

# APPENDIX 31

## Simple theories of the tides in the Middle Ages

THE subject of the tides is of some interest to us, especially since the clock at St. Albans is known from Leland to have had a tidal dial (see vol. ii, p. 366). There is a widespread belief, it seems, that the first tidal dials were of the seventeenth century. H. A. Lloyd, for example, in a discussion of dials by such makers as Thomas Tompion and Samuel Watson, even goes so far as to argue that the tidal dial on the Hampton Court clock cannot be the original one since it was Kepler who first suspected the influence of the Moon's mass on the tides ('Tides and the Time', *Horological Journal*, December 1950). As it happens, Leland was writing on the subject of the St. Albans clock before the Hampton Court clock was erected. But many of the principal tidal phenomena were known in the ancient world; so many, in fact, that it would be impossible to survey the subject in a short space. There are three main periodicities in tidal phenomena, namely the daily ebb and flow, a monthly cycle following the synodical movement of the Moon, and an annual periodicity, producing tide-maxima near the equinoxes. Classical and medieval writers were aware even of the last of the three components. The number of references in classical literature is surprising, bearing in mind the fact that the Greeks, Alexandrians, and Romans, lived on the shores of a sea virtually without tides. The influence of the Moon was appreciated at an early date. Pytheas (fourth century B.C.) mentioned it, and Strabo has it that Seleucus the Babylonian (*c.* 150 B.C.) had correlated periodical inequalities in the tides of the Red Sea with the position of the Moon in the zodiac. This was contrary to the beliefs of Aristotle, who accorded the Sun prime importance. (See Sir Thomas Heath's *Aristarchus of Samos* (1913), p. 307, which follows Schiaparelli's *I precursori*.) At the end of the same century as Seleucus, Posidonius explained the spring and neap tides, introducing the idea of a joint action of the Sun and Moon. (See Gunnar Rudberg, *Forschungen zur Posidonios* (Uppsala, 1918).) The notion of the Moon's importance was never lost to view in the Middle Ages: it is to be found embodied, with greater or lesser acumen, in the writings of St. Basil, Bede, Grosseteste, and Roger Bacon, not to mention a pseudo-Aristotle in the Islamic world (whose work came to be known in the West as *Liber de elementis*), and of many other less influential writers. It is highly probable that

Richard of Wallingford was familiar with the teachings of the Mertonian commentator on Aristotle, Walter Burley, to whom is ascribed an influential treatise *De fluxu et refluxu maris Anglicani*. (Inc.: 'Descriptis his figuris circa . . .' See A. G. Little, *Roger Bacon Essays* (1914), p. 419.)

Discussion of the tides in the Middle Ages was carried on at three quite different levels. On the one hand there was the philosophical problem of tidal causation, which admittedly demanded detailed information about the phenomenon to be explained, but which was not concerned with the mundane problem of calculating future tides. Into this philosophical category we should no doubt place Walter Burley, or whoever was responsible for the treatise *De fluxu et refluxu*, and a number of writers who relied on it, including Jean Buridan. Having something in common with their neo-Aristotelian treatment are the writings of those who followed the astrologers, and especially Albumasar, for which tradition see vol. ii, p. 98. But the third tradition is much more homespun, and one encounters it so often in the manuscripts of the thirteenth to the seventeenth centuries that one need not have serious doubts as to the general character of Richard of Wallingford's tidal dial. This tradition goes one better, as one might say in the words of the treatise of doubtful authorship discussed on p. 384, vol. ii, than a determination of tides by examining shells and seaweed. On the other hand, it does not go as far as the eight-fold cause of the tides discussed in connection with Albumasar and *Exafrenon*. In a nutshell, one may say that it gives flood tides in the form 'a constant number of hours plus the product of 48 minutes and the age of the moon in days, counting in integers from 1 to 30'. The rule is never so explicit, of course, but this statement does justice to the tables for flood tide at London Bridge which are to be found in British Museum, MS. Cotton Jul. D. vii, f. 45$^v$, where the port constant was three hours. This was a manuscript written in the thirteenth century which actually belonged to the abbey of St. Albans, and therefore we may conjecture that the tidal dial there showed the tides not at the Wash, or King's Lynn, or Dover, but at London Bridge. (Some of the manuscript is by John of Wallingford, an abbot who died in 1213.)

Similar rules are to be found on instruments and in tables throughout the following centuries. Dante might write on the natural philosophy of tides, but Chaucer, if we are to accept *Treatise on the Astrolabe*, chapter 46, as his, was satisfied with a rule rather more vague than that given above, but adapted to the astrolabe. In the Museum of the History of Science at Oxford there is an horary instrument with the name of Roger Brechte and the date 1537 (but the instrument might be older) which carries scales of tides. (The instrument is on loan from St. John's College. The 48 minutes parameter is of course unchanged. There are two series of figures, one with a constant of 12 hours, rather than the 3 hours of the

London tables, and the other with a constant of 6 hours 12 minutes. In principle one should be able to guess at the port, from these tidal constants, but local conditions may change drastically over the centuries.) Taking another example, the last item of MS. Ashmole 191, ff. 197ᵛ–211ʳ, written I suspect around 1445; there was a volvelle on the last folio (a parchment leaf, unlike the paper of the remainder), but the moving parts are now missing. Nevertheless, at f. 199ʳ there is a full explanation of the 'rewle of the volvelle', in English. I quote the opening words and the end:

Now folowith here the volvelle that summen clepen a lunarie . . . And thanne loke undir the tunge of the Moone and there ye schulen fynde writen in what signe he sittith, and the Sunne also; and in what tyme of the day thei arisen eny of hem, either goon doon; and what it is of the watir, whether it be floo or eb. For schulen undirstonde that a southwest Moone and a northeest Moone maken an high flood Lundoun brigge, but if caseweltees of the wynde make it otherwhile, and that is whan the wynde blowith so feersly aghens the watir, that it may not have his kyndely cours as it oughte.

The writer goes on with eclipses.

This passage echoes a principle which was exemplified in an Oxford manuscript of the fourteenth century, MS. Jesus College 46. The volume as a whole is filled with material relating to William Rede of Merton College. At f. 2ʳ there is a compass rose, with north to the left and east uppermost, as in early maps, and with 'ebbe' and 'flud' marked on it, ebb being along the directions south-east and north-west, and flood along north-east and south-west. No doubt there are many other comparable tidal volvelles and dials to be found among the host of lunaria which are a commonplace in medieval books. To them we should no doubt look for the outward appearance of the St. Albans tidal dial.

# APPENDIX 32

## *An outline theory of lunar eclipses*

In order to put those parts of *Albion* dealing with eclipses into better perspective, we may adopt a more analytical point of view. In many respects, a modern treatment is little different from Ptolemy's. A lunar eclipse occurs when the Moon enters the shadow cast by the Earth in the light of the Sun. The penumbra was not taken into account until after the Middle Ages, and we ignore it here. A lunar eclipse can obviously only occur near full moon, and would occur every full moon were the orbital planes of the Sun and Moon (with reference to the Earth) inclined at a sufficiently small angle. Their inclination is about 5°. The Moon has an apparent diameter of only about half a degree, which is rather less than a third as great as the diameter of the Earth's shadow at this distance. Opposition must therefore occur when the Moon is sufficiently near a node, if an eclipse is to take place. How near the Moon must be, at true opposition, is not merely a geometrical, but a kinematical, problem. It is a problem which requires not only the geometry of the sphere, but a knowledge of the spatial disposition of Sun, Moon, and Earth. Atmospheric refraction is another important factor in any accurate prediction of an eclipse, but is overlooked here.

As the first result of an eclipse calculation, it is shown whether there will be an eclipse, and if so at what time, whether it will be total or partial, its duration in the different phases, and its magnitude. Ptolemy's definition of eclipse magnitude has been accepted until modern times. The diameter of the luminous disc of the Moon is divided into twelve units ('digits'). The number of digits obscured of a diameter at right angles to the path of the shadow across the Moon's face is then the eclipse magnitude. If we denote the angular semi-diameters of shadow and Moon, respectively, by $u$ and $m$, and if the number of digits eclipsed be $q$ when the angular separation of the centres is $d$, then

$$q = 12(u+m-d)/2m.$$

It is worth noting that Ptolemy preserved the continuity of the magnitude function to beyond the stage where the Moon is fully eclipsed, and did not end his tables of eclipse magnitude at 12 (cf. *Almagest* VI. 8). Ptolemy pointed out that most observers of eclipses tended to give their impression of obscuration in terms of areas. He accordingly offered a second definition of eclipse magnitude, and gave a table for conversion between the two sorts. There is nothing to be gained here by giving both alternatives.

Using the subscripts m, s, and n, for the Moon, Sun, and node ('head or tail of the Dragon'), and our usual notation for longitude ($\lambda$), we begin by assuming that the longitudes of all three are known, and likewise the rates of change of the first two. The rate of change of the longitude of the node is so slight as not to be worth considering during the eclipse itself, the nodes moving round the sky as they do once in about 18·6 years. The angle between the orbits will be denoted by $i$, the centre of the Earth's shadow on the sphere at the Moon's distance by $U$ (moving along the ecliptic at longitude $\lambda_s + 180°$), and $U'$ will be the position of the shadow's centre when the Moon is at the node ($N$). We now require the maximum value of $NU'$ for an eclipse to be possible.

Since $i$ is small, and this value of $NU'$ will not be a very large fraction of $90°$, we can look upon the triangle $MUN$ as plane. The steadily increasing distances $MN$ and $UU'$ are in an almost constant ratio ($\eta$, say), if we regard the orbital velocities of Sun and Moon as constant during the eclipse. They can be written $ht$ and $kt$ respectively. Denoting $NU'$ by $\xi$, and the distance $MU$ by $\theta$, then

$$\theta^2 = (\xi + kt)^2 + h^2 t^2 - 2ht(\xi + kt)\cos i.$$

The condition for a partial eclipse to be possible is that the minimum value of $\theta$ should be less than the sum of the radii of the shadow and Moon, $(m+u)$. By differentiation it is a simple matter to show that

$$\theta_{min} = \xi \sin i \bigg/ \sqrt{\left\{1 - 2\left(\frac{k}{h}\right)\cos i + \left(\frac{k^2}{h^2}\right)\right\}}.$$

An approximate and average value of $(k/h)$—the ratio of the sidereal month to the year—is 3/40. Taking $i = 5°$, we find that lunar eclipses are possible, on Ptolemy's data, when

$$m+u > \xi/10·66.$$

At far apogee, Ptolemy supposed $(m+u)$ to be 56′ 24″, and at far perigee 63′ 36″. With each figure we should use the appropriate value of the ratio $k/h$ in the above equation. The problem of appraising Ptolemy's methods is out of place here, but by way of example we observe that taking an average for $(m+u)$, namely 1° precisely, we find that an eclipse can take place only when $\xi$ is less than 10°·66. If the eclipse is to be total, clearly we must have

$$\theta_{min} < u - m,$$

and therefore, again taking an average value for the quantity $(u-m)$, we find that $\xi$ must be less than $2°\cdot5$. This must serve as a rough illustration of the calculation of eclipse limits. Pursuing the matter further we should find that Ptolemy's lower and upper limits for a partial eclipse to be possible were both about $3°$ too high.

Suppose, now, that the motions of nodes, Sun, and Moon are known with sufficient accuracy for us to be able to say that a particular full moon will fall within the eclipse limits. The next problem is to relate the distance of the Moon from the node, at the moment of maximum obscuration, to the so-called 'circumstances' of the eclipse. Ptolemy took the sine approximation, in effect, and pointed out that the times of the syzygies are not exactly those of mid-eclipse, but that assuming them to be the same involves us in an error of no more than one part in 360. To be concerned over such a small error, he added, would be more a matter of vanity than a love of truth. Putting vanity aside we may use plane diagrams (cf. those of *Albion* I. 22), from which we see that

$$MN = d \operatorname{cosec} i,$$

$$M_1 M = \sqrt{\{(m+u)^2 - d^2\}},$$

and
$$M_1' M = \sqrt{\{(u-m)^2 - d^2\}},$$

to which we add the expression for magnitude, given previously. Ptolemy tabulates these quantities, obtaining $MN$, for instance, by interpolating linearly $MN$ as a function of $q$, a procedure which is good to about a minute of arc. The time of full moon being known, and also the orbital velocity of the Moon with respect to the shadow, it is a simple matter to work out the precise time of commencement of the eclipse, its duration, and the time of total immersion. Another calculation of value involves determining the ecliptic latitude $(\beta)$ of the Moon for a given node distance $(L)$ or longitude difference $(\lambda = |\lambda_m - \lambda_p|)$. Exact formulae are

$$\sin \beta = \sin i \sin L,$$

and
$$\tan \beta = \tan i \sin \lambda.$$

Since $\beta$ cannot be more than about $5°$, there are many reasonable approximations here, involving exchanging the sine of $\beta$ or $i$ for the corresponding tangent, or vice versa, or replacing either by the angle in radian measure.

As usually defined, the moment of full moon is when the Moon's longitude differs from the Sun's by exactly $180°$. If this were the moment of maximum eclipse, which it is not, then any of the formulae mentioned could be used to relate $\beta$, $L$, and $\lambda$ at maximum eclipse. Some modern writers appear to think that there is no error whatsoever in taking instead the great circle through the poles of the Moon's orbit, but the procedure is no more exact than the approximations already mentioned. To discover when the Moon and shadow are closest, the perpendicular from the shadow centre on the Moon's path *relative to that of the shadow* is needed. But all early investigations rely on an approximation of the kind indicated.

# APPENDIX 33

## *Two seventeenth-century eclipse instruments*

AFTER Apian's adaptation of Richard of Wallingford's eclipse instrument, the graphical method seems to have fallen into abeyance, to be replaced by simple methods combining tables (as in *Almagest* VI. 8) and equatorium (for the positions of Sun, Moon, and nodes). Such a combination of methods is in evidence on the equatorium of *c.* 1600 which I described in my 'A Post-Copernican Equatorium', *Physis*, xi (1969), 418–57. There are two somewhat different instruments in existence, which deserve brief notice. One of these is now in the possession of the University of Leiden, and the other, resembling it closely, in the Museum of the History of Science at Oxford. The latter is illustrated in R. T. Gunther's *Astrolabes of the World*, vol. ii (Oxford, 1932), Plate LXXXVII, and described in general terms on pp. 364–5. No early document relating to the use of these instruments is known. Both have astrolabes on one side, and both have plates for latitude 45°. Important towns near this latitude are Bordeaux and Turin, while Lyons, Milan, Genoa, and Venice are all less than a degree of latitude distant. If anything, a French provenance seems more likely than any other. Judging by the Dragon volvelles both incorporate, the dates of the instruments are 1612 (Leiden) and 1613 (Oxford). Both Dragon plates have what are obviously tables (as opposed to scales), with the heading 'TABULA ECLIPSI. [LUNAE] AUX EPIC. PERIGIUM' occurring only on the earlier instrument. ('Table of eclipses of the Moon at the near point of the epicycle at aux.') The Dragon's tongue and tail are presumably auxiliary pointers, to assist in subtraction. The earlier has on the plate an inner table not on the later, while the later instrument has a scale of sines uncovered by a window in the Dragon plate.

These Dragon plates are interesting for their further inscriptions. Engraved alongside the dates there are symbols for Caput the head of the Dragon, together with an 'R', struck through, which Gunther (op. cit.) was probably wrong to take as a symbol for 'Paris'. It surely means *retrogradus*. The inscriptions then run:

| (1612) | S | D | M |
|--------|---|----|----|
| | 2 | 10 | 40 |
| (1613) | S | D | M |
| | 2 | 29 | 59 |

Do the letters indicate *signa, degradus,* and *minuta*? Later French astronomers used the lower-case 'd' for degrees well into the nineteenth century, and the symbols on the instruments could have been intended to stand for either the Latin words given or their French near-equivalents. Leaving this question aside, it seems that the inscriptions must give either the positions of the ascending node at some standard epoch or the epochs at which the node is in some standard position. The second alternative is ruled out by the slowness with which the nodes move— ruled out, that is, unless the maker had some fresh information which made the older instrument *passé*. But the most notable thing about these figures is their difference of $19^d$ $19^m$. This is the approximate annual movement of the nodes, namely 19·38 degrees, which amounts to 19·65 days on an astrolabe calendar scale. An interpretation of the letters above the numerals in terms of time units rather than units of arc is thus very unlikely. But there is still a mystery here. From what point, and in what direction, were the node positions measured? If in the usual way, from the vernal point and in a direct sense, then the 1613 figure should be smaller than that for 1612. Curiously enough, taking a January year beginning, the 1612 position was only about a degree from the truth, whereas the position quoted for 1613 was nearly 40° in error, suggesting that a mistake of increasing rather than diminishing the node's longitude was indeed made. I have found the same mistake often, in inferior medieval writers; but they were not professional makers of eclipsoria.

Both instruments have unusual inner and outer scales, which will be outlined in due course. First, however, we turn to the tables on the earlier instrument, which run as follows, the words 'aux' and 'perigium' referring to position on the epicycle:

| Right-hand side ('Aux') | D | 1 | 2 | 3 | 4 | 5 | 6 | 7 | 8 | 9 | 10 | 11 |
| | P | 19 | 17 | 15 | 13 | 11 | 9 | 7 | 5 | 3 | 2 | 0 |
| | H | 1 | 1 | 1 | 1 | 1 | 1 | 1 | 1 | 0 | 0 | 0 |
| | M | 54 | 53 | 49 | 45 | 40 | 33 | 16 | 3 | 50 | 30 | 10 |

| Left-hand side ('Perigium') | D | 1 | 2 | 3 | 4 | 5 | 6 | 7 | 8 | 9 | 10 | 11 | 12 |
| | P | 20 | 18 | 16 | 15 | 13 | 11 | 10 | 8 | 7 | 5 | 3 | 2 |
| | H | 1 | 1 | 1 | 1 | 1 | 1 | 1 | 1 | 1 | 1 | 0 | 0 |
| | M | 54 | 53 | 50 | 45 | 37 | 33 | 28 | 23 | 12 | 0 | 40 | 16 |

(The later instrument lacks the H–M series, and the P figures occasionally differ by a digit from those above—obviously not an important change.) Taking H and M as standing for *horae* and *minuta*, we observe that the maximum duration listed is $1^h$ $54^m$, which happens to be the approximate maximum half-duration of a (total) lunar eclipse. If, then, we assume that the H–M rows contain these half-durations (or, as would have been said, the sum of the *mora media* and the *tempus casus*), it is not difficult to see that the P-rows contain corresponding eclipse-magnitudes (*puncta ecliptica*). That the figures in these tables then agree fairly closely with those obtained, for example, from Apian's compound instrument, is easily verified. Finally, it may be verified that D stands for the distance of the Moon, at opposition, from a node. This, as we saw in Appendix 32, must be less than about 12° for an eclipse to take place.

The scales engraved on the back of the mater are the most difficult to interpret. There are six on the earlier instrument and eight on the later. The origins of the inner scales on the later instrument are on the diameter through the beginning of 1 January; all other origins are on the vertical. *All* outer scales are alike, within a degree. The six sets of figures have been averaged, and working on the assumption that the maker was aiming at integral marks, the table from which he most probably worked is:

| Outer scale: | 1 | 2 | 3 | 4 | 5 | 6 | 7 | 8 | 9 |
|---|---|---|---|---|---|---|---|---|---|
| Degrees (counting vertical as origin): | 6 | 12 | 19 | 26 | 33 | 41 | 49 | 58 | 69. |

The graph obtained, if these figures are plotted on rectangular axes, is very roughly sinusoidal in shape. If we assume that the instrument is meant for the simple prediction of eclipses from observations made with the astrolabe itself, then it is natural to suppose that this, or one of the other scales, allows for the conversion of an observed lunar latitude into a node distance. The fact that the outer scale, when extrapolated, reaches a maximum at over 9 units, seems to rule out the possibility that this is a latitude scale, but one other curious possibility remains. Let us suppose that the inclination of the Moon's orbit to the ecliptic was taken as 5°. Working now from the exact equations relating latitude and node distance, it is a simple matter to deduce the size of the scale unit ($\psi$) for each pair of values in the above table. A typical calculation leads from $\tan 7\psi = \tan 5° \sin 49°$ to $\psi = 32'\cdot4$. In all cases the result turns out to be within a minute of arc, or so, of this. The figure is not so awkward a number as it may at first appear, for it is very nearly the mean diameter of the Moon. Ptolemy, for example, quoted the Moon's diameter at mean distance as $33' 40''$, al-Bāttanī took $32' 25''$, and today we take $31' 08''$. Can it be that the separation of the Moon from the ecliptic was to be measured in Moon-breadths? This method of measuring small angles was certainly still current in the seventeenth century. Although it is not difficult to guess at possible procedures, none seems plausible enough to be recorded here.

The inner scales are even more inscrutable. They are:

| Scale mark: | | 0 | 1 | 2 | 3 | 4 |
|---|---|---|---|---|---|---|
| Positions of marks (degrees) | | | | | | |
| 1612 | 1st quadrant | 0 | 11 | 24 | 41 | $68\frac{1}{2}$ |
| | 2nd quadrant | 0 | 14 | $32\frac{1}{2}$ | 50 | — |
| | 3rd quadrant | 0 | 17 | 32 | 51 | — |
| | 4th quadrant | 0 | $10\frac{1}{2}$ | 25 | $40\frac{1}{2}$ | $67\frac{1}{2}$ |
| 1613 | 1st quadrant | 0 | 9 | 23 | 33 | 70 |
| | 2nd quadrant | 0 | $10\frac{1}{2}$ | $25\frac{1}{2}$ | 47 | — |
| | 3rd quadrant | 0 | $9\frac{1}{2}$ | 24 | $47\frac{1}{2}$ | — |
| | 4th quadrant | 0 | $10\frac{1}{2}$ | $28\frac{1}{2}$ | $50\frac{1}{2}$ | — |

Were they meant to assist in reducing the difference between the declinations of the Sun and Moon, as obtained from meridian altitudes, to a node distance? This seems unlikely, for a number of reasons. Were the scales totally unconnected with the prediction of eclipses? They were not for determining the equation of time, as their asymmetry might seem to suggest, but are they connected with some other correction term? Unfortunately we cannot take the advice of Francisco Rodrigues, a writer of the fifteenth century, who, after discussing an instrument at length without understanding it, begs the reader to go, 'in case of difficulty, to the person who made it'.

# APPENDIX 34

## *Lewis of Caerleon's use of* Albion

CAMBRIDGE University MS. Ee.3.61 (1017), ostensibly written *c.* 1482, is an autograph manuscript of Lewis of Caerleon, while the Oxford manuscript Digby 178 is glossed by the same 'Lewys'. See P. Kibre, *Isis*, xliii (1952), 100–8, for a short biographical account of Lewis, and a useful conspectus of his extant manuscripts. Whereas the Digby volume contains the *Quadripartitum* of Richard of Wallingford, not to mention the dubious *De corda et arcu* (see vol. ii, pp. 37, 387), the Cambridge manuscript has *Rectangulus*, the short piece beginning 'Fiat triangulus rectangulus . . .' discussed in Appendix 16, and a number of eclipse calculations, tables, and rules. It is the headings to the tables in both these and other manuscripts which I have suggested elsewhere (see p. 386, vol. ii) are responsible for the literary ghosts *De diametris* and *De opimetris*.

Richard of Wallingford is first mentioned in MS. Ee.3.61 in the course of the calculation of an eclipse (28 May 1481). In the midst of this (at f. 13ᵛ, new foliation) we find a reference to *nove tabule* which Lewis calculated 'super quantitate dyametrorum Solis et Lune at umbre ut ponunt Albategni, Gebir, et commentator super Almagesti, et Ricardus de Sancto Albano'. (These 'new tables' will be mentioned again shortly.) At the head of f. 15ʳ, after the signature 'LEWYS CAERLYON', there is a 'Demonstracio geometrica' of the same eclipse, expressly by Lewis himself, in which the same sources are named, together with Arzachel, but now including a specific *Albion* reference (to I. 15). In both cases it is quite clear that *Albion* is the immediate source, and that the other names are simply those of the authorities Richard of Wallingford himself cites.

Another heading in MS. Ee.3.61 to be compared with these notes occurs at f. 107ʳ:

Examinatam componendi(?) tabulam anglorum et diversitatis aspectus ad eclipses per calculacionem Lodowyci Caerlyon in medicinis doctoris, et calculatur latitudinem 51 gra. 50 min., secundum doctrinam Albategni et Ricardi de Sancto Albano libro primo conclusione 12ª, 13ª, 14ª, 15ª, et 16ª, ut patet ibidem ex demonstracionibus eius pro 7 horis ante meridiem . . . etc.

The data which follow, occupying a side and a half, are from such sources as Almeon (the obliquity—23° 33′ 30″) and Simon Bredon, while Alfraganus, Albategni, and Ptolemy are mentioned in addition, but not Richard of Wallingford.

In a further series of references to Richard, Lewis's apparently numerous sources are in reality copied straight out of *Albion* I. 18 sqq. This summary begins at f. 142ʳ with the words (after an erased title, and the name LEWYS, boldly written in his characteristic style) 'Ad opus eclipsium requiritur noticia quantitatum dyametri'. The first figure to be given agrees with that of f. 13ᵛ, and is in fact not that to be found in the best *Albion* manuscripts. (The diameter of the Sun is quoted as 32′ 32″, rather than 32′ 30″, as in *Albion* I. 21.) The ascription now contains the statement that Richard was a Fellow of Merton: 'Istud elicitur ex primo libro Ricardi de Sancto Albano socii domus scolarium de Merton conclusione 21.' The other diameters (which in our commentary were known as *m* and *u*, for two cases) are as printed in our version above. It is worth recalling that Lewis's enthusiastic tailpiece to the *Quadripartitum* is one of the sources of the idea that Richard was a Fellow of Merton. (See Appendix 14.)

After stating initial parameters, Lewis writes on the method of composing eclipse tables. Since the central problem is one of applying the theorem of Pythagoras, it is not surprising to find that Lewis, after writing out the *Albion* data once again (under the heading 'Albategni, Gebir, Arzachel') for the purpose of comparing them with the data from the Toulouse tables ($2s = 32′\ 32″$; (a) $u = 36′\ 0″$, $m = 14′\ 38″$; (b) $u = 44′\ 44″$, $m = 17′\ 18″$), reduces them to seconds of arc and tabulates the squares of these quantities. Immediately preceding this short text there are tables of squares (and cubes) which would be of obvious use in this connection. The verso of f. 142 ends with information as to parallaxes, ostensibly from Alfraganus, Albategni, and Ptolemy; but this trio of references, not to mention the information they are supposed to have provided, was clearly taken (unacknowledged) from *Albion* I. 14.

Instructions follow (f. 143ʳᵛ) for compiling parallax tables 'sine astrolabio'. The tables themselves come next, having been lightly crossed through, like part of the text preceding. Other eclipse tables are added to lighten the working, the manuscript being at this juncture punctuated with the name 'LEWYS'. One of the last of the tables (f. 151ᵛ) is headed *Tabula diversitatis aspectus Lune in latitudine pro Oxonia grosse* [sic] *calculata pro albion*. This refers, of course, to the book and not the instrument. This particular page concludes Lewis's *nove tabule*, and itself ends with an explanation of the cancelling of the tables: 'Nota quod tabulas precedentes cancellavi quia eas non prima precipue calculavi sicut tabulas quas dedi universitatibus Cantebrigie et Oxonie, tamen quasi insensibilis est differencia in calculo experiatur quicunque velit.' It thus appears that Lewis was himself responsible for striking out the tables, for there is no readily detectable change as between the hand of the text and that of the footnote. His gifts to the two universities cannot be identified with any certainty. The Cambridge MS. Ee.3.61 (1017) is a relatively late accession, having been presented by George I in 1715. (It was formerly owned by John Moore, Bishop of Norwich, shelf-mark 820.) Worth investigating would be St. John's College, Cambridge, MS. 41 (B.19) (f. 1ʳ), and British Museum, MSS. Royal 12.G.1 (f. 1ʳ) and Sloane 1697 (f. 25ʳ), for all have some tables by Lewis adjusted for the latitude of Cambridge (52° 20'). None is seemingly in Lewis's own hand, however, and this fact might be thought to tell against them. The Oxford book cannot have been that recorded in the *Registrum Annalium Collegii Mertonensis* (ed. H. Salter, Oxford Hist. Soc., 1923, p. 139; cf. p. 215) as having been donated by Lewis. This gift, no longer identifiable, is of course of some interest, but was not made until 1490 (it is recorded under 26 October).

After two more tables (distinguished from what has gone before, one of them being copied from a manuscript by the friar John Somer), and an eclipse canon by Lewis, MS. Ee.3.61 has a further note relating to Richard of Wallingford (f. 153ᵛ): 'Secundum Albategni, Arzachelem, et Ricardum de Sancto Albano, si distancia a nodo sit ultra 12 gra. et 15 min. non erit eclipsis Lune. Secundum alios, si distet per 14 gra. vel plus non erit eclipsis.' As before, this comes in all probability from one work alone, namely *Albion*.

The copies of Lewis's tables in the British Museum and Cambridge manuscripts referred to above are all given long titles, in which Richard's name occurs in such a way as to have made a later cataloguer believe that he wrote a work *De diametris*:

Hic incipit tabula eclipsis lunaris secundum dyametros Ricardi abbatis de S. Albano libro suo primo de composicione albionis conclusione 18, 19, et 21 ad longitudinem longiorem . . . noviter facta et expansa ad singula minuta argumenti latitudinis Lune . . . per me, lodowycum, Anno Christi 1482°. Et huic tabule finaliter adhereo ut in principio huius operis premisi. (MS. Royal 12.G.1. f. 1ʳ; the table follows.)

At the foot of the first page of each copy of the table is a note of biographical importance, explaining, perhaps, why Lewis was so addicted to the composition of tables. 'Per expoliationem regis Ricardi [III]', Lewis was passing his time incarcerated in the Tower of London. His misfortune did not cause him to overlook the need to calculate eclipses, and he seems to have been sufficiently resigned to his fate to calculate them according to his new latitude. After the first four pages of lunar eclipse tables we find, in the same three manuscripts, 'Tabula eclipsis Solis secundum dyametros Ricardi de Sancto Albano ad longitudinem longiorem . . . per me Lodowicum anno Christi inperfecto 1482° apud

London'. (Pearl Kibre transcribed this, in the article cited above, but reading 'Londoniensis' for 'longiorem'.) It is noted at a later point (MS. Royal 12.G.1, f. 4$^r$) that although the diameters are those of Richard of Wallingford, yet these differ from Albategni's; and later still there occur two separate tables of diameters themselves, one of them comparable with the table of MS. Ee.3.61, and the other contrasting Albategni's diameters with those of *Albion*. The tables as a whole in this trio of volumes bear dates ranging over the period 1482–5, 1485 being the date of a solar eclipse observed by Lewis (16 March) from the Tower.

# APPENDIX 35

*A brief description of the contents of MSS. Savile 100 (Bodleian Library) and Radcliffe 74 (Museum of the History of Science, Oxford)*

MS. SAVILE 100
Dimensions: 26 × 28 cm.

f. 1. A blank end-leaf.

f. 2$^r$. An instrument almost certainly copied from the 1551 edition of Schöner's works (*Opera mathematica*, Nuremberg, 2nd edition 1561), called there a meteoroscope, for the illustration of which see f. clv$^r$, second section of foliation. The instrument is for determining the separation of points on a sphere whose coordinates in latitude and longitude are known. It is dated 1550 in the printed book, and in the manuscript 1552.

f. 2$^v$. Tables of true motus of the Sun and Moon, in hours and minutes only.

ff. 3$^r$–4$^v$. See vol. ii, p. 277. These are instruments related to the albion.

f. 5$^r$. A Rojas universal astrolabe projection, with fifty stars marked, but no auxiliary rules attached. The book by Rojas from which this was presumably taken was published in 1550 and again in 1551.

f. 5$^v$. A solar calendar, giving the Sun's position in the zodiac throughout the year, together with a subsidiary table giving corrections to be applied during years between 1516 and 1590. (Notice that the mean of these years is 1553; cf. MS. Radcliffe 74, f. 2$^v$.)

f. 6$^r$. Equatorium of card volvelles on which are pasted printed pieces cut from an unidentified work resembling in some respects that by Willem van Wissekerke (printed 1494 and 1495), Sarzosus (printed 1526), and Oronce Fine (printed 1526 also). The Savile instrument covers all planets and auges, and radix dates between 1500 and 1600 are marked.

ff. 6$^v$ and 7$^r$. Printed constellation maps of the northern and southern hemispheres, and carrying the title 'INCLYTO SENATUM CORONENSI DD', being dated 1532. This is reprinted in a 1551 (Basle) edition of Ptolemy, and it seems that 'I. H. C.', whose initials are engraved on the blocks, was Johannes Honter von Kronstadt (1498–1549), who first published it as a broadsheet star map in 1532, hence the date engraved.

f. 7ᵛ. Blank.

f. 8ʳ. The Sarzosus (manuscript) equatorium discussed in the article by E. Poulle and F. R. Maddison, for which see vol. ii, p. 261.

f. 8ᵛ. A device for evaluating planetary aspects, of unknown origin but relatively trivial as an instrument. It is of interest for some of the unusual symbols marked on it (clover leaf, various crosses, etc.).

f. 9ʳ (a modern end-leaf). This carries four pasted pieces, one being cut from a printed broadsheet (or book), giving Arabic and Latin star-name equivalents; another, probably from the same source, naming further stars and constellations, and ending 'BASILEAE, M.D.XXXIX' (not from the astronomical works of that year and town, by J. J. Pontanus, but in the same type-fount); one 'Tabella possibilitatis Ecclipsium' (Schöner's spelling!) in the hand of the main manuscript; and the last an alien slip (bookmark?) of the following century, referring to 'Cl[arissimus] Richardus' and 'D[ominus] Barrow', the latter probably Isaac Barrow.

f. 9ᵛ. Blank.

The manuscript, like MS. Radcliffe 74, is bound in soft leather, and this was originally coloured green in both cases. Both have pieces of parchment in the bindings which appear to be from late fifteenth-century legal documents written in French, while the end sheets are from canon law incunabula. Perhaps this evidence, together with the Basle printings mentioned, points to a provenance in that city, or near by. The impression is that the two manuscripts are in the same hand, although there are slight differences in style—which are easily reconciled, on the other hand, with the four or five years apparently separating the two documents.

## MS. RADCLIFFE 74
Dimensions: 37 × 39·5 cm.

Loose enclosure no. 1. A complete manuscript equatorium in paper for the planet Mercury only, and of a simple kind.

f. 1ʳ. Astrolabe, with rete (now loose), plate of horizons, and equal/unequal hour conversion chart only. There is one small flourish in the form of Greek lettering for the name of the star Procyon. Two stars are named in Greek on the astrolabe of MS. Savile 100, f. 4ᵛ. See above.

f. 1ᵛ. Volvelle with a central band characteristic of the Rojas projection, the disc moving within a zodiac scale which is squared at the corners in a curious way.

f. 2ʳ. Astrological volvelle (now loose) showing planetary aspects and certain astro-medical information. A pointer on the volvelle moves over sixteen different scales carrying further astrological information (domiciles, terms, faces, decans, etc.).

f. 2ᵛ. A series of scales deriving almost certainly from Apian's *Astronomicum Caesareum*, and entitled 'Motus horarius planetarum'. These scales, and Apian's indebtedness to *Albion*, are discussed at pp. 278–85, vol. ii. This set of scales in the manuscript is surrounded by a solar calendar scale, totally unrelated. The calendar scale occupies two turns of a spiral almost square in general form, the corners only being rounded. A marginal table 'Emendatio loci solis' runs by years from 1520 to 1600. On the calendar scale, the Sun's noonday position is marked for 1 January 1556 and 1 January 1557, and we may therefore suppose that this instrument at least was drawn during the year 1556.

f. 3ʳ. A quadrantal diagram with no explanatory script, containing two distinct sets of lines the purpose of which can only be conjectured. One set is a series of parallels, as parallels of latitude on a sphere, orthogonally projected. The other set might be regarded as originating in a projection of the *saphea Arzachelis* type. It is possible that the two sets of lines were to be used separately.

f. 3ᵛ. A paper volvelle marked out twice over, as recommended in *Albion*, with projections of the whole sphere as in the *saphea Arzachelis*. The two sets of coordinate lines are for ecliptic and equatorial coordinates. The disc moves within a normal scale of degrees, and is traversed by a diametral zodiacal rule as in the universal astrolabe for which the projection was devised. See p. 36 above, for a reproduction of the volvelle.

f. 4ʳ and ᵛ. Blank.

# APPENDIX 36

## *The will of John of Gmunden*

THE will by which John of Gmunden bequeathed books and instruments to the new library of the Faculty of Arts in the University of Vienna (the library was founded in 1415) was drawn up in 1435 and proved in 1443. These and other circumstances are to be found in Rudolf Kink, *Geschichte der kaiserlichen Universität zu Wien*, 2 vols. (Vienna, 1854), where also is printed the greater part of the will (i, part 2, pp. 108–11). The following excerpt concerns the instruments, which bear comparison with those of the fourteenth and fifteenth centuries in the possession of Merton College (see Appendix 15):

Item de instrumentis placet, quod sphera solida reponatur in armario ad hoc deputato et quantum pro honore facultatis . . ., raro tamen extra Librariam concedatur. Item instrumenta Campani de equacionibus planetarum cum figuris extractis ex Albione de eclipsibus in ladula custodiatur et rarissime extra librariam concedatur. Item figure communes in theorica planetarum sub custodia vigili in sua ladula teneantur. Item astrolabium ligneum, duo quadrantes, sphera materialis, unum chilindrum magnum, quatuor theorice lignee simili modo custodiantur.

# APPENDIX 37

## *Some manuscript notes of the early fifteenth century for the design of an astronomical clock with epicyclic trains*

In a volume of astronomical tracts in the Bibliothèque Royale, Brussels (MS. 10117–26, a single volume, foliated consecutively), composed for the most part in the thirteenth and fourteenth centuries, but copied between around 1410 and 1445 (see the last section of this appendix), the scribe at first left a number of pages unused. Judging from the hand, he or a near-contemporary then filled some of the blank pages with notes which were clearly intended to assist, in principle at least, in the design of an astronomical clock. The folio references are: 22$^v$, 23$^r$, 23$^v$, 79$^v$ (bottom quarter), 80$^r$, 80$^v$, 88$^v$, 117$^v$–118$^r$, 121$^v$, and 147$^{rv}$. There is no rubric: the first page of notes simply begins 'Pro Saturno', and follows with the symbolism characteristic of the notes as a whole. The symbolism is possibly unique, and hence this first entry may be used to explain it:

There follows a note: 'Hec excrescunt 2 min. et 47 sec. unius hore.' The first pattern of words and figures is obviously to be interpreted thus: 'A wheel moving through a full circle in 6 days, with 8 teeth, drives a wheel of 37 teeth (37 d[entium]) to the arbor of which is fixed a second wheel of 8, this driving the wheel of Saturn of 109 teeth.' With the notation used for trains of wheels in *Horologium*, above, we may write

$$\frac{(6^d)}{8} \cdot \frac{37}{8} \cdot \frac{109}{(\hbar)},$$

where the symbol for the planet Saturn here denotes the period of the wheel of Saturn. Since, as explained formerly, the numerical value of the product so expressed must be equal to unity, it is easily verified that the wheel of Saturn will have a periodicity of 378 days 2 hours 15 minutes. This is close to the mean synodic period of Saturn, rather than its mean sidereal period, a fact which perhaps suggests (since the same will be found to hold good of the other planets) that the symbolism for the device was to be interpreted as relating all movements to that of the Sun; or that the Sun was actually fixed in position, with a moving zodiac, and moving planets; or that epicyclic motions were envisaged (since they involve a synodic period).

Disjointed though the notes may be, they are worth investigating, if only because they tell us something of the good intentions, if not the mechanical activity, of an early fifteenth-century astronomer. They are of interest for the way in which he exploits his notation, and because it is possible to form a rough idea of his general plan (especially from some connected paragraphs f. 147$^{rv}$); but over and above all this they are of interest simply because they are for the most part notes, and nothing more. Their author was involved in repeated calculation, trying out one idea after another, evaluating the error (as in the example given above, which was held to lead to a period 2$^m$ 47$^s$ in excess of the true (synodic) period), often writing down the same idea two or three times, and occasionally crossing out his schemes. Such semi-informal writing tends not to have survived from the Middle Ages, for obvious reasons, and although this specimen does not contain any record of the associated arithmetical computations—which would probably in any case have been done on a slate—it is an unusual preliminary draft of what might have become in due course a design for a monumental astronomical clock. Not all astronomical clocks were large, even at this relatively early date, but I know of no evidence for a small *planetary* clock as early as 1430. A reference to a small striking astrolabe clock of 1379, with several latitude plates, alarm mechanism, and lunar indicator, is to be found mentioned in a letter from Pere III of Aragon to his daughter, quoted by A. Rubio i Lluch, *Documents per l'historia de la cultura catalana mig-eval*, 2 vols. (Barcelona, 1921–8), i. 265.

There is no clear guide to the identity of the author of the notes. Elsewhere (f. 79$^r$) probably the same writer alludes to a 'master' who calculated the movement of the eighth sphere for 1400: 'Anno Domini 1400 completo, stella antecedens in cornu Arietis octave spere erat in nona spera habens o in signis 25 gr. 22 min. 21 sec. 10 ter. 50 quar., si bene calculat magister.' An unsigned personal note at the foot of f. 116$^v$, which seems to have been written later, and which refers to a date of 1445, mentions a place seemingly in the Low Countries, 'Leermich iuxta Halsenberch in Zoma' (if this is a correct transcription). The year 1444 occurs in another such note on f. 117$^v$.

The vocabulary in the notes themselves is sparse and conventional enough. *Opus* is a common term for a single train of wheels, and *vector* for an arm carrying an epicycle, as in 'vector per quam Luna movetur in epiciclo', while *dens* is never written out beyond the first letter. There are other technical terms in the passages on f. 147, which will be noted in due course.

Although it would not be unnatural to begin with Saturn, as happens on the first page of notes in the way already explained, one might equally regard f. 80ʳ as the first page, beginning 'Pro Sole'. The main part of f. 79ᵛ opposite, which does not belong to any previous text, comprises a summary of planetary parameters and, rather more unusual, a table of synodic periods. The entries in the table have been verified, seemingly by a later writer, and the word 'bene' written against all but the last (for the head of the Dragon), which has been altered. The table has also been partially extended in the same less formal hand to include data 'secundum Campanum'. (The style of these additions is identifiable with that of some glosses, and, owing to the writer's predilection for Campanus, it may be assigned to 'hand C'.) The synodic periods are highly relevant to the planetary *opera* which follow, and for this reason the entire page is transcribed here at the outset, with the exception of the horological notes of the last quarter. These end-notes, which were in any case probably a continuation of those on f. 80ʳ, will here be treated like the contents of the other six pages of notes, and completely rearranged by planet, duplicated trains being noted, but written out only once.

The trains were certainly envisaged as being incorporated into a time-piece, as opposed to a manual planetarium. The passage at f. 147 makes this clear, but the only statement which does not concern gear ratios or periodicities in the earlier notes runs as follows: 'Ad inveniendum horas inequales diei, divide emisperium nostrum in limbo matris in 12 partes seu lineam Capricorni et Cancri similiter(?) in 6 et provenient in limbo circa 20 et 21 gradus, ipsaque linea secabit lineam Arietis vel Libre, ut patet practicanti.' (I am indebted to Mons. Emmanuel Poulle for drawing my attention to the later pages of the text, and verifying a number of readings from the original manuscript, which were unclear on microfilm.) These notes were obviously written by someone familiar with astrolabe practice, and were probably only intended for his own use. They are in hand C, and their precise meaning is doubtful enough.

The planetary data of f. 79ᵛ are as follows:

|                                       | Gra. | Min. |
| ------------------------------------- | ---- | ---- |
| Centrum Solis a centro Terre          | 2    | 30   |
| Centrum Lune a centro Terre           | 12   | 28   |
| Semidiameter epicicli Lune            | 6    | 20   |
| Deferens Saturni a centro Terre       | 3    | 25   |
| Equans eius                           | 6    | 50   |
| Semidyameter epicicli eius            | 6    | 30   |
| Deferens Jovis a centro Terre         | 2    | 45   |

| | Gra. | Min. |
|---|---|---|
| Equans eius | 5 | 30 |
| Semidiameter epicicli eius | 11 | 30 |
| Deferens Martis a centro Terre | 6 | 30 |
| Equans eius | 13 | 0 |
| Semidyameter epicicli eius | 39 | 30 |
| Deferens Veneris a centro Terre | 1 | 15 |
| Equans eius | 2 | 30 |
| Semidyameter epicicli eius | 43 | 10 |
| Deferens Mercurii a centro Terre | 9 | 0 |
| Equans eius | 3 | 0 |
| Semidyameter epicicli eius | 22 | 30 |

[A note above the table in hand C points out that the units are such that the semidiameter of the deferent circle is of 60 parts.]

| Auges planetarum | Sig. | Gra. | Min. |
|---|---|---|---|
| Solis et Veneris | 3 | 0 | 30 |
| Saturni | 8 | 12 | 38 |
| Jovis | 5 | 22 | 41 |
| Martis | 4 | 16 | 17 |
| Mercurii | 6 | 29 | 44 |

Notandum quod Caput et Cauda Draconis moventur cum firmamento, sequuntur cursum suum modo firmamenti; unde a Leone transeunt in Cancrum, et a Cancro in Geminos, et deinceps. Et semper sunt Caput et Cauda signis et gradibus oppositis, et venter in medio, ut si Caput est in Cancro venter est in Ariete et Cauda in Capricorno, et si Caput fuerit in Leone Venter erit in Tauro, Cauda in Aquario.

| | Dieb. | Hor. | Min. | Sec. |
|---|---|---|---|---|
| Sol percurrit zodyacum in | 365 | 5 | 49 | |
| Saturnus coniungitur Soli in | 378 | 2 | 12 | 13 |
| Jupiter coniungitur Soli in | 398 | 21 | 12 | 8 |
| Mars coniungitur Soli in | 779 | 22 | 22 | 36 |
| Mercurius circuit suum epiciclum in | 115 | 21 | 5 | 6 |
| Venus circuit suum epiciclum in | 583 | 22 | 18 | 48 |
| Luna coniungitur Soli in | 29 | 12 | 44 | 3 |
| Caput Draconis coniungitur Soli in | 345 | 15 | 11 | 11 |

[Columns for days, hours, and minutes 'secundum Campanum' are added in hand C, but only the first column is filled in, and then not in the first or in the last two rows. All such entries take the earlier figures to the next higher figure for the day. The Venus figure has been subsequently altered in the last two places to 14$^m$ 26$^s$. The data for Caput Draconis were later struck out and replaced by 346$^d$ 14$^h$ 53$^m$ 16$^s$. All this is done by hand C, which is also responsible for the following note:]

Notandum secundum Campanum quod ambitus convexe superficiei spere Saturni, et ipse est ambitus concave spere stellarum, erunt 461254410 miliaria et 440·|660^me unius miliarum [*sic*]. Et quoniam semidiameter a centro Terre usque ad celum stellatum dividitur in 60 equales partes tunc cuilibet parti correspondet 1223129 miliaria et 71·660^me miliaris unius. Et ma[?] 40·660^me indivise prout miliarie constat ex 4000 cubitis.

## Comment on the data

With the exception of the first two sets for Mars, the initial figures could have been taken from Ptolemy's *Almagest*. The data for Mars could, like the rest, have come from the Toledan tables, although a more immediate source might have been the writings of John of Linières, whose treatise on the equatorium was printed from this very manuscript (f. 142^v–) by D. J. Price, *The Equatorie of the Planetis* (Cambridge, 1955), pp. 188–96.

The aux positions quoted, since they are not given beyond minutes of arc, are difficult to place with certainty, but I suggest that they are arrived at by rounding off Alfonsine data for 1420, and then adding 5′ to each, as appropriate to about 10 years. (Strictly speaking, 5′ 56″ corresponds to 10 years.) Rounding off to the next lower minute before performing the addition, one finds the aux positions to be, in the order of the text, 3$^s$ 0° 30′, 18$^s$ 12° 28′, 5$^s$ 22° 42′, 4$^s$ 14° 17′, 6$^s$ 29° 44′. There are only three small discrepancies, easily explained on the grounds of careless calculation or miscopying. At all events, it does seem that 1430 is a more suitable date to assign to this part of the manuscript than the 1410 suggested by the text on the recto of the same folio.

The passage on 'Caput et Cauda Draconis' (the ascending and descending lunar nodes) is commonplace, announcing that the Dragon moves contrary to the natural order of signs, that is, in the same sense as the daily rotation. The reference to the belly of the dragon is very curious, and it has no astronomical counterpart.

The synodic periods listed are those of Alfonsine astronomy (being the periodicities in mean argument), except that the 6$^s$ for Mercury should be 2$^s$, while the Venus figures should end in 14$^m$ 26$^s$, rather than 18$^m$ 48$^s$. Modern figures for mean synodic periods, different from these, are as follows:

|         | d   | h  | m  | s  |
|---------|-----|----|----|----|
| Saturn  | 378 | 2  | 8  | 8  |
| Jupiter | 398 | 21 | 15 | 45 |
| Mars    | 779 | 22 | 28 | 26 |
| Mercury | 115 | 21 | 3  | 22 |
| Venus   | 583 | 22 | 7  | 6  |
| Moon    | 29  | 12 | 44 | 3  |

We observe that the manuscript has the Moon's period correct, not unnaturally, since no calculation is involved. The Sun's sidereal period could have been given more precisely without difficulty ($365^{d}\,5^{h}\,48^{m}\,48^{s}$). All the planetary periods are at least fully accurate in the first two columns.

Finally, the quoted actual sizes of the spheres of Saturn and the Moon are part of a long tradition now known to go back at least as far as Ptolemy himself. (See B. R. Goldstein, 'The Arabic Version of Ptolemy's *Planetary Hypotheses*', *Trans. American Philosophical Society*, N.S. lvii, part 4 (1967).) They are of no apparent relevance to the notes which follow. The references to Campanus are of some interest since the writer of the notes has added the comments (on ff. 88$^{v}$ and 23$^{v}$) that the Alfonsine tables are being followed, while a gloss has been added to the former statement 'et in theoricam planetarum'. This, being written in the same hand as the notes referring by name to Campanus, is apparently a (gratuitous) reference to his *Theorica*.

## The planetary gear-trains

|  |  |  | d | h | m | s | Δ |
|---|---|---|---|---|---|---|---|
| *Pro Sole* (etc.) | | | | | | | |
| I | $\dfrac{(21^{d}\,16^{h})}{7}\cdot\dfrac{118}{\text{(zod.)}}$ | 88$^{v}$, 80$^{r}$, 23$^{r}$ | 365 | 5 | 42 | 50 | $-7^{m}$ |
| 2 | $\dfrac{(20^{d})}{12}\cdot\dfrac{26}{7}\cdot\dfrac{59}{\text{(zod.)}}$ | 23$^{r}$, 88$^{v}$ | 365 | 5 | 42 | 51 | $-7^{m}$ |
| 3 | $\dfrac{(15^{d})}{9}\cdot\dfrac{26}{7}\cdot\dfrac{59}{\text{(zod.)}}$ | 23$^{r}$ | 365 | 5 | 42 | 51 | $-7^{m}$ |
| 4 | $\dfrac{(9^{d})}{12}\cdot\dfrac{487}{\text{(zod.)}}$ | 23$^{r}$ | 365 | 6 | 0 | 0 | $+11^{m}$ |
| 5 | $\dfrac{(\text{☽})}{35}\cdot\dfrac{29}{17}\cdot\dfrac{29}{8}\cdot\dfrac{70}{\text{(zod.)}}$ | 23$^{r}$, 80$^{r}$, 88$^{r}$ | 365 | 5 | 34 | 11 | A |
| 6 | $\dfrac{(\text{☽})}{45}\cdot\dfrac{27}{17}\cdot\dfrac{29}{12}\cdot\dfrac{145}{\text{(zod.)}}$ | 23$^{r}$, 80$^{r}$, 88$^{r}$, 88$^{v}$ | 365 | 5 | 34 | 11 | A |
| 7 | $\dfrac{(\text{☽})}{45}\cdot\dfrac{16}{7}\cdot\dfrac{37(?)}{9}\cdot\dfrac{59}{\text{(zod.)}}$ | 23$^{v}$ | | | | | |
| 8 | $\dfrac{(\text{☽})}{45}\cdot\dfrac{16}{21}\cdot\dfrac{40}{12}\cdot\dfrac{220(?)}{\text{(zod.)}}$ | 23$^{v}$ | | | | | |
| 9 | $\dfrac{(\text{☽})}{45}\cdot\dfrac{16}{7}\cdot\dfrac{20}{9}\cdot\dfrac{26}{14}\cdot\dfrac{59}{\text{(zod.)}}$ | 23$^{v}$ | 365 | 5 | 38 | 24 | |
| 10 | $\dfrac{(26^{d}\,6^{h})}{8}\cdot\dfrac{27}{17}\cdot\dfrac{29}{12}\cdot\dfrac{29(?)}{\text{(zod.)}}$ | 88$^{v}$ | | | | | |
| 11 | $\dfrac{(3^{d}\,10^{h})}{10}\cdot\dfrac{1069}{\text{(zod.)}}$ | 23$^{r}$ | 365 | 5 | 48 | 0 | |

| | | d | h | m | s | Δ |
|---|---|---|---|---|---|---|

*Pro uno circulo septem dierum*

12   $\dfrac{(7^{\mathrm{d}})}{24}\cdot\dfrac{18}{8}\cdot\dfrac{45}{(\mathbb{D})}$     $23^{\mathrm{v}}$     29   12   45   0

*Pro motore epicicli Lune*

| | | d | h | m | s | Δ |
|---|---|---|---|---|---|---|
| 13 | $\dfrac{(7^{\mathrm{d}})}{12}\cdot\dfrac{43}{12}\cdot\dfrac{197}{(\mathrm{M})}$ | $23^{\mathrm{v}}, 80^{\mathrm{v}}$ | 411 | 18 49 57 $-2^{\mathrm{m}}$ | | |
| 14 | $\dfrac{(7^{\mathrm{d}})}{12}\cdot\dfrac{197}{12}\cdot\dfrac{43}{(\mathrm{M})}$ | $23^{\mathrm{v}}, 80^{\mathrm{v}}, 88^{\mathrm{v}}$ | 411 | 18 49 57 $-2^{\mathrm{m}}$ | | |
| 15 | $\dfrac{(15^{\mathrm{d}})}{12}\cdot\dfrac{67}{12}\cdot\dfrac{59}{(\mathrm{M})}$ | $23^{\mathrm{v}}, 80^{\mathrm{v}}, 88^{\mathrm{v}}$ | 411 | 18 29 57 $-22^{\mathrm{m}}$ | | |
| 16 | $\dfrac{(6^{\mathrm{d}})}{12}\cdot\dfrac{241}{12}\cdot\dfrac{41}{(\mathrm{M})}$ | $79^{\mathrm{v}}$ | 411 | 17 0 0 0 | | |
| 17 | $\dfrac{(4^{\mathrm{d}})}{12}\cdot\dfrac{97}{15}\cdot\dfrac{191}{(\mathrm{M})}$ | $79^{\mathrm{v}}$ | 411 | 17 4 0 $+3^{\mathrm{m}}$ etc. | | |
| 18 | $\dfrac{(22^{\mathrm{d}}\,12^{\mathrm{h}})}{12}\cdot\dfrac{31}{12}\cdot\dfrac{85}{(\mathrm{M})}$ | $80^{\mathrm{r}}$ | 411 | 17 15 0 $+15^{\mathrm{m}}$ | | |

*Pro Dracone*

| | | d | h | m | s | Δ |
|---|---|---|---|---|---|---|
| 19 | $\dfrac{(13^{\mathrm{d}}\,12^{\mathrm{h}})}{12}\cdot\dfrac{79}{10}\cdot\dfrac{39}{(\mathrm{D})}$ | $23^{\mathrm{v}}, 88^{\mathrm{v}}$ | 346 | 14 42 0 $-11^{\mathrm{m}}$ | | |
| 20 | $\dfrac{(5^{\mathrm{d}})}{4}\cdot\dfrac{79}{12}\cdot\dfrac{42}{(\mathrm{D})}$ | $23^{\mathrm{v}}\,80^{\mathrm{v}}$ | 345 | 15 0 0 $-11^{\mathrm{m}}\,11^{\mathrm{s}}$ | | |
| 21 | $\dfrac{(9^{\mathrm{d}}\,16^{\mathrm{h}})}{7}\cdot\dfrac{251}{(\mathrm{D})}$ | $23^{\mathrm{v}}$ | 346 | 14 51 25 A | | |
| 22 | $\dfrac{(5^{\mathrm{d}})}{4}\cdot\dfrac{8}{12}\cdot\dfrac{79}{12}\cdot\dfrac{63}{(\mathrm{D})}$ | $80^{\mathrm{v}}$ | 345 | 15 0 0 $-11^{\mathrm{m}}\,11^{\mathrm{s}}$ | | |
| 23 | $\dfrac{(47^{\mathrm{d}})}{8}\cdot\dfrac{59}{(\mathrm{D})}$ | $88^{\mathrm{v}}$ | 346 | 15 0 0 $-7^{\mathrm{m}}$ | | |

*Pro Saturno*

| | | d | h | m | s | Δ |
|---|---|---|---|---|---|---|
| 24 | $\dfrac{(3^{\mathrm{d}})}{9}\cdot\dfrac{53}{5}\cdot\dfrac{107}{(\hbar)}$ | $80^{\mathrm{r}}$ | 378 | 1 36 0 $-36^{\mathrm{m}}$ | | |
| 25 | $\dfrac{(8^{\mathrm{d}})}{12}\cdot\dfrac{53}{10}\cdot\dfrac{107}{(\hbar)}$ | $80^{\mathrm{r}}$ | 378 | 1 36 0 $-36^{\mathrm{m}}$ | | |
| 26 | $\dfrac{(13^{\mathrm{d}})}{12}\cdot\dfrac{349}{(\hbar)}$ | $80^{\mathrm{r}}$ | 378 | 2 0 0 $-12^{\mathrm{m}}$ | | |
| 27 | $\dfrac{(6^{\mathrm{d}})}{8}\cdot\dfrac{37}{8}\cdot\dfrac{109}{(\hbar)}$ | $22^{\mathrm{v}}, 80^{\mathrm{r}}$ | 378 | 2 15 0 $+2^{\mathrm{m}}\,47^{\mathrm{s}}$ | | |

*Pro Iove*

| | | d | h | m | s | Δ |
|---|---|---|---|---|---|---|
| 28 | $\dfrac{(7^{\mathrm{d}})}{12}\cdot\dfrac{39}{15}\cdot\dfrac{263}{(\math04)}$ | $22^{\mathrm{v}}, 80^{\mathrm{r}}$ | 398 | 21 12 0 $-8^{\mathrm{s}}$ | | |

|  |  | d | h | m | s | Δ |
|---|---|---|---|---|---|---|
| **Pro Iove** (cont.) | | | | | | |
| 29   $\dfrac{(6^d)}{9}\cdot\dfrac{31}{10}\cdot\dfrac{193}{(♃)}$ | $22^v, 80^r$ | 398 | 20 | 48 | 0 | $-24^m$ |
| 30   $\dfrac{(15^d)}{8}\cdot\dfrac{111}{12}\cdot\dfrac{23}{(♃)}$ | $22^v, 80^v$ | 398 | 21 | 45 | 0 | $+33^m$ |
| **Pro Marte** | | | | | | |
| 31   $\dfrac{(9^d\,4^h)}{12}\cdot\dfrac{1021}{(♂)}$ | $22^v$ | 779 | 22 | 20 | 0 | $-2^m$ |
| 32   $\dfrac{(6^d\,21^h)}{9}\cdot\dfrac{1021}{(♂)}$ | $22^v$ | 779 | 22 | 30 | 0 | $-2^m$ |
| 33   $\dfrac{(49^d)}{12}\cdot\dfrac{191}{(♂)}$ | $22^v, 80^r$ | 779 | 22 | 0 | 0 | $-22^m$ |
| 34   $\dfrac{(14^d)}{8}\cdot\dfrac{28}{12}\cdot\dfrac{191}{(♂)}$ | $22^v, 80^r$ | 779 | 22 | 0 | 0 | $-22^m$ |
| 35   $\dfrac{(4^d)}{12}\cdot\dfrac{49}{16}\cdot\dfrac{48}{12}\cdot\dfrac{191}{(♂)}$ | $22^v, 80^r$ | 779 | 22 | 0 | 0 | $-22^m$ |
| 36   $\dfrac{(14^d)}{12}\cdot\dfrac{26}{16}\cdot\dfrac{34}{10}\cdot\dfrac{121}{(♂)}$ | $22^v, 80^r$ | 779 | 22 | 42 | 0 | $+20^m$ |
| 37   $\dfrac{(5^d\,12^h)}{12}\cdot\dfrac{55}{10}\cdot\dfrac{34}{10}\cdot\dfrac{91}{(♂)}$ | $22^v, 80^v$ | 779 | 22 | 42 | 0 | $+20^m$ |
| 38   $\dfrac{(5^d\,12^h)}{12}\cdot\dfrac{22}{10}\cdot\dfrac{85}{10}\cdot\dfrac{91}{(♂)}$ | $80^v$ | 779 | 22 | 42 | 0 | $+20^m$ |
| 39   $\dfrac{(5^d\,12^h)}{12}\cdot\dfrac{22}{10}\cdot\dfrac{25}{10}\cdot\dfrac{34}{10}\cdot\dfrac{91}{(♂)}$ | $80^v$ | 779 | 22 | 42 | 0 | $+20^m$ |
| 40   $\dfrac{(5^d\,12^h)}{12}\cdot\dfrac{91}{10}\cdot\dfrac{25}{10}\cdot\dfrac{34}{10}\cdot\dfrac{22}{(♂)}$ | $80^v$ | 779 | 22 | 42 | 0 | $+20^m$ |
| **Pro Venere** | | | | | | |
| 41   $\dfrac{(12^d\,6^h)}{6}\cdot\dfrac{44}{10}\cdot\dfrac{65}{(♀)}$ | $23^r, 80^v$ | 583 | 22 | 0 | 0 | $-18^m$ |
| 42   $\dfrac{(22^d)}{12}\cdot\dfrac{49}{6}\cdot\dfrac{39}{(♀)}$ | $23^r, 80^v$ | 583 | 22 | 0 | 0 | $-18^m$ |
| **Pro Mercurie** | | | | | | |
| 43   $\dfrac{(23^h)}{12}\cdot\dfrac{1451}{(☿)}$ | $23^r$ | 115 | 21 | 5 | 0 | $-6^s(?)$ |
| 44   $\dfrac{(25^d\,18^h)}{8}\cdot\dfrac{36}{(☿)}$ | $23^r, 80^v$ | 115 | 21 | 0 | 0 | $-5^m$ |

## Comment on the ratios

Following each ratio, there is a reference to the folio(s) on which it is
to be found, an evaluation (in days, hours, minutes, and seconds) of the

periodicity given to the planet by the gear train in question, and finally, in the column headed 'Δ', a statement of error *as cited in the manuscript*. Thus ratio 26 provides a period for Saturn of $378^d 2^h$. Since the error stated is $-12^m$, we should expect, if the author's calculation be supposed correct, that he believed Saturn's synodic period to be $378^d 2^h 12^m$. Reference to the table transcribed above from f. $79^v$ shows that, ignoring seconds, this was so. Seconds were in fact only occasionally mentioned in the statements of error, but there are many signs that the writer(s) generally took the calculation as far as seconds. On three occasions, indicated by an 'A' in the Δ-column, the resulting period is stated, rather than its deviation from some norm. That the norm is indeed the table of f. $79^v$ is easily verified by comparing this table with my calculations combined with the stated errors. And here some comment might be made on the general accuracy of the chosen ratios, and the arithmetical accuracy of the 44 calculations. The main problem was that of choosing the ratios, and there is no indication of the way in which this problem was solved; but to evaluate the error, it was still necessary to perform a long division (sexagesimal), often by a four-figure number.

There is no doubt that the author, or principal author, was fully aware of the axiomatics, as it were, of gear ratios (cf. *Horologium*, Part I). As may easily be seen, he recognizes the possibility of alone altering the number of teeth and the period of a single wheel simultaneously (cf. 2 and 3); he interchanges two wheels separated by one other (cf. 13 and 14); and he will even interchange a wheel number with the period of the driving wheel (cf. trains 33 and 35, with a wheel of 49 days in the first and a wheel of 49 teeth in the second). This might all seem very obvious in the notation used here, but it cannot have been so at the time the manuscript was written, and judging by the verbal cavortings of many a modern writer on the choice of gear ratios, it is not always obvious today.

One of the unexpected results of the above analysis is that three of the five trains for the Dragon and three of the six trains for the lunar epicycle are the work of hand C, that is, of the person who altered the figure for the Dragon in the table of f. $79^v$, having verified the remainder of the table. It should be emphasized that this may simply be the work of the original writer at a later time. Ratios 21 and 23 agree with the altered figure for the synodic period of the Dragon. Unfortunately there is no quoted figure *pro motore epicicli Lune* except in the later hand: on f. $79^v$, beneath the notes mentioning Campanus and above the *deleted* trains 16 and 17, which are nevertheless incompatible with this later addition, we find 'Luna circuit suum epiciclum in $27^d 13^h 18^m 33^s 56^t$. Vector per quam Luna movetur in epyciclo coniungitur Soli in $411^d 18^h 52^m$.' It is the fact that the data here given in the later hand disagree with ratios 16, 17, and 18 in the same hand, and yet agree with 13, 14, and 15 in the more formal hand, which suggests that the entire notes are the work of one man.

Ratios 1 to 4, 10, and 11 require no comment, but the remaining solar ratios involve the lunar period. Fortunately there is a note telling us that a lunar synodic period of $29^d\ 12^h\ 45^m$ was assumed in the calculation. This does not accord with the $29^d\ 12^h\ 44^m\ 3^s$ of the table, but there is no doubt that the figure was chosen to simplify the calculation. (Even then it introduces the factor 945/32.) The 'wheel of seven days' makes it seem almost certain that the device was a timepiece rather than merely a geared planetarium. A dial showing the day of the week is not unusual in elaborate clocks of the sort envisaged, and we notice that the drive for the week-day indicator could have been taken off any of the trains 6 to 9. There were other trains mentioned in the manuscript, not recorded above, in connection with 5 and 6, namely,

$$\frac{(6^d\ 18^h)}{8}\cdot\frac{16}{(13^d\ 12^h)}, \quad \frac{(21^d)}{32}\cdot\frac{8}{(5^d\ 6^h)} \quad \text{and} \quad \frac{(5^d\ 6^h)}{8}\cdot\frac{16}{(10^d\ 12^h)}.$$

The first connects with 5 thus:

$$\frac{(13^d\ 12^h)}{16}\cdot\frac{35}{(\mathrm{☽})},$$

and the others with 6 or 10 typically thus:

$$\frac{(5^d\ 6^h)}{8}\cdot\frac{45}{(\mathrm{☽})}.$$

The figures queried in 7, 8, and 10 are suggested readings of figures illegible or hidden in the binding of the manuscript.

The letter 'M' used in the notation for trains 13 to 18 signifies the synodic period of the radius of the Moon's epicycle. The period in question is not commonly encountered, and like the synodic periods of all the planets is relatively tedious to calculate. The two different figures apparently calculated by the writer are not, perhaps, as accurate as they might have been, but it is impossible to make a specific criticism in ignorance of the data from which he worked. The calculation may be done as follows. The angle between the rotating vector and the line from the Earth to the mean Sun is equal to the difference between the true argument of the planet in the epicycle and the mean anomaly of the Moon, that is to say, its mean motus reduced by the mean motus of the Sun. In a complete cycle, the change in the mean argument is the same as the change in the true argument, and therefore the angle in question may be considered as compounded from three angles whose mean rates of change are known. Working with Alfonsine data, for example, we find that the resultant of the three mean angles changes in a year (365 days) by $319° 5' 51''$. In 411 days, therefore, the change is a little under $5'$ in excess of a complete revolution. There are so many factors which could be taken into account to affect the result that there is little point in saying more than that the writer was broadly correct in his analysis.

Thus far it is not possible to form a very clear picture of the device which might have resulted from these manuscript notes. We shall have a better idea after considering the text of f. 147, but for the moment we observe that trains 9 (Moon and Sun), 12 (day indicator), 13 or 14 (lunar epicycle), and 28 (Jupiter) might have been very easily fitted together to provide an accurate and economical representation of these five motions. (The wheel of seven days plays a central role, in this example.) Any one of 37 to 40 for Mars would fit conveniently with 42 for Venus (interposing a ratio of four to one), and 27 for Saturn is easily added to almost any gear sequence, with its integral number of days for the period of the wheel of 8 teeth. The same could be said of the simplest, and at the same time most accurate, train (23) for the Dragon (following the later intentions of the writer), and we are then left with only Mercury to accommodate. The more accurate Mercury ratio is number 43, but as in the case of the most accurate train for Mars, this requires a wheel with well over 1,000 teeth.

### The tract 'Ad laudem et gloriam . . .'

Although there is a certain formality in the opening words to the text of f. $147^{rv}$, the finished result has all the marks of a preliminary essay, or perhaps one only for relatively private use. The phraseology is clipped and the writing is hastily done, whilst the mechanism described—or advocated—is hardly intelligible from the text alone. This begins with an elaborate prayer to the Creator of the world, from the lips of a man who perhaps saw something in common between his work and His. The author has spent a great deal of trouble calculating with the help of ancient tables, and his calculations can only work to the greater glory of the Creator of the things they are concerned to represent. But the writer avers that he was more usefully employed in designing a 'mechanical instrument' with a diurnal rotation, on which the moving Sun marks out the time of day and night. The instrument shows the place of all that moves in the heavens, and by a single motion; that is, presumably, all motions are driven from a common wheel.

There follows a prayer to the glory of the Virgin, by whose intervention human affairs run smoothly. And then the writer gives us his own prescription for a well-ordered universe. First there is to be a wheel (*runtellus*), going round once in 24 hours, and on another instrument, that is, a timepiece. Unspecified wheels link the driving journal wheel to the astrolabe face, and a word which occurs very frequently is *fiala* or *fyala*, which probably is the equivalent of Richard of Wallingford's *nux*, although *phiala* in classical Latin referred to a shallow dish, or saucer, while *MLWL2* lists only the two meanings, 'phial' and 'cruet (eccl.)'. Whatever its meaning, there is a *fiala* fixed, it seems, to an astrolabe plate for the local horizon, under a rete. There is another *fiala* taking a drive to a large *tabula* which contains all the wheels of the Dragon train, and which seems to have wheels for the lunar epicycle. There is also a device for rotating the centre of the lunar eccentric, and another train for the lunar mean motus. The epicyclic train is attached by a pin (*clavis*) to the deferent. The waxing and waning of the Moon is also arranged *ad placitum*. It is at this point that we begin to doubt whether the writer was doing more than close his eyes to envisage the Ptolemaic system as it might be one day turned into metal. The style is not that of a man who has faced difficult practical problems. With the same abandon he goes on to describe the movement of Mercury, with *fiale* and *runtelli*, missing none of the Ptolemaic subtleties, but giving none of the details. And so for Venus; but that is all. The superior planets are not included in the exercise for the time being.

   If the scheme is to be supposed consistent, or even to have materialized, then from the previous notes we have it that the zodiac moved, while there was no mention of a moving Sun. From the design *Pro Sole* we might have supposed a fixed Sun, which would require, if there were an astrolabe dial, both a moving zodiac plate and a moving rete of local coordinate lines (horizon, meridian, etc.). The latter would have been necessarily pivoted eccentrically if the zodiac plate had been pivoted about the geometrical centre of the ecliptic, rather than about the pole. A simpler arrangement would have been to adopt the usual polar pivot, and to constrain the Sun and planets to move along the ecliptic by some sort of sliding joint (as in the Prague clock), if at all. Some such arrangement was probably envisaged for movement in the lunar epicycle. To conjecture further would be as easy as it would be valueless. There are lists on f. 121$^v$ of the wheels which are needed for the representation of the different planetary motions, this time with the superior planets included, but they add nothing of significance to what we may learn from f. 147, of which it seems to be an extended précis.

*On the dates in the manuscript*

   It has already been seen that planetary auges for 1430 are quoted in the notes, and that other notes relate to 1444 and 1445. The more formal, and undoubtedly earlier, sections are such, however, that 1410 might otherwise have been suggested for their approximate date. This would have been on the basis of a series of summary statements on ff. 78$^v$–79$^r$, beginning 'Universe igitur stelle que sunt in parte meridiei sunt 316, quarum in magnitudine prima sunt 7, in secunda 18 . . .' There is in Bodleian MS. Laud misc. 674, at ff. 77$^r$–78$^v$ a text beginning in the same way (with the exception of 'ergo' for 'igitur'), but the resemblance ends after the opening paragraph. The Oxford text is not, as is the Brussels text, a compilation of loosely related remarks, but both draw heavily, in all probability, on material introducing a version of the star catalogue either of Ptolemy, or of ʿAbd al-Raḥmān (better known as al-Ṣūfī). The latter was essentially a revision of Ptolemy's catalogue, hence the similarity in the totals of stars in each magnitude category. It was the basis for the four books on the fixed stars to be found among the Alfonsine books. Immediately preceding these notes in the Brussels MS. comes what is ostensibly a revised version of Ptolemy's catalogue, with fine illustrations of the constellations (ff. 54$^r$–78$^r$). The rubric to some lists which precede it is *Tabula stellarum fixarum Parsius* [sic] *inventa Anno Domini Iesu Christi 1320, festo beati Mathie apostoli* (f. 50$^r$). Returning to the two pages of extracts beginning 'Universe igitur . . .', we find Gulielmus Anglicus of Marseilles cited; and then, after the first five paragraphs, 'Explicit Anno Christi ut supra', presumably referring back to the date of 1320. (A verse from Priscian and an Aratos fragment intervene, introducing the neo-Ptolemaic lists!) Despite the 'explicit', however, the notes continue with five more paragraphs: 'Nota quod motus stellarum fixarum est in anno 38 sec. 49 ter. 25 quar., unde nota quod Anno Christi 1344 incompleto circa nona, diem Martii . . .', and so on, the main object being to arrive at the correct figure for the movement of the eighth sphere, first for the year 1400, basing the calculation on 1344 (1334 occurs once in error), and thence for 1410. Since these paragraphs are all in the same hand as the horological notes (somewhat less formal than the hand of the star lists, although not necessarily that of a different person), 1410 might have been thought close to the date of composition of those notes, had planetary auges for 1430 (as it seems) not been presented on the verso of the same folio.

# APPENDIX 38

## *Walter of Odington and traditions concerning the movement of the eighth sphere*

I⊤ is currently *de rigueur* to maintain, with J. L. E. Dreyer, that it is greatly to the credit of a medieval astronomer that he was not led astray by the 'imaginary phenomenon' of trepidation. The fact is that most of the best astronomers of the time believed that Arabic writers had established the reality of the phenomenon beyond doubt. They were not sufficiently careful readers of Ptolemy's *Almagest* to appreciate the source of the fallacy, but were content to write minuscule histories of the path to the truth. Richard of Wallingford nowhere shows that he was in any sense critical of current ideas, and in the works considered we have seen a number of statements on the subject which are hard to reconcile (vol. ii, pp. 242–4). The astronomical problem is complicated by the extreme slowness of precessional motions. The evidence of this appendix is that the textual problem, as seen through the eyes of an early-fourteenth-century astronomer, must have seemed no less daunting. At the same time, the appendix will serve to exhibit some of the parameters in circulation at the time, and to sort out some of the confusion which exists over the work of Walter of Odington (who, among other things, has been named as the author of *Exafrenon*), John Maudith (on whose star positions Richard of Wallingford seems to have placed some reliance), John Ashenden, and other Oxford contemporaries. Walter of Odington is an interesting case, because it is doubtful whether he belongs to the university tradition which was beginning to dominate the astronomical scene in England by the end of the thirteenth century.

Confusion about Walter of Odington begins over the time at which he is supposed to have lived. All who have written on him have been misled as to his dates, and it will be shown that one of the errors goes back to the end of the fourteenth century. On the fact that he was a monk of Evesham, Thomas Werkworth and William of Worcester, whose texts are reproduced below, are agreed, and there is an almanac with calendar beginning in 1301 which was evidently compiled by Walter of Odington for Evesham Abbey. Since he was, in this case, a Benedictine monk, we can safely refuse to identify him with that Walter of Evesham who, as a member of Merton College, is mentioned in a record of *c.* 1330 as having required a new lock for his door. (See the biography by Henry Davey in the *DNB*.) Following Henry Davey we may also dismiss the Walter of Einesham (i.e. Eynsham, near Oxford) who was chosen, but not elected, Archbishop of Canterbury in 1228. This second confusion originated with a mistranscription by Bale.

It will be seen later that Walter of Odington put the Pleiades at 17° 22′ of Taurus in 1301, which is a fortunate confirmation of the date of the almanac cited, and at first sight independent of it. It is of some interest to try to add another date to this in order to decide, roughly, on the generation to which Walter of Odington belonged. The one date which every writer on Walter of Odington seems to accept without question is 1316, in which year he is said to have been making astronomical observations. This date stems from the title of William of Worcester's abbreviated and modified version of Walter of Odington's text, reproduced here below. William of Worcester was no scholar, and he was writing in 1463. We know the source text through two other versions, one by Thomas Werkworth, also given below, and the other the version which we shall conjecture to have been the original, or a substantial part of it. The former (only) contains the statement that 'in 1316, the Masters of Oxford added 16° 40′ to the places of the fixed stars of Ptolemy's *Almagest*' (Bodleian MS. Laud misc. 674, f. 75ʳ. The earlier occurrence of the same statement, at f. 7, is clearly derivative. A statement on f. 101ʳ, which follows a star catalogue, and which is mentioned again below as deriving from John Ashenden, might also be William of Worcester's.) There is here no mention of Walter of Odington. Moreover, since Thomas Werkworth, writing in 1396, refers elsewhere to 'this year of 1350', he was either blending his basic text with another, or using an already revised version of it. But more important still is the fact that John Maudith (see Appendix 25) was the author of a star table for 1316, which was in small part based on original observations, and in part formed by adding 16° 40′ to some of Ptolemy's stellar longitudes. What is more, this figure was not an easy one for the medieval dabbler in astronomy to have calculated, for it was based on Thebit's theory of trepidation. It seems fairly clear that the 'Oxford Masters' was an allusion to John Maudith, and perhaps his colleagues, and that some writer, perhaps William of Worcester himself, made a careless identification. It is not improbable that he did so, having been misled by a statement in the full text by Walter of Odington concerning the date at which the true equinox would occur in 1316, this being a date in the future, as is evidenced by the tense of the associated verb.

Walter of Odington's second substantial work was his *Icosahedron*, a book on alchemy, in twenty chapters. It has been edited by P. D. Thomas under the title 'David Ragor's Transcription of Walter of Odington's

*"Icocedron"'*, *Wichita State University Bulletin*, University Studies no. 76 (August 1968). The main work ends in at least one copy (MS. Digby 119, f. 147$^v$) with words confirming the Evesham connection. In this work comes a suggestion that certain virtues (heat, dryness, etc.) may be measured quantitatively, and use is made of the concept of intension and remission of qualities. A part of his astronomical treatise, not quoted below, has words which are reminiscent of this fact: 'Sed aliud operantur directi et aliud retrogradi, hoc est intensius vel remissius.'

In our discussion of the authorship of *Exafrenon*, Walter of Odington was introduced as Lynn Thorndike's candidate. We have there given reasons for dismissing the suggestion, originally made on no better grounds than that *Exafrenon*, like *Icosahedron*, is formed from a Greek number-word, and that Walter is said to have written a work *De mutacione aeris*.

It goes without saying that it is necessary to exercise care in identifying every fourteenth-century 'Walter of Evesham' with 'Walter of Odington, monk of Evesham'. The Cambridge manuscript (University Library MS. Ii.1.13), which contains the newly identified text under discussion, has three other works ascribed to Walter of Evesham, none of which appears to have survived elsewhere:

(i) 44$^v$–51$^r$. *De multiplicatione specierum in visu secundum omnem modum.* Inc.: 'Hic investigantur condiciones . . .'

(ii) 51$^v$–55$^v$. *Ars metrica.* Inc.: 'De proprietatibus numerorum secundum Boecium . . .'

(iii) 55$^v$–56$^v$. *Liber quintus geometrie* . . . (Euclid on proportion). Inc.: 'Est prima questio quinti: *Si fiant . . .*'

The manuscript appears to date from the second half of the fourteenth century. The first of the above items is ascribed to *Magister Walter of Evesham* [sic] by the original scribe, while the explicit of the second is joined to the rubric of the third in this way: 'Explicit 60 proprietatibus numerorum, incipit liber quintus geometrie . . . compilatus per M. Walterum de Evesham.'

It seems fairly clear that Walter of Odington, monk of Evesham, was the author of all the works mentioned here. It is not necessary to attach much weight to the 'M[agister]', which was a title apt to find its way to the fore of any scholar's name after a period of time.

Without giving any reason, the sixteenth-century antiquary John Leland suggested that Walter of Odington flourished in 1280. This date is generally ignored, but should be reinstated. In the text reproduced at length below, Walter of Odington is stated to have put the star *Alaioc* (α Aur) at 11° 20' of Gemini, *Alabor* (α CMa) at 4° 0' of Cancer, and *Aldebaran* (α Tau) at 29° 0' of Taurus. Each of these three positions is

precisely 16° 20′ in excess of the corresponding position given by Ptolemy, for the year A.D. 138. The text also makes it clear that Walter of
Odington rejected trepidation in favour of a steady precession of 1° in
70 years precisely. This being so, the figure of 16° 20′ was accumulated
over a period of 1143⅓ years, and therefore the quoted star positions
would have been believed correct around the year 1281. It is not impossible that Leland arrived at his date in this way, although those familiar
with his work will agree that this is unlikely. On the whole, the evidence
is that Walter of Odington flourished during the last two decades of the
thirteenth century. Allowing that the system of monastic education
would not have encouraged competence in handling a turketum—
which he is said, in our text, to have used—by a monk under the age
of 20 or 25, we may reasonably suppose that he was not born much after
1260 and may have been born long before.

It may be added that he was using the turketum skilfully, if we are
to believe that his star positions derive from it. But there is room for
doubt about this. The figure of 1° per 70 years is, however, closer than
most astronomers were wont to come to the correct figure. It probably
comes, as we shall see, from al-Ṣūfī.

## Walter of Odington's editors

We shall not reproduce more than a small fraction of the contents of
the calendrical, chronological, and astronomical composition identified
below as Walter of Odington's, and that will be the part which deals
with the eighth sphere. The two works to be considered, apart from his,
are by Thomas Werkworth and William of Worcester. They were put
together, respectively, in 1396 (or a little after) and 21 June 1453. So far
as I know, only one copy of each survives. Out of them a composite text
will be assembled, divided into roughly three parts:

(i) (In Werkworth's version only.) A general history of the theory
of the movement of the equinoxes, mentioning Ptolemy, Albategni, Thebit, Almeon, Arzachel, Alfonso, 'the moderns' (associated with 1350), 'many Oxford astronomers' (perhaps the
same, also associated with 1350), 'the masters of Oxford' (see
above, for the possible confusion with John Maudith), Simon
Bredon (*circa* 1340 is mentioned, and a calculation of 1357, using
his figures).

(ii) (The second part of Werkworth and the first part of Worcester
are virtually the same.) A more elementary history than (i), for it
is mostly concerned with actual star positions and their changing coordinates as recorded down the ages. The astronomers

mentioned are Hermes, Hipparchus, Ptolemy, Albumasar, Thebit, Arzachel, Albategni, Grosseteste, Walter of Odington. In addition, Werkworth mentions an observation (or calculation?) of his own, for 1396, which Worcester omits.

This common part begins half-way to turn into a discussion of the limitations set on judicial astrology by the fact of changing star coordinates.

(iiia)  Werkworth adds some of Walter of Odington's data, with a remark which might be an allusion to trepidation.

(iiib)  Worcester continues, trivially, on the dilemma in judicial astrology, and the way to avoid mistakes. This section will be shown to have been drawn from Walter of Odington's text, as I originally conjectured was the case with (ii), before finding the corresponding passage in the Cambridge manuscript.

More difficult to decide is the authorship of (i), which does not have the support of the Cambridge manuscript. Several quotations will be given below, suggesting sources for the writer's facts, as well as for his 'historical' inspiration. A missing part of Walter of Odington's text could have mentioned the astronomers up to Arzachel, and Worcester might still have left the subject alone, if trepidation was beyond the compass of his intellect. Not only is Walter of Odington not mentioned by name, but judging from (ii) he was himself vague about trepidation. Perhaps that is why he stuck to the uniform motion of the eighth sphere. On the whole, Werkworth's remarks about Simon Bredon, and his examples for 1350 and 1357, suggest that he was using some such source material as the book by John Ashenden (see below).

Of the lives of the two editors, Worcester's is by far the better known. Apart from referring the reader to K. B. McFarlane's 'William Worcester; a Preliminary Survey',[1] it is worth adding that Worcester was a prolific collector of books and papers, with an antiquarian rather than a scholarly interest in them. Born in 1415, he was a scholar at Oxford at the charge of Sir John Fastolf, whose ill-paid secretary he was until Fastolf's death in 1459. As an antiquarian he might be looked on as one of the first of a line which included Leland, Bale, Twyne, Wood, and Hearne. Although he had an interest in humanism, his Latin, as McFarlane points out, was unpolished.

Of Thomas Werkworth, nothing whatsoever appears to be known, apart from the fact that he compiled the treatise on the eighth sphere. There were later members of Oxford and Cambridge universities with his surname (sometimes spelt Warkworth), but apparently none called Thomas at this time.

---

[1] In *Studies presented to Sir Hilary Jenkinson* (Oxford, 1957), pp. 196–221.

*The manuscripts*

(D) The Werkworth text occurs only in Bodleian MS. Digby 97, f. 143$^{rv}$. Its style suggests that it was copied fairly early in the fifteenth century. The heading 'Tractatus 8 sphere Thome Werkwoothe' is in a librarian's hand of the seventeenth or eighteenth century, but is justified by the contents.

(L) The Worcester text occurs only in Bodleian MS. Laud misc. 674, ff. 75$^r$–76a$^v$. This is an unusually small codex, with proportionately compressed writing. Its style suggests that it was copied in the last half of the fifteenth century, not long after it was composed.

(C) The great bulk of the Walter of Odington material, as already outlined, is to be found in Cambridge University, MS. Ii.1.13. Further details as to its contents will be given below. The crucial 'Stelle fixe . . .' passage begins on f. 180$^r$.

Since there is a substantial section which is virtually common to the end of the Werkworth text and the beginning of Worcester's, and since the intrinsic interest of the second half of Worcester's text (still derivative, of course) is negligible, the Werkworth text is taken as basic, and Worcester's variant readings are noted in footnotes, unless they are used to correct it. In this way, the two plagiarized versions are presented, leaving the much lengthier original to be dealt with in outline at the end. Footnotes concerned with textual matters are given letters of the alphabet. Superscript numerals refer to the sections of the commentary which follows. *Not all sections of the commentary are noted against the text.* The only other person to have discussed either of the texts in print, to the best of my knowledge, is Lynn Thorndike: 'Thomas Werkwoth on the Motion of the Eighth Sphere', *Isis*, xxxix (1948), 212–15. Thorndike does little more than give a translation of the one text, together with a useful collection of references to articles discussing the significance of the name 'Almeon'. When Thorndike's readings (inferred from his translation) differ from mine, the fact is noted.

⟨*Tractatus de motu octave spere*⟩

(i) NOTA pro motu octave spere quod Ptholomeus posuit eam moveri uniformiter ab occidente in oriente in 100 annis gradu uno[1]; et post eum Albategni[6] fuit primus qui invenit errorem in posicione illa. Eo quod invenit polum octave spere accedere sensibiliter ad polum none spere, sed non fecit consideracionem super hoc. Postea venit Thebit[4] et sui sequaces, scilicet Almeon[5] et Arzachel[4], qui posuerunt motum accessus et recessus octave spere secundum motum parvi circuli descripti a capite Arietis octave spere circa caput Arietis none spere; sed posuerunt centrum illius circuli continue quiescere in capite Arietis none spere. Et sic posuerunt motum octave spere esse difformem. Postea Alfonsus[7] voluit concordare cum utroque Ptholomeo scilicet et Thebit. Posuit utrumque, sed posuit centrum predicti circuli parvi continue moveri uniformiter versus orientem. Posuit etiam motum augium planetarum esse uniformem, et propter hoc totum motum octave spere difformem, propter eius accessionem et recessionem. Sed ista posicio contradictionem includit. Nam cum auges omnium planetarum moventur ad modum octave spere, contradictio est quod auges planetarum moverentur uniformiter et spera octava difformiter.[8] Et propter istam causam multi reputant tabulas Alfonsi magis falsas tabulis Arzachelis; tamen utreque fient false per processum temporis, sed adhuc tabule Alfonsi veriores fient tamen magis false per longum processum temporis.

Unde tenent moderni quod polus octave spere accedit continue polo mundi in quinque annos minuto uno; et quod octava spera movetur versus orientem modo scilicet anno Christi 1350 adquirendo in 60 annis gradum unum. Cum tamen tempore Ptholomei in 100 annis quesivit[a] gradum unum; unde tam iste motus octave spere versus orientem quam motus accessus poli eius ad polum mundi continue est velocior et velocior.[9] Nam octava spera adquesivit[b] per motum eius versus orientem a tempore Ptholomei usque in hunc annum Christi 1350 18 grad. et 10 min., sicut expertum est Oxonie per multos valentes; et polus octave spere adquesivit[c] a tempore Indorum usque ad tempus Ptholomei nona minuta; et a tempore Ptholomei usque ad tempus Albategni 15 min.; a tempore vero Albategni usque ad Arzachelem 4 min.; et ab Arzachele usque modo anno supradicto 5 min. Nam maxima Solis declinacio modo est 23 grad. et 27 min., cum tamen fuerat tempore Indorum 24 grad.

Et Ptholomeus fecit consideraciones suas per 138 annos post Christum, et anno Christi 1316 addiderunt Magistri Oxonie super loca stellarum fixarum in *Almagesti* Ptholomei 16 grad. 40 min., quia tantum processit octava spera a tempore Ptholomei. Nota etiam quod 21 grad. Cancri in spera nona est modo in directo earundem stellarum fixarum atque ymaginum in quarum directo fuit tercius gradus Cancri tempore Ptholomei. Nam octava spera a tempore Ptholomei usque in annum Christi 1357 pertransivit[d] 18 grad. 17 min., quia modo in quolibet anno pertransit unum minutum. Ista patent secundum Simonem de Bredone,[e] qui circa annum Christi 1340 equavit motum octave spere cum maxima diligencia.[f']

(ii) STELLE fixe alium habent motum preter cotidianum, sicut patet per observaciones[f"] notas. Nam stellam Aldebaran invenit Hermes 25 grad.

<sup></sup>  a D quesunt.         b  D adquisunt.          c  D adquisunt.          d  D pertransunt.
  e  There is a final flourish on the *n*.                                       f'  L +Memorandum quod.
  f"  L consideraciones.

17 min. habere Arietis, scilicet a nodo habendo respectum ad nodum equinoxii; quam quidem invenit Abrachim[g] habere 10 grad. Tauri; et Ptholomeus 12 grad. 40 min. Tauri; et Albumasar[h] 19 grad. 15 min. Tauri; Thebit 21 grad. 17 min. Tauri; Arzachel 23 grad. 20 min. Tauri; Albategni 24 grad. 30 min. Tauri; Dominus Robertus[i] Lincolnie Episcopus 28 grad. 40 min. Tauri; et nos, ut dicit Odyngton,[j] per observacionem turketi et per compotum, inveniemus eam habere 29 graduum Tauri complete. Et ego, Thomas Werkwooth, in anno Christi 1396[k] perfecto, inveni eam in Geminis 0 grad. 24 min.[l] Unde, ut[m] dicit Odyngton, sunt motus stellarum fixarum et augium planetarum in 70 annis gradus unus,[11] ut patet in brevi tabula precedenti, in folio precedenti.[n] Unde[o] etiam patet quod motus stellarum fixarum est semper progressivus, et[p] non reductivus. Quapropter iudicia referenda sunt ad motum earum, hoc est ad[q] progressionem. Quare enim maius[r] debemus inniti radicibus Thebit vel Arzachelis qui ponunt ad aliquos[s] gradus spere reductionem quam radicibus antiquorum, cum experiencia instrumentorum agat[t] pro antiquis in progressione spere, et cum[u] per viam mundus fit finem adepturus, quando, scilicet, spera illa alium cursum complevit,[v] secundum opinionem dico antiquorum.[w]

Quod autem stelle[x] fixe et planete habent[y] referri ad coniunctionem circulorum, et non econverso, patet per hoc, quod horas accepimus domus[z] a Sole qui dirigit equinoctium. Et sic componunt instrumenta. Et domos accepimus secundum horas. Aliter Sol in meridie existens diceretur una hora post meridiem horam facere meridianam. Diceretur-que quod nona domus esset decima. Et sic vacillarent iudicia.

(iii a) NOTANDUM etiam quod secundum Walterum Odyngton mona-chum de Evesham, qui composuit hanc tabulam precedentem, quod Pliades fuerunt anno Domini 1301 in 17 grad. 22 min. Tauri; et Alaiok in 11 grad. 20 min. Geminorum; et Alabor in 4 grad. Cancri;[12] accipiendo longitudinem respectu equinoxii non plus figura est ad talem figuram: 8.[zz]

---

[g] C Abracam; L Abrachas.                   [h] L Albumazar.                   [i] L+Grostet.
[j] C naturally omits 'ut dicit Odyngton', and the sentence begins 'Et ego ...'
[k] Thorndike, op. cit., claims that Macray misread this figure, but it was his own '1395' which was wrong. (W. D. Macray compiled the catalogue of Digby manuscripts, Oxford, 1833.)
[l] L omits sentence.                                                        [m] L 3λ.
[n] Neither D nor L has such a table, but it is present in C.                [o] L 3λ.
[p] L +quod non est aliquis motus earum; C +et quod non est motus earum.
[q] D λ.                            [r] C magnus; D 2λ.                    [s] CL aliquot.
[t] L 3λ+ qui agant per antiquos.          [u] L λ.              [v] L complementer.
[w] C aliquorum.                            [x] L λ.                        [y] L habeant.
[z] D omits 'a Sole ... et domos accepimus'. C omits 'domus' only.
[zz] Thorndike, in a footnote, shows that he reads the ending as 'non plus figura est ad talem signum 8'. He offers two translations: 'Taking the longitude with respect to the equinox, the

(iii b) SOLEM non minus cum divertit ad septentrionem . . .

[*This section, omitted here, ends as follows:*] Explicit abbreviacio declarata de motu octave spere per Walterum ef Evesham compilata per Willelmum Wyrcestre, die Martis 21 Iunii anno Christi 1463, in Norwici civitate in vico de Pokethorp.

Addicio equacionum motus stellarum fixarum a tempore Ptolomei, videlicet usque annum Christi 1415, quo anno W. Wygorniensis natus fuit, videlicet 18 grad. 50 min. 32 sec. 34 ter. 27 quar.

## On the reconstruction of the original work by Walter of Odington

Apart from minor additions and a few variant readings, sections (ii) and (iiib) are to be found in manuscript C, in association with an almanac which the catalogue of Cambridge manuscripts ascribes to Walter of Evesham. The ascription is almost certainly correct, but was equally certainly made for a reason which by itself is insufficient: the difference in longitude between Evesham and Marseilles is stated at the side of it. This nevertheless adds weight to our identification based on the overlap of texts, and on some remarks by Ashenden referred to later. There is reason for thinking that 24 folios, beginning with the almanac, form some sort of unity, and are all by the one author, and a survey of this part of the manuscript is therefore made here. Folio numbers are different from those in the catalogue, which is in any case very misleading, and will not be referred to again, beyond mentioning that the old folio numbers are 10 or 11 less than the new.

f. 160$^r$. 'Cum loca planetarum . . .' The text is placed alongside the almanac, which resembles that of Profatius both in style and in having commenced in 1301, but not in the planetary positions it predicts. The differences are of the order of a few degrees at most, but a first impression is that the Evesham almanac is the more accurate of the two. It is curious that there is no readily apparent connection between the two, since the text goes on to compare the longitude of Evesham with that of Marseilles. The two in any case probably share their source, namely the almanac of Arzachel, of which more is said below. There is only occasional glossing, until f. 170$^v$.

f. 170$^v$. The *Kalendarium* begins. As in the almanac, the year begins with 1 January. There is no distinctive heading, which suggests continuity with the first item. The central column of kalends, saints' days, etc., is flanked by a table of conjunctions of the Sun and Moon, the first year of the (19-year) cycle being 1292, and a table of the Sun's true

figure to such a sign is not more than 8.' 'Taking . . . the figure to such a sign is no longer 8.' I can only guess at one further interpretation: 'Taking the longitude with respect to the equinox, longitudes of the stars never vary by more than 8 degrees from some mean position.' The section is not in C, but is compiled from a star catalogue in C (see below).

position, beginning with the year 1300. To the right of the latter comes a table giving the motus and argument of the Moon in days. The pattern is repeated for the twelve months, each being given a single side. Further tables are given under the calendar and following it—hourly motions of Sun and Moon, radices, tables of equations, eclipse limits, movement of the Moon's nodes, and so on. Some are lifted from *Almagest*, including that for solar declination (= *Almagest* I. 15). Folio 172$^v$ has a note about the devastation wrought by the plague, but in a different hand. The tables are of interest in that they include tables for latitude, which many medieval texts are happy to ignore (178$^v$–). Aux and node positions are listed on the first page of the latitude tables. On f. 179$^r$ is a canon for their use: 'Ad habendum latitudines planetarum . . .'

f. 179$^v$. Star catalogue, listing the coordinates (longitude and 'latitude'—in reality declination) of 35 stars. This number is suggestive of a revised version of the Arzachel list (Kunitzsch, op. cit., Typ. XIII), which in many forms has longitude, latitude, declination, and mediation. The present list is emphatically not a rearrangement of the earlier catalogue, let alone a trivial one, nor do the longitudes of stars common to both simply incorporate a precessional constant added to the earlier figures. The claim of the text that calculation and the use of a turketum were resorted to seems to be borne out by the coordinates cited, although the matter bears closer investigation. Those longitudes which differ from Ptolemy's by a constant amount are 16° 20′ greater, as in the case of the star much cited in the text, Aldebaran. The longitudes are all of the order of 20′ less than those given by John Maudith in his catalogue of 1316, which is another reason for thinking it a mistake to associate Walter of Odington's catalogue and tables with the year 1316.

f. 180$^r$. Table giving the dominical letter. Beneath it is the text of (ii) above: 'Stelle fixe alium habent . . .' That part of the text which continues as (iiib) above suggests that the author is familiar with Ptolemy's *Almagest* and *Centilogium*. There are six lines over and above those reproduced by William Worcester, before the next section commences, and these seem to refer back to the lunar and eclipse tables, and to form a natural introduction to the next section. This fact, together with the scribe's habit of using the word 'explicit' strictly at the end of a work, and a wavy line elsewhere, is one of our reasons for thinking *all* of manuscript C considered here to have been composed by Walter of Odington as a single work, although not necessarily complete as we have it.

f. 180$^v$. 'Luna 4 habet circulos . . .' A section dealing with the Ptolemaic theory of the Moon's motion, competently written, and perhaps from Campanus.

f. 181$^v$. 'Cum motus planetarum possit salvari absque ecentricis et epiciclis . . .' This section seems to relate to the Evesham almanac with which the work began. It is a statement of broad principles underlying the construction of an almanac, and although not very explicit, predisposes us to think that the earlier almanac was composed on an arithmetical, rather than on a trigonometrical, basis. This short section, less than a page long, mentions only one authority, namely Humenuz. This reference is highly significant. Before considering the identity of this 'compiler of the first almanac', we may quote from the Evesham text: '. . . Secundum predicta accidencia componuntur tabule manuales quas almanach vocant, cuius primo fecit Humenuz philosophus rogatum filie Ptholomei ad quarum similitudinem propter promtidunem [promptitudinem?] istas elaboravit.'

This curious statement is typical of many to be found in medieval texts concerning Humenuz, and has led some to suppose that he was to be identified with some Hellenistic astronomer. (For a summary of the discussion, see *Estudios sobre Azarquiel* by J. M. Millás Vallicrosa (Madrid, 1943–50), *passim.*) The fame of Humenuz extended at least until the late fourteenth century, as witnessed by the following quotation from an anonymous text of the time, written in English (Trinity College, Cambridge, MS. O.5.26, f. 112$^v$): '. . . and another olde man that was seyde Humenidis, which calculed perpetuel tables of the almanak to Cleopatre, the doughter of Ptholomei.' This is obviously a corrupt version of an already corrupt account to be found in the only Latin text of which I am aware purporting to be by Humenuz, namely that reproduced in the work by Millás already referred to (pp. 379–92), which text begins:

Sciendum quod Humeniz philosophus summus Egipciorum, magister filie Ptholomei, composuit istas tabulas equacionum planetarum super annos Egipciorum quas Azachelus Grecorum philosophus de annis Egipciorum ad annos Alexandri magni mutavit. Post hoc magister Iohannes Papiensis eas transtulit ad annos Christi.

Now in the Oxford manuscript Laud Misc. 644, of which Millás made no use, the canon beginning with these words is ascribed to Humenuz (the ascription cannot be strictly correct), and is adjacent to an almanac which it is usual to assign to Arzachel. This almanac has been examined in some detail by M. Boutelle ('The almanac of Azarquiel', *Centaurus*, xii (1967), 12–19), using an electronic computer to assist in the analysis. In attempting to discover the identity of the original compiler of the tables, an attempt was made to find a date on which the initial entry in the first column of each of the planetary tables was correct, but 'no set of positions reasonably close to this has been found' (op. cit., p. 15). My impression had been that the initial positions corresponded with reasonable accuracy to those obtaining on 1 September (the beginning of the year of the tables) A.D. 994. The proximity of the entries can be judged from the following table:

|                       | Saturn | Jupiter | Mars | Venus | Mercury |
|-----------------------|--------|---------|------|-------|---------|
| (Almanac)             | 342    | 118     | 162  | 149   | 173     |
| (Tuckerman's tables)  | 343    | 123     | 162  | 150   | 185     |

Between 300 B.C. and A.D. 1300 this was the only occasion for which I had found any measure of agreement. In A.D. 994 there was an astronomer very much concerned with the royal house of Egypt, namely Ibn Yūnus. In identifying him with Humenuz, one might consider as good evidence not only the similarity of names, but the similarity of 'Humenides' to 'Johannides', the transcription used by the seventeenth-century scholar Edward Bernard, in copying out some star positions from an Arabic manuscript which he fails to name, of the Hakemite tables. (See *Phil. Trans.*, no. 158 (1684), 567-76.) The Hakemite tables are usually said to have been compiled over the period 990 to 1007 by Ibn Yūnus, which fits well with the hypothesis that he is to be identified with Humenuz.

There is one strong objection to the simple account of the compilation of the Humenuz/Arzachel almanac presented here, which may well explain why others have refused to consider the resemblance between the initial almanac figures and those corresponding to the actual planetary positions as being significant. The reason is that although the correspondence is apparently much closer than can be explained by chance, in some cases the column chosen to come first is ostensibly the wrong column of the cycle. This could nevertheless be explained by careless rearrangement of some still earlier tables, and even the fact of a single planet's being at year 1 in 994 would seem to me to be corroborating evidence for the correctness of that choice. No casual survey of the almanac can do justice to the situation. Humenuz was apparently not alone in making the mistake of identifying the wrong point of the cycle, for it seems from a casual examination that Arzachel did so in the case of Jupiter, with which 994 is correctly column 1, and 1088 is column 19, and not column 66, as he suggested. (If this is a correct diagnosis, it explains why the sum of the squared deviations accumulated so rapidly in Miss Boutelle's computation.) Despite these difficulties, it should be possible to distinguish between the three sorts of adjustment involved in an analysis of the almanac: that incorporated by the original compiler, to allow for the movement of the apse lines of the planets from the time of Ptolemy, that recommended by the author of the canon, first Arzachel and second the Latin reviser of the Humenuz canon, and last the correction which we know to have been necessary to bring us to the actual position of a planet. The first was calculated by Miss Boutelle, and for the three outer planets was about 10°. It was a natural inference from this that, since astronomers were not inclined to make precession faster than 1° in 66 years, the date of composition must be later than about

A.D. 800. This presupposes that the method of compiling the almanac has been correctly determined. (In the study by Miss Boutelle it was assumed that an epicyclic model with eccentrics was the correct one.)

The second adjustments, namely those recommended by Arzachel, were quite insufficient to bring the mean paths of the planets into their correct positions, and even after applying the suggested correction, all positions were found to fall short by between 7° and 13°. Where, as with Arzachel, the suggested corrections are considerably less than this, it is natural to suggest two possible explanations. First, it may well have been intended that longitudes be measured from some star, presumably about 8° from the equinox at the time of compilation. The star α Arietis would only be a candidate for this honour in the first century B.C., or there-abouts. The second suggestion is that Arzachel was uncritical of the almanac, and that the errors in his longitudes are due to a *true* movement of the respective apse-lines. This would allow a very rough estimate of the date of composition, not dependent on the beliefs of the original author. Clearly one cannot decide between these alternatives without a careful search of relevant texts. It is noteworthy that the later Latin 'Humenuz' text gives quite reasonable correction terms, to allow, as it claims, for the movement of the eighth sphere.

That it is imperative to look carefully into the matter of texts, especially those dealing with the *construction* of almanacs, should be obvious. Most historians of astronomy seem to believe that medieval almanacs were obtained by repeated application of the ordinary Ptole-maic methods of evaluating planetary longitudes, perhaps with a few final modifications to yield the necessary cyclical relations. This is indeed the impression given by some of the texts themselves. Walter of Oding-ton, however, seems to be hinting in the present section that a much more thorough application of arithmetical methods was in use, at least in his own case. If he is at the end of a long tradition of almanac com-puting, the lines of transmission are, for the time being, obscure. Broadly speaking, in this all too short section, he seems to believe the following sorts of information essential (taking Mercury as an example):

(i) the number of synodic periods, and the number of revolutions round the deferent, in the number of years making up the cycle;
(ii) the number of degrees and the number of days in each of three retrogradations—maximum, minimum, and mean;
(iii) the same also for the three types of forward motion;
(iv) the period in the epicycle; and
(v) the number of degrees and minutes between one maximum elongation and the next in the same sense.

He also gives the gradient at either side of the stationary point (1° in five days). It is this last information which makes it seem likely that Walter of Odington did not contemplate the construction of his almanac by strictly trigonometrical methods, but by an arithmetical procedure. Thus in the case of Mercury, it is possible that the intention was to divide the 46 years of the cycle into 348-day periods, and an examination of the almanac suggests that these were to be made up as follows:

$$95 \,(D) \quad 25 \,(R) \qquad 91 \,(D) \quad 21 \,(R) \qquad 93 \,(D) \quad 23 \,(R) \quad (D\text{—direct, R—retrograde})$$

with    129        15        122        8        126        12

as the corresponding movement of the planet in degrees of longitude. According to Ptolemy (*Almagest* IX. 3), 145 cycles of Mercury's epicycle are performed at the same time as $1°$ in excess of 46 cycles of the deferent. In one cycle, therefore, the epicycle covers on average $114° \, 12' \cdot 8$ of longitude (cf. $129° - 15°$, etc., above). Bearing in mind the fact that the epicycle centre moves with the mean solar motion, we look for the time in which the mean Sun covers this distance in longitude, and find that it is only two or three hours short of 116 days (cf. one-third part of the 348-day period above). In other words, the almanac fits very well with Ptolemaic data at the level of averages, even though there are many respects in which it falls short of a precisely calculated Ptolemaic almanac.

We are still faced with the problem of deciding how the intervening values were filled in. Very tentatively I suggest that a few cycles were fully calculated on the basis of a Ptolemaic model, and an arithmetical procedure was determined which gave a good approximation to it. The algorithm would go something like '$1°$ over the first five days, $2°$ over the next five, $10°$ over the next ten, . . .', and so on, the motions being uniform over greater or longer periods, as in the Babylonian texts. It would be difficult to decide between the claims of such a scheme and those of the strictly trigonometrical scheme—with rounding off, of course. The only satisfactory solution is to make a thorough search for further texts of the sort found in the Cambridge manuscript, and perhaps have an electronic computer set in motion along arithmetical rails.

f. 181ᵛ. The last section of what we believe to be essentially a single text, or at least the work of a single man, is a chronological work beginning 'Ait Solinus . . .' This work mentions some precession constants, and it is clear that the author, as we know was the case with Walter of Odington, believes in a movement of $1°$ in 70 years and an Aldebaran position of $29°$ Taurus. But more significant is the mention of the year (then in the future) 1316, which would explain the misunderstanding, if such it was, by the two abstractors of Walter of Odington's work; and still more significant are the many references by John Ashenden to a work on chronology by Walter of Odington. (See Comm., Note 10.)

Moreover, at the end of the section (184ʳ) the word 'Explicit' occurs, for
the first time since the beginning of the Evesham almanac. It is for these
reasons, together with the internal evidence presented above, that it is
possible to say with some confidence that all the intervening work is by
Walter of Odington, although it might represent only a part of an even
longer work.

## COMMENTARY

(Where there is no reference from the Latin text to one of the following
numbered sections, the number will be given an asterisk.)

1. Our most reliable knowledge of the early history of precession comes
from Ptolemy's *Almagest* VII. 2. According to Ptolemy, Hipparchus discovered the precession of the equinoxes, by comparing his own measurements of stellar longitude—using lunar eclipses—with those of Timocharis, about 150 years earlier. Hipparchus was said to have written about
this in his *On the Precession of Solstitial and Equinoctial Points*. The figures
quoted show that he measured Spica as 6° west of the autumnal
equinox, whereas Timocharis had recorded it as nearly 8° west. From
this Hipparchus would have argued for a precessional movement of 1°
in a little over 75 years. Ptolemy went on, moreover, to quote another
work by Hipparchus, *On the Magnitude of the Solar Year*: 'If for this reason
the tropics and equinoxes move westwards not less than 1/100° per
annum, then they must needs move not less than 3° in 300 years.' This
quotation, with its phrasing in terms of an inequality, is perfectly con-
sonant with the earlier figures. Ptolemy, however, deduced his own
figure. Having made observations with an 'astrolabe' (not the plane
astrolabe, of course) of Regulus (α Leo), which he placed at $32\frac{1}{2}°$ from
the Summer solstice, he compared this position with that given by
Hipparchus ($29\frac{5}{6}°$ from the solstice) for a date 265 years earlier. The
eastward movement of the stars, using this date, is almost exactly 1° in
100 years ($2\frac{2}{3}°$ in 265 years)—'as Hipparchus seems to have guessed',
says Ptolemy, somewhat disingenuously.

In no section of our Latin text is there evidence that the writer was
familiar with these facts. Turning to the next section of the *Almagest*
(VII. 3), we come to Ptolemy's proof that the eastward drift of the stars
takes place about the pole of the ecliptic, as opposed to the poles of the
equator. (The point was not discussed by Werkworth or Walter of
Odington.) There we find equatorial latitudes quoted for (amongst
others) 'the bright star in the Hyades', namely Aldebaran. (Timocharis
gave $8\frac{3}{4}°$ N; Hipparchus $9\frac{3}{4}°$ N; Ptolemy 11° N.) The drift of the argument is to show that these latitude changes are what would be expected
from a stellar movement of 1° per 100 years centred on the pole of the

ecliptic. Neither the longitude of Aldebaran nor those of any other stars are adduced to verify directly the figure of 1° per century. Where, then, did Walter of Odington get his Aldebaran data for Hermes and Hipparchus? We can only assume that he—or someone before him—took the longitude from Ptolemy's catalogue, subtracted 2⅔° to arrive at a Hipparchan figure in keeping with the data for Spica, and worked on the assumption of a movement of 1° per century in deducing a figure for the mythical Hermes. In this case we should expect Hermes to have been supposed alive *c*. 1600 B.C.

2.* Ptolemy's figure of 1° per century was accepted by Theon of Alexandria, but nevertheless from Theon we have the first intimation of a theory of trepidation:

According to some opinions, ancient astrologers believe that from a certain time the solsticial signs have a movement of 8° in the order of the signs, after which they return to the same distance. Ptolemy, however, is not of this opinion; for dispensing with the idea in the calculations, when these are made from the tables they invariably agree with the [star-]positions observed. We also, therefore, recommend that the correction be not used, although we shall explain it.

It is assumed that 128 years before the reign of Augustus [i.e. 155 B.C.], the greatest movement (8°) having occurred in a forward sense, the stars begin to move back. We add to the 128 years before Augustus 313 years to Diocletian and 77 years since his time. Since in 80 years the motion is 1°, we take an eightieth part of the sum. The result [6° 28′ 30″ precisely], subtracted from 8°, will give the amount by which the solsticial points will be in advance of what the tables predict.

(Translated from Halma's ed. of the *Commentaires de Théon . . . sur les Tables manuelles*, vol. I (1822), p. 53.)

3.* The 'ancient astrologers' mentioned by Theon (see 2* above) are apparently the 'antiqui' of the text (halfway through section (ii)): 'We should be more willing to take the radices of Thebit and Arzachel, who suppose that the sphere turns back at some point, than the radices of the ancients [antiqui].'

It is unlikely that Theon's work was known to Walter of Odington. He is more likely to have obtained his information from an intermediate source, perhaps one of those excerpted here (see 4, 5 below). Of the possible sources, first there is this passage from the Pseudo-Aristotelian *De proprietatibus elementorum* (1496 edition of Aristotle in Latin, Venice, published by Gregorius de Gregoriis, f. 367ᵛ):

Aut est illud propter rem quam dixerunt auctores Atalasimet orbi signorum est accessio septem partium et recessio octo partium in omnibus octoginta annis gradu . . . aut est illud propter mutationem orbis stellarum fixarum: et ipse permutatur in omnibus centum annis gradu . . . et hoc est ultimum rerum omnium super quam imittuntur auctore recidivationis et sententia quam inducunt.

This is clearly a muddled passage which, in addition, has suffered much in the process of copying. It differs at no fewer than 12 points, for example, from the late thirteenth-century text in Bodleian MS. Auct. F.5. 28, f. 217$^{rv}$. The most important characteristic is that the manuscript makes the access equal to the recess, that is to say, 8 parts. It therefore agrees exactly with section [9] of the *Declaraciones supra Kalendarium*, which was possibly by Richard of Wallingford (vol. i, p. 561). On the meaning of 'Atalasimet', see Note 5 below. The manuscript renders this word very obscurely as 'Achata philosophi mech'.

The pseudo-Aristotelian passage (on which Albertus Magnus commented in his *Liber de causis proprietatum elementorum*, I. ii. iii) seems to be a reference to the ancient theory of trepidation. This seems fairly certain, the more so when the 'antiqui' are compared with those whom Abraham ben Ezra (Note 15 below) called *magistri ymaginum*, and with those whom Massahalla had, four centuries earlier, called 'the first authors to make *imagines* according to Altasamec', who 'said that there are nine spheres'. (*De elementis*, Nuremberg edition of 1549, cap. XX, beginning 'Et orbis secundus ab orbe magno, est orbis signorum, et est orbis non stellatus iterum, et motus eius est ab oriente sicut motus orbis magni. Et dixerunt autores primi facientes imagines secundum Astronomiam Altasamec, primum quod ipse est unus ex circulis orbis magni, et orbes omnes sunt novem.')

Plato of Tivoli(?) used words reminiscent of these when translating Albategni's *De scientia stellarum*, cap. 52. Albategni was clearly referring to the same authors as Theon, and in claiming to have taken his information from Ptolemy, he may well have been making a simple mistake. The following passage is from the printed edition of 1645 (Bologna). (The manuscripts differ from it at several points, only one of which seems to be significant here. The opening words in Bodleian MS. Canon. misc. 61 (*c.* 1300) are 'Quod ymaginum auctores celum . . .' and *ymaginum*, rather than *imaginaverunt*, is obviously correct. In this manuscript, Ptolemy becomes 'Batholomeus'.)

Quod imagina verunt authores caelum in longinquitate temporum in 80 scilicet annorum spacio unius gradus habere motum alteracionis olim asserere videbantur, Ptolomaeus manifeste in suo libro declarat. Hunc autem motum usque ad 8 gradus in anteriori parte cresecere, et post illo eodem tramite ad posteriora redire dicebant . . .

Albategni goes on to explain the difficulty of ascribing two motions to a body. He quotes Ptolemy's figure for precession ($1°/100$ years) and his own ($1°/66$ years).

It is generally held that Massahalla's work was translated by Gerard of Cremona, or his school, and even if we assume that it is correct to ascribe the Albategni translation to Plato of Tivoli, roughly contemporary with Gerard, only a very close study of the translations is likely to decide which came first. When Albertus Magnus used the very same terminology, however, he clearly had the Massahalla work at his elbow, as is evident from the following passage (*De caelo et mundo*, ed. Borgnet, iv. 196):

> ... circulus declivus primus habens duos motus, quorum unus est diurnus ab Oriente in Occidentem, et alter declivus ab Occidente in Orientem tardissimus, qui est in omnibus centum annis gradu uno. Auctores autem imaginum ponunt etiam istum motum, et ad ipsum potissime referunt effectum circuli obliqui in motibus planetarum omnium: et hanc opinionem judicamus veram, et ipsam sequitur Messalach in libro de *Sphaera mota*.

Al-Biṭrūjī was more explicit (see Note 16 below), in speaking (Michael Scot's translation) of 'antiqui sicut Hermes et illi qui fuerunt post ipsum sicut componentes imagines'. In the Calonymos translation this becomes 'ut Hermes et posteriores eo qui dant operam [*sic*] imaginibus'. The explanation of the phrase 'masters of images' is to be found in B. R. Goldstein's modern translation of al-Biṭrūjī, where it becomes 'Masters of the Talisman'. See Note 16, below. The phrase certainly appears to refer to those astronomers to whom Theon alluded, since they were also said to give 'istis stellis motum aliquando secundum signum aliquando contra signa . . .'

Here then are six writers from whom Walter of Odington could have gleaned his very incomplete information about these particular 'ancients', and two of them, as we shall see, were possible sources for almost all his early information. There are fewer comparable sources, at least in the thirteenth century, than might be supposed. Thus although Grosseteste gives in his *Compendium sperae* a straightforward (and reasonably good) outline of the rival theories of Ptolemy and Thebit, it is not historically slanted. Much the same may be said of Gulielmus Anglicus, and Campanus of Novara, neither of whom, however, wrote very clearly on the subject, and both of whom indulged rather more than Grosseteste in name-dropping. Gerard of Cremona's *Theorica Planetarum* is concerned only with precession, quoting Alfraganus and Albategni. Sacrobosco has nothing comparable to Walter of Odington's treatment, and the same goes for his commentators.

Finally, although there are some obvious problems raised by the quotation, evidently coming in the last resort from Theon, they do not concern us here. One of these relates to the choice of the year 155 B.C. as the epoch at which the change in direction of the motion of the equinoxes occurred. Was that the year in which Hipparchus was supposed to have deduced the movement of the eighth sphere? Were the 'antiqui' trying to go one better than Hipparchus, without actually contradicting him? It seems more likely that some inferior astronomer got hold of a set of inconsistent positions of the equinox, as stated by different writers down the ages, and tried to reconcile them. (For further details see J. L. E. Dreyer, *History of the Planetary system from Thales to Kepler* (1905), p. 205.)

4. The theory of trepidation as it was most widely known in the Middle Ages was, of course, that of Thebit. An edition of his *On the Motion of the Eighth Sphere*, which survives only in Latin translation, has been twice published by F. J. Carmody (Berkeley, 1941 and 1960), as well as by J. M. Millás Vallicrosa (see especially *Estudios sobre Azarquiel* (1950), p. 496 sqq.). An English translation with excellent commentary, by O. Neugebauer, is in *Proc. Amer. Phil. Soc.* cvi (1962), 290 sqq., while B. R. Goldstein has discussed Thebit's theory further, and contrasted it with Arzachel's in *Centaurus*, x (1964), 232 sqq. Goldstein analyses the latter theory, basing himself on a *Treatise on the Fixed Stars* which is preserved only in Hebrew. The theory which was usually accepted as Arzachel's in the Latin West was nevertheless Thebit's pure and simple, although perhaps with slightly different parameters. Part (i) of our text is perfectly typical in conflating the two. It would be of interest to find any Latin writer who showed a knowledge of the somewhat different procedures which Goldstein considers. It is, incidentally, Goldstein's contention that al-Biṭrūjī's homocentric planetary model was based on trepidation theory, rather than on the Eudoxan model.

5. Thorndike, in *Isis*, xxxix (1948), 213—his article on the Werkworth text—equated 'Almeon' with the Caliph al-Ma'mūn. This caliph lived earlier than Thebit, and therefore the text is said to have been in error. In a footnote, Thorndike at some length allows a hearing to Steinschneider's case for identifying Almeon with the son of Albumasar. Since the whole matter is inconclusive, with the evidence if anything in Steinschneider's favour, perhaps the author of (i) might be given the benefit of the doubt. What remains to be done is to find his reason for coupling the name of Almeon with those of Arzachel and Thebit. Had Thorndike known of a thesis later to be put forward by Henri Michel, on the origin of the theory of trepidation, he might well have looked on this as evidence that al-Ma'mūn was indeed intended. (See 'Sur l'origine de la théorie de la trépidation', *Ciel et Terre*, 9–10 (1950).) One of the

outstanding problems of Thebit's theory concerns the source of the seemingly arbitrary parameter 10° 45'. Michel points out the remarkable fact that a point of the ecliptic at longitude 23° 33' rises and sets within the range of azimuth (measured from east and west respectively) ±10° 45', for the geographical latitude of Baghdad (33° 19'). Was the theory of trepidation arrived at by false analogy? If so, we can see a connection with al-Ma'mūn, for it was his astronomers who settled on the figure of 23° 33' for the obliquity of the ecliptic, on which the analogy rests.

Michel's reasons for thinking his own idea to be close to the truth are the references to 'Altasimec' in medieval astronomy. (See Note 3 above.) This word he derives from *Khatt al-azimah* (line of azimuth). Millás had in fact earlier noted the similarity in the use of the phrase 'autores altasimec' and the 'ancient astrologers' in Theon. Would it be confusing the issue to remark that Thebit's 'azimuthal astronomers' had no connection with either Thebit's brand of trepidation or with Baghdad? Doing so does not make the numerical coincidence to which he draws attention any the less remarkable.

6. Albategni made observations of great accuracy, during the last two decades of the ninth century in particular. (He died in A.D. 929.) He found that the longitude of the Sun's apogee had increased by 16° 47' from the time of Ptolemy. Although he was a younger contemporary of Thebit (the text is wrong in making Thebit the later), he accepted the simple theory of precession with a movement of 1° in 66 years. This figure he derived from a comparison of his own observations with some by Menelaus. It is not the figure one would deduce from his value for the motion of the solar apogee (roughly 1° in 44 years). For a general discussion of Albategni, see C. A. Nallino's biography of him in *Encyclopedia of Islam*, i (1911), 680. The accuracy of Albategni's observations was proverbial in the Middle Ages, although few were in a position to judge of it.

7. In the four centuries between Thebit and the astronomers of Alfonso X of Castile, the equinoxes were observed to have moved further than Thebit's theory of trepidation would allow, with the parameters he had provided. The mean equinoxes about which trepidation was said to take place were now themselves given a steady precessional movement of one revolution in 49,000 years, while the trepidation period was increased to 7,000 years, from Thebit's figure of about 4,181·5 mean Arabic years. Exactly as the recognition of precession or simple trepidation was deemed to require the hypothesis of a ninth sphere between the *primum mobile* and the sphere of Saturn (the sphere of course being now the

'eighth' by name), so the Alfonsine theory required ninth and tenth spheres. What was now the 'eighth sphere' took care of the trepidational component of the resulting motion of the equinoxes, whilst the new 'ninth sphere' took care of the steady precession.

One of the common mistakes of the later Middle Ages was to regard the Alfonsine books as the work of a single man, or at least as having inner consistency. The (incomplete) edition by Manuel Rico y Sinobas (Madrid, 1863–7) shows how mistaken was this idea; and we notice in particular that the four books giving the positions of the stars (which are said to be carried on the eighth sphere!), comprising essentially al-Ṣūfī's catalogue, incorporate Albategni's figure for precession. The 'Alfonsine' theory of precession is found in the tables and the canon for their use, prepared under Alfonso's direction by Judah ben Moses and Isaac ben Said (sometimes called Hassan). The canon in the original Castilian is in volume 4 of Rico's edition. There are many Latin manuscripts of the tables and canon, and two incunabulum editions (1483, Venice, Ratdolt; 1492, Venice, Hamann), but no modern edition. The finer details of the Alfonsine theory were not given much currency in northern Europe.

8. The author of (i) cannot see how the auges of the planets can move in any other way than with the equinoxes, since he assumes that both are fixed in the eighth sphere. This involves a misunderstanding of the Alfonsine view, which involved putting the auges of the planets in the ninth sphere, thus advancing with a steady precessional motion. If, as the writer goes on to say, 'many' thought the Alfonsine tables deficient for this reason, then their arguments were probably expressed in terms too convoluted for our author to appreciate. In fact one of the most difficult problems facing medieval astronomers was that of deciding whether the auges of the planets moved in the same way as the equinoxes. Ptolemy fixed the solar apogee, that is, he made the tropical and anomalistic years agree. However, a careful determination of the solar apogee by Thebit suggested that it was moving at a rate which was fairly close to the precessional constant to which he adhered at the time (1° in 66 years, the figure later adopted by Albategni), and led Thebit to postulate the exact equality of the motions. This he did on the grounds that hypotheses were not to be multiplied without necessity. (For this determination, and the work of al-Bīrūnī mentioned below, see W. Hartner and M. Schramm, 'Al-Bīrūnī and the Theory of the Solar Apogee', in *Scientific Change*, ed. A. C. Crombie (1963), pp. 206–18.)

One of the great paradoxes of the history of astronomy is the fact that al-Bīrūnī, one of the greatest astronomers of all time, should have been almost, or perhaps entirely, unknown to the Latin West in the Middle Ages, so that there is no established Latin form of his name. His work on precession is mentioned briefly here, merely to indicate how inferior was the astronomy we are discussing, by comparison. As Hartner and Schramm indicate (op. cit.), he used the observations of Timocharis, in preference to Ptolemy's. More precisely, Hartner and Schramm show that he used the longitude of Spica, deduced from an occultation of 294 B.C., first accepting the longitude deduced by Ptolemy, but eventually deducing the longitude with the help of his own lunar theory. He corrected for parallax and for the difference in longitude between Alexandria (Timocharis) and Ghazna (his own residence). His own observations of Spica, taken together with the deduced longitude for the earlier epoch, yielded a precessional motion of 1° in 68 years 11 months. This is of course quite close to the modern figure (1° in 71·6 years). Like any other competent astronomer, he now deducted the daily motion of the apogee (taken as equal to the precessional motion) from the daily motion of the Sun (as derived from his own determination of the tropical year), obtaining the Sun's mean (anomalistic) daily motion. Unlike other astronomers, he remarked, first, that this was strictly improper, since the two motions were about different centres; and second, that the equality of the motions of the auges and the equinoxes was something which had to be accepted as a hypothesis, which for the time being could be neither proved nor refuted. We should here remember that the two motions are nearly, but not exactly, equal, and that their difference is one of the things a satisfactory theory of gravitation must explain. Arzachel eventually proved the motion of the solar apogee.

For a detailed study of Thābit's solar theory and several relevant parameters of the solar year, see the forthcoming 'Thābit ibn Qurra between Ptolemy and Copernicus' by Kristian Peder Moesgaard, in *Archive for the History of Exact Sciences*.

9. This passage, with the remainder of (i), contains some interesting but confused material. It is unusual in that it resolves the movement of the eighth sphere into two parts, and in keeping with his remarks about Alfonsine precession, the author does not use two spheres for his purposes. He explicitly accepts an accelerated motion for the changing obliquity of the ecliptic, as well as for the movement of the equinoxes. Neither would have been unusual in an astronomer accepting a theory of trepidation, and yet the figures quoted for the changes in obliquity between the epochs of different astronomers show that the author of (i), or someone before him, was merely stating the differences between the obliquities found by the authorities named. This may be verified by comparing the differences with the differences between the 24° of the 'Indians', Ptolemy's 23° 51′ 20″, Albategni's 23° 35′, Arzachel's 23° 33′, and the stated 23° 27′ for 1350. The differences given, which are apparently rounded off, look rather as though a steady change of 2′ per

century had been assumed by the astronomer responsible for the last figure. In short, it looks as though it was calculated, rather than obtained from new observations. It is in error by about 4′, whereas the earlier observations were within 1′, accepting Newcomb's expression for the obliquity ($23° 27' 08'' - 46'' \cdot 85 T$, $T$ being the time in Julian centuries, measured from 1900·0). Since the calculation does not seem to be based on any theory of trepidation, it remains an open question as to whether the dual accelerated motion is meant to be the same as trepidation. If not, it is pretty worthless, since the values of the accelerations are not stated.

10*. We notice three values for the movement of the equinoxes since Ptolemy:

for 1316 (the 'Oxford Masters' means John Maudith; see
    Appendix 25) .   .   .   .   .   .   .   . 16° 40′;
for 1350 ('as was proved at Oxford by many worthies') . 18° 10′;
for 1357 ('because it now covers a minute per annum') . 18° 17′.

Here are the clear marks of a feeble compilation, with the writer ignorant even of the names of his authorities, and, since the year of writing is around 1395, making use of a writer who was himself a compiler, but one with the ability to bring the story up to date. I shall now show that this man was very probably John Ashenden, and the work to which reference was made his *Summa iudicialis de accentibus mundi*. (Notwithstanding claims which have been made to the contrary, the 1469 Venice edition of this work is very inaccurate, and Oriel College, Oxford, MS. 23 (entire volume, 224 folios) is used here. The work is in at least three other Oxford MSS., namely Digby 225, Bodley 369, and Bodley 714.)

The *Summa* was completed, by Ashenden's own statement, in two parts, on 20 July 1347 and 18 December 1348 respectively. In the Bodleian MS. Ashmole 209, f. 148ᵛ, a seventeenth-century writer (Richard Foster) lists the names of those authors mentioned in the *Summa*, nine in all, and not mentioning Walter of Odington. In Bodleian MS. Rawlinson D.1290, f. 105ᵛ, there is, however, this passage by Archbishop Ussher:

Ex libro astrologico Jo. Eschenden (qui claruit anno 1345) MS. in Bibliotheca Bodleiana, maxima Solis declinacio per Oddington anno 1340 inventa est grad. 23 min. 29 (without respect to parallax and refraction). Loca stell[arum] fix[arum] per multos sapientes 18 grad. distantia a locis illis quibus positae sunt a Ptolemaeo (all the error that Ptolemye hath in the place of the fixed starres, ariseth only from his error in the motion of the sunne seu[?] loco Solis). The place of the starres assigned by Oddington agreeth with ['Ptolemies hypotheses' is here struck out] Tychoes positions . . .

Turning to MS. Oriel 23, we first notice that Ussher was no less careless than Werkworth and Worcester before him, for on f. 34$^v$, ll. 22 sqq., occurs the passage

Nota motum stellarum fixarum a tempore Ptolemei usque ad annum Christi 1340 nam anno Christi 1340 verificate sunt *per quosdam famosos* secundum magnas experiencias, et tunc addebantur super loca earum temporum Ptholomei 18 gradus quam quidem addicionem reputo multum appropinquare veritati . . .

Nowhere in the text is the year 1340 associated with Odington's name. It is now obvious, however, that *quidam famosi* may be a veiled reference to Bredon. Or is the occurrence of the year 1340 in (i) and here a simple coincidence? Compare, then, two further quotations. The first is a gloss to a Bodleian copy of the printed text of Ashenden's *Summa* (Ashmole MS. 576, f. 37$^r$): 'Simon Bredon 1340 equavit motum octave sphere cum maxima diligentia 1837 [= 18° 37′?] a tempore Ptolomei.' The second is from another of Ashenden's works, *De significatione coniunctionis Saturni et Martis . . . isto anno 1357 . . . et coniunctionis . . . que erit . . . 1365*, and is here taken from Bodleian MS. Digby 176, f. 45$^r$: 'Ista patent secundum Magistrum Simonem de Bredon qui circa annum Christi 1340 equavit motum octave spere cum maxima diligencia.' Perhaps this was the source of Ussher's quotation. These *quidam famosi* of 1340 turn up again in the very manuscript whence comes the Worcester text (MS. Laud misc. 674, f. 101$^r$): 'Nota quod anno Christi 1340 verificate fuerunt stelle fixe per quosdam famosos secundum magnas experiencias; et tunc addebantur super loca Ptolomei 18 gradus quam addicionem multum reputo apropinquare veritati . . .' There is no reason for thinking that Worcester copied out these lines from Ashenden's *Summa*, but the coincidence is striking. Where, then, did 'Oddington's' name come from? If Ussher was familiar with the *Summa*—and, since it is as much concerned with chronology as with astrology, it is extremely unlikely that he was not—then Walter of Odington's name would be very naturally associated with Ashenden's, for in the first two chapters of the *Summa* alone, Odington's name occurs forty times! He is mentioned as 'frater Walterus de Odyngton, monachus de Evesham', for example (notice that he is never 'Magister'), his dates do not appear to be given anywhere, and in so far as he is cited as an authority on any subject, that subject is chronology. Thus on f. 4$^r$ of the Oriel manuscript 'Et Odynton monachus de Evesham in suo tractatu *De etate mundi* dicit sic: Dicunt namque philosophi et precipue Indi omnes planetas in capite Arietis et in principio creationis eorum esse constitutos . . .'

It seems that Walter of Odington must have written at some length on chronology, and this being so is it not likely that his writings on the eighth sphere are contained in it? There are two reasons why it would

have been appropriate to include a discussion of precession and trepidation in a work on the age of the world: first, it was widely thought that
vicissitudes of the world's history were linked with the periodicities of
trepidation—notice how Alfonsine trepidation had a turning-point near
the time of Incarnation. Second, in order to evaluate the length of the
tropical year accurately, the movement of the equinoxes must be known.
Thus the thirteenth-century English computists Grosseteste and Bacon
had set an example for the Evesham monk to follow. (See Notes 13 and
17 below.) It seems not unlikely that an anonymous work on chronology
will one day turn up with the at present insignificant first half of text (ii)
in its midst, thus identifying itself as Walter of Odington's *De etate mundi*.

We finally notice that the '1357' of the text (i) coincides, perhaps
fortuitously, with the year of Ashenden's calculations relating to the
conjunction of Saturn and Mars. And Ussher's '18°' as a precession
constant 'per multos sapientes', which was shown to have come from
Ashenden, coincides with none quoted in any of our texts.

11. If Walter of Odington used this list of star coordinates, apart from
the last, to show that precession was constant at one degree per seventy
years, then let us hope that he took the epochs of the eight different
observations into account. That he did so seems likely from the existence
of a table, hinted at in the manuscripts, which presumably related the
movement of the eighth sphere (for auges as well as for the stars, be it
noticed) to the epoch. That Werkworth's work was worth little should
be evident from the clear contradictions between this section and (i),
which advocates an accelerated movement of the equinoxes, and which
settles for a movement of 1' per annum between 1350 and 1357. If
Werkworth did indeed observe Aldebaran himself, his observation was
accurate enough; but the passage reads as though the figure was calculated, in which case the calculation contained a mistake, unless the
figure was wrongly copied.

12. Walter of Odington, 'by means of observation with a turketum',
and (perhaps this was rather more honest, since the coordinates we know
of his are Ptolemy's plus 16° 20') 'by calculation', has left us four sets
of star coordinates. (Notice that in modern terms Aldebaran = α Tauri,
Alaioc = α Aurigae, Alabor = α Canis Maioris.) Alabor's position is
given in the correct sign: some editors of the *Almagest* table would have
had him put Alabor in Leo, it seems. The Pleiades are not commonly
given in star lists, since their coordinates are various. (See P. Kunitzsch,
however, *Typen von Sternverzeichnissen* . . . (1966), p. 65, and my 'Kalenderes enlumyned ben they: Part III', *Review of English Studies*, xx (1969)
421.) Although the year for which the star positions are quoted is 1301,

they do not justify Walter of Odington's stated choice of a movement of 1°/70 years. (With Ptolemy's they give 1°/71·22 years, which if adhered to would have been perhaps the most accurate value available to the Middle Ages: compare with Newcomb's figure for annual precession of 50″·2564+0″·0222$T$, $T$ as before. The error would have been about 2 parts in 1000.) It seems more probable that the '1301' is a confusion with the date on Walter of Odington's calendar. (See p. 246 above.) If this is not so, then the date deduced for his observations (1281) is in doubt. But then Leland's '1280' must be explained away.

Here it is worth remarking that Henry Bate, who wrote a treatise on the equatorium which was mentioned in connection with *Albion* (p. 259, vol. ii), there, in his prologue, referred to the 'masters of proofs' who had argued for a figure of 1°/70 years: 'Est etiam advertendum quia movet octava spera secundum posteros magistros probationum uniformiter in 70 annis gradu uno; et hunc motum sequuntur auges planetarum . . .'

13*. It seems unlikely that, even if Walter of Odington's full text were found, it would prove to be of especial merit, however interesting it may be to the history of astronomy. It is, however, tempting to speculate, if not on its contents, at least on the likelihood of its having been in a certain tradition of brief quasi-historical discussions of the eighth sphere, some examples of which are given in these remaining notes. This is an Arabic and Hebrew, rather than a Latin, tradition; and it is not very much in evidence in the best European works of theoretical astronomy of the thirteenth century, although fairly common in all writings on a related subject, namely the length of the year. Grosseteste's *Compendium spere* (see, for instance, f. 160 of the 1531 edition) is a straightforward—one might even say polished—external account of the two theories, Ptolemy's and Thebit's; but it is not overtly historical. If Sacrobosco was acquainted with Thebit's theory, his *De sphera* gives no hint of the fact. (See L. Thorndike's *The Sphere of Sacrobosco and its commentators* (Chicago, 1949).) There are commentaries on Sacrobosco's work which do bring in trepidation, in late thirteenth- and early fourteenth-century manuscripts, while Robertus Anglicus in his commentary cites Thebit's *De motu octave spere* by name, and even introduces the rudiments of the theory. But this is not historical, by comparison with the passages from Abraham bar Ḥiyya, Abraham ben Ezra, al-Biṭrūjī, and Bacon, which are given below. All that is here suggested is that the list of authorities cited in what remains of Walter of Odington's text makes it likely that he borrowed more extensively from one or more of these authors—all but the first being known in Latin translation in England.

When Sacrobosco and Grosseteste are mentioned together in connection with their works on the sphere, it is inevitable that the question of priority be raised. Thorndike's argument (op. cit., p. 5) that Grosse-teste's was the later, since Grosseteste mentions Thebit whereas Sacro-bosco does not, is clearly gratuitous; for Thebit's text was available to both authors, to use or to overlook. It is highly probable that Sacrobosco could not appreciate the ideas involved. Again, Sarton's quoted state-ment that Grosseteste's *De sphera* contains 'the first mention of the trepidation of the equinoxes in a non-Muslim work', with which Thorn-dike apparently concurs, needs amending in the light of the two initial passages given here below. And did not Gerard of Cremona's translation of Arzachel cause anyone to mention the idea?

That Michael Scot's(?) commentary on Sacrobosco's *De sphera* con-tains no reference to trepidation is another reason given by Thorndike for putting both before Grosseteste's work (op. cit., pp. 22–3). Thorndike points out that the commentary belongs to one or other of the periods 1210–15 and after 1231. (Michael died *c.* 1235.) He decided in favour of the second period. Michael Scot translated the *De motibus celorum* of Alpetragius (see note below) in 1217, and this work does mention trepi-dation—in fact the theory is explicitly rejected. If the commentary is his, which for a number of reasons seems doubtful, then by Thorndike's method of argument the earlier period is the more probable. For other arguments which seem to favour this view, Thorndike's book must be consulted.

From outside England we may briefly mention Albertus Magnus, who in astronomy did little more than list the titles of books; Campanus of Novara, whose *De sphera* (printed Venice, 1518) has an insignificant chapter (XI) mentioning Ptolemy, Arzachel, and Thebit in the same breath (the tenth chapter of his *Computus maior*, in the same printed volume, being no better); Gulielmus Anglicus, who wrongly interpreted Thebit's theory (Duhem, *Système du monde*, iii. 290); and, earliest of all, the author of the *Theorica planetarum* commonly ascribed to Gerard of Cremona. The *Theorica planetarum* had a continued popularity from the time it was written in the thirteenth century until at least the sixteenth. (There are four printed editions.) What he has to say on the movement of the eighth sphere is trivial enough, but not uninteresting, considering its early date:

Nota etiam quod auges dicuntur moveri versus orientem 7 gradibus in 900 annis: et totidem versus occidentem in aliis 900 annis. Item dicuntur moveri ab Albategni in 60 annis et quatuor mensibus uno gradu semper ad orientem. Alfraganus narrat eas moveri in 100 annis uno gradu versus orientem . . .

A corrupt version indeed, but how can the translator of *Almagest* have given Alfraganus priority over Ptolemy, and have even overlooked

Ptolemy? Following Duhem, perhaps, it is usual to say that Gerard
failed to mention trepidation; but clearly he does not, although he has
only the faintest idea of what it involves. Or is this eastward and west-
ward movement he mentions a fossil of the idea referred to by Theon?
Did Gerard not translate Thebit's treatise on the eighth sphere? Or was
the *Theorica planetarum* a work of his youth? Or are we to look elsewhere
for its author?

Finally, although I have not searched systematically for competent
reiterations of the full mathematical theory of trepidation—whether
Thebit's, Arzachel's, or the Alfonsine—it is my impression that these are
not to be found in the Latin West before the end of the thirteenth century and the beginning of the fourteenth. That it was then taken seriously
by some astronomers should be evident from its use in the calculation of
star positions.

14*. The following account of the history of theories of the eighth sphere
is by Abraham bar Ḥiyya, and was not translated until this version was
prepared by Sebastian Münster (1489–1552). It is taken from *Sphaera
mundi . . . autore Rabi Abraham Hispano filio R. Haijae* (Basle, 1546), p. 196.
(The Hebrew text is in parallel with the Latin.)

Agit hic ultimus tractatus de motu octavae seu potius nonae sphaerae, qui
est motus stellarum fixarum, qui secundum veterum opinionem fit ab
occidente in orientem, absolvitque singulis centum annis unum gradum: et
sic completur in 36000 annis. At posteriores astronomi communiter sentiunt
hunc motum non amplius quam 44 minuta et 4 secunda absolvere in centum
annis, nec completur nisi post 49000 annos, vocaturque motus remissionis et
fit super polis Zodiaci, atque ob id stellae semper habent eandem latitudinem
ab ecliptica. Veteres vero Indi, Aegyptii et Chaldaei, Graeci et Latini,
fuerunt primo in hac opinione, ut putarint polum zodiaci revolvi in parvo
circulo cuius diameter esset 8 graduum, unde fiebat quod stellae octo
gradibus moverentur in orientem, ac deinde rursum octo retraherentur in
occidentem. Et hunc motum dixerunt compleri in 1600 annis. In 800 vero
annis semicirculus revolvebatur de puncto occidentali in orientem. Et
quoniam is motus fiebat in polo zodiaci, accedebat quoque stellis ipsis
quaedam latitudo ab ecliptica.

Hunc motum vocabant progressionem et retrogressionem: item abi et redi.
Cum hunc conatum Indorum Ptolemaeus vidisset, non improbavit nec
probavit, Albategni vero Saracenus ille confutavit opinionem eorum.

[198] Scripserunt Indi: ante Christum natum Ptolemaeus autem post anno
circiter 140 et hic quoque invenit coelum motum in 420 annis 4 gradibus.
Venit tandem Albategni, annis scilicet 720 post Ptolemaeum, et sudans
quoque in hac difficultate invenit calculum suum superare calculum Ptol.
fere 11 gradibus, quapropter tribuit motui stellarum singulis centum annis
graduum unum et dimidium, fuitque differentia calculi sui et calculi veterum

15 gradus et fere una quinta in 1140 annis, cum tamen secundum veterum supputationem tantum 11 gradus 36 minuta esse deberent, si initium calculandi fiat a puncto orientali, aut 3 gradus et 36 minuta si a puncto posteriori numeratio incipiat. Est enim diameter circuli octo graduum, quibus si 3 addantur, habentur 11 gradus. Sin supputatio fiat a medio puncto, locus stellarum mutatus est 7 gradibus. Nec refert an medium punctum accipias secundum punctum priorem aut posteriorem. Sed facile convincuntur veteres erroris cum calculus Albategni iam 15 gradibus et non octo tantum auctus fuerit.

15*. Abraham ben Ezra was a late contemporary of Abraham bar Ḥiyya (sometimes known as Savasorda), from whose writings the last passage comes. Both were Jewish savants who started life in Spain, and not surprisingly they are often confused in medieval ascriptions. Until a little over twenty years ago, the consensus of opinion was that the younger Abraham had composed about eight distinct treatises relating to astronomy, translated into Latin no earlier than the late thirteenth century. (See Thorndike's survey in *Isis*, xxxv (1944), 293.) A translation deriving from the twelfth century, however, had frequently been said to be from a work by Abraham ben Ezra, on the basis of its opening words: 'Dixit Abraham Iudeus. Cognitum est corpus solare . . .' José M. Millás Vallicrosa edited the text of this in *El libro de los fundamentos de las Tablas astronómicas* (Madrid–Barcelona, 1947), and there defended Abraham ben Ezra's authorship, seemingly without convincing Thorndike. To agree entirely with his arguments for Abraham ben Ezra's authorship, it is not necessary to go so far as Millás, who holds that the text was written by Abraham in 1154 at Angers *in Latin*. These arguments are not weakened by a comparison of the text below, taken from Millás's edition (with the critical apparatus omitted; see p. 77), with the somewhat similar text of note 14.

We notice in particular that the second paragraph is a very strong candidate for the source of that part of (ii) which seems to be Walter of Odington's. We observe also that on the preceding page of Millás's edition, Thebit is described as a Christian—'christianorum summus philosophorum fuit'. This may be compared with the opening words of Roger Bacon's chapter on the same subject (see Note 17 below): 'Thebith vero maximus Christianorum astronomus . . .'

Antiqui omnes et Hermes et Indi et doctores ymaginum omnes in hoc consentiunt, quod in circulo firmamenti duo motus sunt, ascendendi in septentrionem et descendendi in austrum; inter hos tamen est aliqua discordia, nam magistri ymaginum dicunt eos motus esse polorum, Indi vero duorum circulorum qui sunt in capite arietis et libre; omnes tamen in proximo dicti in hoc consentiunt quod gradus horum motuum sunt 8, Azarchel vero asseruit 10 gradus et duas tertias gradus. Sed Ptholomeus et omnes magistri probationum sententias predictorum de motibus deriserunt supradictis, preter solum Abencine.

Similiter discordia est in declinatione solis, nam Indi dicunt 24 graduum integrorum declinationem solis esse, sed Abrachis et Ptholomeus dixerunt 23 graduum 51 minutorum, secundum horum sententiam arcus declinationis sic se habebit ad totum circulum ut 11 ad 83. Omnes vero alii magistri probationum dixerunt declinationem esse 23 graduum et 35 minutorum, exceptis Abnebimezor et Azarchel qui dixerunt eam esse 23 graduum et 33 minutorum.

16*. The next passage, taken from Michael Scot's Latin translation of a work by al-Biṭrūjī (Alpetragius) has affinities with those of the last two notes. The translation was edited by F. J. Carmody (*Al-Biṭrūjī: De motibus celorum* (Berkeley, 1952); see pp. 100–1), from whose edition the passage is taken, without the critical apparatus. In 1528 there was another translation made of the same work, this time from the Hebrew by Calonymos, a Neapolitan Jew: *Alpetragii Arabi planetarum theorica physicis rationibus probata, . . . a Calo Calonymos . . .* (Venice, 1531). The wording of this translation is occasionally more interesting than that of Michael Scot's, but it is here worth mentioning only his equivalent of the clause 'sicut Hermes . . . componentes imagines', near the beginning of our passage. This becomes: 'ut Hermes et posteriores eo qui dant operam [*sic*] imaginibus'. See Note 3 above, for phrases comparable with these.

In a passage not quoted here (Carmody's edition, p. 85) the author speaks as though he knew a work on the trepidation of the eighth sphere by Arzachel. When Carmody wrote, he was apparently not aware of a Hebrew manuscript containing the translation of such a work, which had been published in facsimile by Millás, together with a translation into Spanish (*Estudios sobre Azarquiel* (Madrid–Granada, 1943–50), ch. 5). B. R. Goldstein has recently put the crucial part of this into English (*Centaurus*, x (1964), 238–41), and indeed has edited the whole from Arabic and Hebrew versions, translated it, and written an illuminating introduction (*Al-Biṭrūjī: On the Principles of Astronomy*, 2 vols. (New Haven and London, 1971)). The translation offered for the phrase corresponding to the Latin 'illi qui fuerunt post ipsum sicut componentes imagines' is 'Masters of the Talisman'. Goldstein's note referring to an extensive treatment of the relationship between astrology and talismans is: *Picatrix*, ed. H. Ritter and M. Plessner, Studies of the Warburg Institute no. 27 (London, 1962), pp. lix–lxxv, and *passim*.

Et quia non verificaverunt antiqui rem firmam in salvatione huius motus, ideo est magna ambiguitas in isto motu; quia antiqui (sicut Hermes et illi qui fuerunt post ipsum sicut componentes imagines) dant istis stellis motum aliquando secundum signa aliquando contra signa; et videtur quod ista res fuit scita sive concessa eis. Et illi qui venerunt post illos, ut Alkaldemein (de illis qui inspexerunt has stellas ante tempus Nabugodonosor, ut salvent quod dixerunt antiqui), non invenerunt eis motum, sed abnegaverunt illum motum quem fecerunt antiqui primo, quia non induxerunt super illas neque tabulas neque astrologiam quibus salvent possibilitatem essendi illum motum. Et crediderunt quod celum stellarum est motor diurnus, et quod celum signorum (et est celum inclinatum) secat equatorem diei super duo puncta (et unum nominatur punctum vernale et aliud autumpnale) que sunt initia Arietis et Libre; et iste sectiones salvantur semper. Et illi qui venerunt post istos per magnum tempus ante tempus Alexandri, cum eo quod induxit Abrachis de considerationibus Timocharis et Arsatilis in anno 450$^{mo}$ a tempore Nabugodonosor, et post consideravit Mileus geometer in anno 845$^{to}$ a tempore Nabugodonosor, et postea consideravit Abrachis per se post mortem Alexandri fere anno 400$^{mo}$ [sic]. Et de considerationibus hominum qui erant illo tempore inventum fuit quod erat stellis motus secundum signa, et sententiaverunt illud quod adepti sunt in suo motu et firmaverunt secundum quod est iste motus solum secundum signa.

Et post Tholomeus aspexit post Abrachis in anno 265$^{to}$, et fuit motus stellarum fixarum semper secundum signa; et iam Abrachis numeravit istum motum secundum signa, et dixit quod est in 100 annis 1°. Et postquam invenit Tholomeus progressum stellarum in uniformitate numerationi Abrachis, firmavit et sententiavit istum motum secundum illud. Et post illi qui venerunt postremo post Tholomeum, quando aspexerunt istas stellas, invenerunt loca sua per visum et loca sua per computationem diversa et non convenientia, et non confisi sunt in salvatione illorum motuum; et credidit Theon Alexandri (de illis qui venerunt post Tholomeum) quod stelle fixe habent motum accessus et recessus, et quilibet illorum motuum est 8°, et cum hoc habent motum secundum signa 1° in 100 annis. Et illi qui venerunt post expulerunt istud, qui non invenerunt sua loca per visum alia a suis locis per equationem computationis quam fecerunt antiqui aliquando diversa per additionem aliquando diversa per diminutionem. Et post Albategni explanavit quod stelle in temporibus equalibus moventur per spatia inequalia a puncto equalitatis vernalis; et divisit rem suam in hoc.

Et quando inspexit Abu Isac Azarkel post, invenit artem coniungendi istos motus secundum quod apparuit illi; sed adhuc non verificata est eius res complemento complete. Et posuit eis astrologiam et tabulas secundum quod duo poli huius celi moventur super duos circulos equidistantes equatori diei, quia isti duo poli moventur super duos circulos, et erit motus stellarum sequens motum polorum. Et expergefecit nos cum eo quod induxit ad inveniendum illud super quod cecidimus quod latuit ipsum, quod facit hoc secundum veritatem, et est motus huius celi super suos polos transmissus ad complendum istum motum quem incurtabat, ut distinguatur ab illo superiori; quiescat ergo motus secundum quod dixit Abu Isac Azarkel, quod illud scilicet quod apparet de diversitate motus stellarum est motus accessus et recessus. Sed res adversatur secundum eos, quia quod est accessus secundum eos est recessus, et recessus secundum eos est accessus; quia accessus secundum illos est econtrario motui generali, et recessus secundum illos est motus ad partem motus generalis; et est secundum veritatem econtrario, ut explanabitur. Et cum hoc illud quod numeravit Tholomeus, quod motus contrarius motui generali est salvus cum accessu et recessu, sicut est modus in motu stellarum erraticarum; et hoc sicut est cum motu eorum retrogradatio et progressio, qui motus contra motum generalem; sed non adhuc verificata est quantitas huius motus.

17\*. The first English writer to echo what was said by the two Abrahams and Alpetragius was probably Roger Bacon. The following passage comes from Steele's edition of the *Opera hactenus inedita Rogeri Baconi*, reminiscent of a passage of Bacon's *Compotus* (Fasc. VI, 1926, Cap. 3, pp. 16–17). It is from the *Communia naturalia* (Fasc. IV, 1913, pp. 454–5), cap. 19.

Thebith vero maximus Christianorum astronomus addidit aliquid operi Ptholomei in motu octave spere. Nam Ptholomeus posuit celum stellatum moveri motu continuo per successionem signorum, hoc est ab occidente in centum annis gradu uno secundum motum declivum. Posuit auges planetarum preterquam Solis moveri hoc motu unde posuit stellam in medio celi octavi moveri secundum circulum signorum, et alias stellas moveri secundum circulos equidistantes. Sed tempore suo accidit celum stellatum sic fortasse moveri in orientem, cuius contrarium secundum ea que apparent in celis perceperunt posteriores. Sed Albategni hoc vidit in universali, et quod motus mutabatur secundum apparenciam. Non autem dedit consideracionem per tabulas et canones in hoc motu.

Thebith vero hunc motum figuravit et ordinavit canones et tabulas et docuit quod stella non semper apparet moveri in equidistante respectu zodiaci, sed aliquando magis elongatur, aliquando minus, tam a zodiaco quam equinocciali in orbe stellato; considerans ut quod stella que est in capite Arietis visibilis describat circulum parvum cuius diameter est 8 graduum et 37 minutorum et 26 secundorum, et similiter Libre visibilis, et sic per consequens caput Cancri et Capricorni; et totum celum aliquando movetur progrediendo secundum successionem signorum, et regrediendo contra successionem. Et progressus vocatur accessus vel accessio et regressus vocatur recessus.

Ptholomeus autem estimatur fuisse in tempore accessus et ideo non percepit regressum. Centrum autem circuli istius, quem stella in capite Arietis mobilis describit, est caput Arietis in celo nono intersecans circulum rectum ad modum equinoccialis in eo. Et hic circulus vocatur zodiacus fixus et immobilis; non quia non moveatur, set fixus est respectu zodiaci in celo stellato, quoniam ille zodiacus movetur respectu alterius, qui est in celo non . . .

A passage reminiscent of this is to be found in Bacon's *Opus tertium*, at pp. 110–13 of Duhem's *Un Fragment inédit de l'Opus Tertium de Roger Bacon* (Florence, 1909). After mentioning Ptolemy, Timocharis, Hipparchus,

Thebit, and Alpetragius, Bacon went so far as to state the trepidation period on Thebit's theory. It is quite unusual to find this periodicity mentioned. Bacon quotes it as one revolution of the small auxiliary circle 'in 4181 annis lunaribus', that is, 4,181 Arabic years. Thebit does not himself quote a specific periodicity, but leaves it implicit in his tables of access and recess. It must be presumed that Bacon, or someone before him, deduced this figure from the tables, with which it agrees very closely, as was seen on p. 157 above (Appendix 25). The common Arab year contains 354 days. Eleven years out of a 30-year cycle are of 355 days, and are described as 'embolismic'. The mean value of the year is thus $354^d\ 8^h\ 48^m$, and it is clear that this is what Bacon, or the man he was ultimately quoting, had in mind.

# APPENDIX 39

## *Evidence as to the size of the St. Albans clock*

Of the different sorts of evidence available, none is very convincing in isolation. The only manuscript illustration of the finished clock, namely that in MS. Cotton *Nero* D.vii, shows the clock in a finialled wooden case, mounted above the level of an arch, which may be either a window or an arch in a colonnade. The very dimensions of church architecture require that, for the dial to be easily visible at this height, it be of the order of five feet square or more. This is in keeping with later monumental clocks in the transepts of churches, the front of the case being often of the order of nine feet square. We recall that the dimensions of the clock are given throughout in terms of a unit—the size of a certain tooth—of unspecified absolute length. As to the dial, we are told that we may decide for ourselves, but the 'plate for the natural hours' is to be $47^t$ 20′ by $42^t$ 20′. This is marginally less than the front of the frame, which is approximately $48^t$ by $43^t$. (The frame was probably $49^t$ deep or thereabouts.)

Now the planks of wood which go to make up the clock case are $3^t$ in breadth, and since it seems unlikely that these would be anything but typical wooden planks of the time, we may suppose that the unit tooth was *at least* two inches. In fact a nine-inch plank (making $1^t = 3$ inches) can be considered far more typical of the fourteenth century—as of any century before large mechanical saws became available. (The modern plank is usually of four to six inches.) Taking our unit as 3 inches, the clock case was roughly twelve feet high and ten feet six inches across the front. This is not at all an improbable size. At Durham, where the original clock no longer survives, the clock case (of the period 1494–1519) is only a little narrower, but is higher. It is, furthermore, above a false colonnade in the south transept, being mounted on corbels in very much the same way as the manuscript illustration suggests for the St. Albans clock. (At Durham, the clock is under the south window, as I think was that at St. Albans.)

We may suppose that the drop for the weights was of the order of twenty feet, or more. In one day, the barrel of the going train turned 40 times (or perhaps 30). We are not told anything about the diameters of the barrels, but naturally they would be as large as was compatible with the design, so as to reduce the weight necessary to drive the clock,

and at the same time to minimize the effects of friction. With the simple
pulley system for effectively doubling the drop, for which we admittedly
have no historical evidence one way or another, and a barrel turning
40 times daily, the circumference of the barrel would be about a foot.
This is perfectly compatible with the design; that is to say, it does not
imply interference with the adjacent wheels. A similar argument holds
for the striking barrel, but in neither case are we offered much assistance
in deducing over-all dimensions.

The dimensions of the bell were 14$^t$ by 12$^t$, which on the hypothesis
of a 3-inch unit would mean a bell 42 inches (maximum diameter) by
36 inches (in height). These measurements are perfectly reasonable, and
indeed, the bell which now hangs in the Curfew Tower in the town of
St. Albans, serving the striking clock below, is 47 inches by 39 inches.
On a modern wooden plaque on the outer wall of the tower there is a
statement, the source of which I have been unable to find, to the effect
that this bell dates from the year 1335—the year before Richard of
Wallingford's death. Since the Curfew Tower was built in 1403–12,
acceptance of the earlier date for the bell would make it natural to
conjecture that the bell was removed from the abbey clock after the
Dissolution. This would mean that the unit 1$^t$ was equal to about
3$\frac{1}{4}$ inches. Unfortunately, the style of the lettering on the bell does not
ring as true as the inscription itself: *Missi De Celis habeo Nomen Gabrielis*.
I am inclined to place the lettering a century or more after 1335, and
to do so whilst preserving the idea that the bell comes from the abbey
clock it is necessary to postulate a recasting of the original. Although this
is not out of the question, the existing bell can hardly be counted as
evidence in any evaluation of the size of the clock. Its very existence,
however, suggests that a 3-inch unit might be a conservative estimate.

We have hitherto been considering the outer case, rather than the
frame holding the mechanism. From the text of II. 7. 2, where the size
quoted for one of the wheels in the astronomical section of the clock is
said to be 'nearly two feet', together with the relative dimensions of the
figure (on f. 176$^r$) on which the wheel is drawn, we deduce that the
frame was approximately 10 feet square, viewed from the front. Since
this frame is presumably a natural extension of the frame carrying the
going and striking trains, we may deduce a value for the accepted unit,
which is now required to be approximately 2·8 inches.

Although nothing by way of a precise argument has been offered, the
evidence collected here all points to a large clock-frame. The fabric of
the abbey church is now so altered from its fourteenth-century state as to
offer virtually no guidance over dimensions. Assuming the clock to have
been beneath the south window, we have no means of knowing the
height of the sill, for the oldest plans available depict a window put in

after Abbot Richard's time. (The sill of this, however, was rather more than twenty feet above the ground.) On the east wall of the south transept, and even more so on the west, there would have been ample room for a clock of the suggested dimensions, but even the south wall was not unsuitable. Part of the depth of the clock—say four or five feet— could have been carried on a gallery, or within the wall, access being obtained by the usual recessed spiral staircase.

In summary, we may imagine the unit to have fallen somewhere between 2·5 and 3·25 inches, making for an iron frame between 10·2 by 8·96 by 10 feet and 13·27 by 11·64 by 13 feet (depth by width by height) and a case very little larger.

Finally, we may turn to the information we have as to the Norwich clock (vol. ii, p. 362), built by the same craftsmen. Although there are no actual dimensions given in the sacrist's rolls, we know that the iron plate for the dial weighed no less than 87 lb. Its volume was therefore about 312 cubic inches, but more we cannot say, unless we hazard a guess as to the thickness of the plate. It was presumably hammered out of slag-bearing malleable wrought iron, and might have been reduced to 0·05 inches in thickness, in which case it would have been nearly 80 inches square. This conjecture is in rough agreement with the figures previously deduced from the Ashmole manuscript for the St. Albans clock. Increasing the thickness to 0·1 inches, the plate would have been about 56 inches square. The correct solution probably lies between the two.

# ADDENDA

References in square brackets are to volume and page

[Plates XXIII-XXIV] The only surviving (but incomplete) medieval albion is that illustrated here. Now in the Rome Astronomical Observatory (Monte Mario), it was first recognized as an albion by M. Emmanuel Poulle, and for information and photographs I have to thank him, Mr. and Mrs. R. S. Webster, and Signor Giorgio Buonvino. The second face of the albion, not shown here, is a constellation map in the Alfonsine (and ultimately al-Sūfi) tradition. It carries a set of ecliptic coordinates and a series of horizon-lines. The second disc is now lost. The first is engraved correctly on both sides. The albion appears from photographs to be of the fourteenth or early fifteenth centuries, and in its execution few instruments from the late Middle Ages can equal it.

[i. 6] An earlier English writer on trigonometry than John Maudith, but in the same style, was a certain Roger de Cotum. The evidence—too involved to be given briefly—is that he worked in London in the latter half of the thirteenth century.

[i. 531] Research into the history of the Rule of St. Benedict has now reached such proportions that a new annual publication has appeared, devoted to this subject alone: *Regulae Benedicti Studia* (1972–   ).

[ii. 1, 15] The statement that Richard was adjudged worthy to lecture on the *Sentences* means no more than that he was given leave to take the degree of Bachelor of Theology. See H. Rashdall, *The Universities of Europe in the Middle Ages*, 2nd ed., Oxford, 1936, vol. iii, p. 70. According to Wood (*Athenae*, fasti, *item* Musgrave), the formula was the same in the sixteenth century.

[ii. 71-3, 243, 303; iii. 60, 87] While not abandoning the assumption that the column of *gradus cum quo stella mediat celum* in the *Albion* star list was meant to give mediations, I now recognize the strong ambiguity of the expression (q.v. in the Latin Glossary), originating no doubt in part from the fact that the two concepts are side by side in the fundamental source, *Almagest* VIII. 5. Tract II might well have been meant to give right ascension (assuming that *circulus divisus*, vol. i, p. 435, line 22, means 'equator'), or even that and mediation.

It has been suggested recently by more than one writer that a column with the given heading, with star coordinates rounded to degrees, is an indication of *new observations* rather than of the readjustment of old lists. Medieval advocacy of graphical methods (as in Tract II), and our evidence for the difficulty Richard had with calculation (i.e. the only alternative), argue against the suggestion.

[ii. folding table] There is another copy of Albion, or rather a fragment of such, in Bodleian Library, MS. Canon. misc. 162, ff. 130ᵛ–141ʳ. This is a late John of Gmunden version, and is without textual value.

[ii. 137] Dr. R. P. Lorch notes that Aleksander Birkenmajer (writing in 1919) ascribed *Almagestum Parvum* to Walter of Lille. See *Studia Copernicana*, vol. i, Warsaw, 1970, p. 215, for a reprint of the article in which the statement is made.

[ii. 183] Another astrolabe (with Arabic inscription) having a rete with a pierced 12-degree zodiacal band is illustrated in the December 1928 issue of *International Studio*.

[ii. 255, 272] M. Poulle finds that the treatise 'Quia nobilissima sciencia . . .', which opens with the same words as the treatise by John of Linières, does in fact describe an equatorium of a type of which the Merton instrument is an example.

[ii. 271] The albion is named as one of several instruments introduced to the Benedictine monastery of Reichenbach by its abbot, Engelhardt. See D. B. Durand, *The Vienna–Klosterneuburg Map Corpus of the Fifteenth Century*, Leiden, 1952, App. 2.

[ii. 275] A good biography of John of Gmunden (by Kurt Vogel) in the *Dictionary of Scientific Biography*, vol. vii, New York, 1973, pp. 117–22, rules out the family names of Nyder and Schindel, and settles the question of birthplace in favour of Gmunden am Traunsee (Austria).

[ii. 277] Sebastian Münster's *Organa Planetarum*, Basle, 1536 (the second edition, and the only one with the volvelles, shown to me in the rare Harvard copy by Professor Owen Gingerich), has not only 'polar coordinate curves' like Apian's but also (between D2ᵛ and E1ᵛ) Richard of Wallingford's solar and lunar eclipse instruments.

[ii. 303; iii. 133] Ptolemy's *Analemma*, preserved only in William of Moerbecke's translation, advocates the graphical method. The analemma was drawn 'on a circular table on which the lines commonly used were painted, while others could be drawn in a thin coat of wax and easily erased again' (O. Pedersen, *A Survey of the Almagest*, Odense, 1974, pp. 403–4).

[ii. 314; iii. 195] It was J. L. E. Dreyer's thesis that Rede's version (*c.* 1340) of the Alfonsine tables contained evidence for the survival of an earlier form than that known from the 'continental' versions. (The thesis influenced what I wrote here only to the extent that I used Rede's version to represent the Alfonsine tables, which they do well enough.) I now wholly reject the idea, having found it possible to trace the transmission of the tables from Paris to London (*c.* 1321) and thence to Oxford. See the reference to my chapter in the Hartner Festschrift, in the Bibliography.

[ii. 329–39; iii. 64] The publication of *The Madrid Codices of Leonardo da Vinci*, New York, 1974, following the chance discovery of the originals in 1965, adds drawings of relevance to the *strob* mechanism, as noted at p. 332, vol. ii. It does seem likely, however, that there is also an affinity between the St. Albans striking mechanism and Leonardo's large (double) screw, with its stops suitably placed in the grooves (Madrid, I, f. 15$^r$). The 'screw revolves freely until the corresponding hour is rung'.

In letters published in *Antiquarian Horology* in December 1971 and March 1974, Signor G. Brusa discusses the escapement, and points to affinities with that by Pierre De Baufre. In the later issue, at p. 627, there is a photograph of Mr. Haward's model, now in the British Museum.

[ii. 333] The late P. G. Coole had apparently followed a reference given by E. Zinner to Cracow, MS. 551, ff. 44$^v$–49$^r$ ('In compositione horologii premittenda sunt . . .'). Two other copies are listed in T.K. M. Poulle has made a transcript from the Yale copy, and it is clear that the text, of about 4,000 words, is an important one for the history of basic horology. (The astronomical content is rather elementary.) Wislocki's catalogue gave Zinner the date of 1388, but where he got the idea that there was an Erfurt association is not established.

[ii. 341, 350, 363] The publication by D. J. de S. Price of his full analysis of the Antikythera device (see Bibliography) shows that this incorporated the idea of the transported train of wheels. It is now natural to ask for intermediaries between this mechanism and Richard of Wallingford's. In this connection, note his awareness of *instrumenta manualia* (*Horologium* II. 4. 2).

[ii. 355] That teeth on the common run of iron clocks continued to be made in a form little different from a simple isosceles triangle well into the fifteenth century is evident from the clock preserved in Cotehele House, in Cornwall, a medieval manor acquired by the National Trust in 1947, and formerly the home of the Mount Edgcumbe family.

[iii. 163] The use of one word to mean both 'north' and 'left', and another both 'south' and 'right', hinted at in connection with the treatise by Abraham ben Ezra, is apparently common to many languages (e.g. Sanscrit, Chaldaean, Syriac, Hebrew, Arabic, Mongolian, Latin, Greek, and Irish). See Guilelmus Gesenius, *Thesaurus Philologicus Criticus Linguae Hebraeae et Chaldaeae Veteris Testamenti*, Leipzig, 1835, pp. 72*b*, 599*b*, 1332*b*, where several sources are quoted. Note that the convention agrees with that used in the orientation of medieval maps.

# GLOSSARY OF LATIN WORDS

THE following list is meant to supplement such standard dictionaries as *MLWL2* and that by Lewis and Short. It will also serve to some extent as an index of minor themes. Many words are included which, while being classical (orthography apart), do not have classical meanings. Many are the subject of explanation, and discussion in footnotes and commentary to the text, and reference is made to this in the following terms: [A: 366 (22); i. 353 n. 1; ii. 219]. This is to be understood as recording a typical or interesting use of the word, in the given sense, in *Albion* (A—see the key below) at p. 366, line 22 (of necessity in vol. i) and that its etymology or its meaning in the context is discussed in n. 1, p. 353, vol. i and p. 219, vol. ii. Line numbers include all lines other than those containing *treatise* titles (running headlines, etc.).

A few words of common medieval occurrence, especially from arithmetical passages, are included on the grounds that they are currently unfamiliar and potentially ambiguous. No attempt is made to record all orthographic variants of the words listed, and the fluid conventions outlined in vol. i are followed here.

A—*Albion*
D—Dubious & spurious works
E—*Exafrenon*
H—*Horologium*
K—*Kalendarium*
M—*Canones* to Maudith's tables, & *Tractatus*
Q—*Quadripartitum*
R—*Rectangulus*
S—Spiritual and Ecclesiastical writings
T—Tracts concerning coordinates

**abicere** [Q: 128 (34)], to discard (in the operation of rounding off a numerical value). (The word commonly means to subtract, deduct.)

**absis,** *f.* [E: 214 (5)], apse, aux, apogee, far point of planet's deferent. (The architectural uses of the word are earlier than the astronomical, but Pliny uses it for a star's orbit.)

**accipere,** to find (as of a result, whether mathematical or observational). (Thus **acceptus, -a, -um,** measured.)

**actonus**—*erroneous form of* **atomus** [(D) ii. 379], atom.

**adequare** [E: 196 (21)], to equate, calculate. (Used of the calculation of the astrological houses. Cognate with **equare** as used of the calculation of a planetary position, when a mean position is combined with planetary equations to provide a **locus equatus** [A: 350 (8)]. See **equacio.**)

**albion, albeon, albyon,** *n.* [A: 248 (1); ii. 137–8, ii. 272; Plates XXII, XXIII], albion, (Richard of Wallingford's) universal astronomical instrument, 'all by one' (*al-bi-on*, Mid. English).

**allidada, alidada,** *f.* [R: 416 (24); iii. 57], alidade, sighting rule (on the rectangulus as on a conventional astrolabe). (Arabic al-'iḍāda.)

almanac(h), n. [E: 212 (4); A: 248 (1); ii. 314–15], almanac, collection of solar, lunar, or planetary ephemerides. (These may be tabulated by days, as in an ecclesiastical calendar, or by groups of days. The approximate periodicity of the ephemerides makes possible the almanac perpetuum. M. Steinschneider, 'Über das Wort Almanach', Bibl. Math. (1888), pp. 13–16, gives examples of the use of the word, which is puzzling because it is known from earlier Latin texts rather than Arabic.)

almicantar, almucantar (usual pl. -at, -ath, -az, or -a; but note Richard of Wallingford's use of pl. form as an invariable sing.), n. [A: 380 (10); H: 488 (17); iii. 45, 165], parallel of altitude (as on an astrolabe plate). (Arabic al-muqanṭara, (pl.) al-muqan-ṭarāt.)

almuri, n., indecl. [A: 366 (22); i. 353 n. 1; ii. 219; iii. 41], index, ostensor, bead (on thread, used as a marker). (Arabic al-murī, the little hand.)

altitudo accepta, f. [Q: 134 (28)], measured altitude.

angulus, m., angle. (But note angulus Terre, the lowest point of the equator on the meridian [Q: 116 (15)], echoing astrological use for the cardinal points of a division of the houses [(trans. of E): 199 (32); E: 230 (24); ii. 121]. Note also [H: 476 (5)], where angulus denotes one of the sharp edges of the semicirculus, q.v.)

anni collecti, m. [iii. 176–7], groups of years (used in the tabulation of uniformly changing quantities).

anni expansi, m. [iii. 176–7], years tabulated consecutively.

applicare [H: 514 (12)], to apply, mesh (as of one gear with another). (Generally in passive.)

apsis—v. absis.

aptare [H: 502 (4)], to mesh, engage (as of one gear wheel with another). (Generally in passive.)

arcuare [M: 10 (30); A: 256 (9)], to find the arc corresponding to a chord (as in phrase arcuare chordam).

argumentum, n. [A: 256 (29); A: 276 (30); iii. 175], argument (of a planet or of the Moon, being then the movement of the planet on the epicycle). (Note phrase argumentum epicicli; note that for the Sun the word is synonymous with motus. For medius motus argumenti as a synonym of medium argumentum, see motus.)

arsmetricus, -a, -um [E: 192 (12); A: 248 (1)], pertaining to arithmetic. (The word arsmetrica for arithmetic is an etymologically erroneous medieval derivative, not of the Greek ἡ ἀριθμητική, but of μέτρον compounded with Latin ars. Thus in MS. Digby 147: 'Dicitur arsmetrica de ars artis et metros . . .' For some of the literature on the word see D. E. Smith, History of Mathematics (1925), vol. ii, pp. 7–8.)

ascendens [E: 194 (18); A: 382 (13); iii. 146], the ascendent, (gradus ascendens, m.) the point of the ecliptic rising above the horizon.

ascendere—v. ascendens, but note also the use [E: 212 (33), etc.] to describe the increasing planetary distance from the centre of the universe.

ascensio in circulo directo, f. [A: 400 (6); iii. 147], ascension on the direct circle, right ascension.

ascensio in circulo obliquo, f. [A: 400 (17); iii. 147], ascension on the oblique circle, oblique ascension.

aspectus, -ūs, m. [E: 226 (8); K: 562 (22); ii. 120–1], aspect (planetary). (The aspects named are five: coniunctus (or coniunctio), sextilis, quartilis, trinus, oppositus (or opposicio), these corresponding to zodiacal separations of 0, 60, 90, 120, 180 degrees respectively. Each is qualified as either bonus or malus [E: 226 (1)].)

assimetrus, -a, -um [Q: 46 (17); Q: 48 (31); i. 47 n. 3], incommensurable. (For possible derivation, cf. arsmetrica; or possibly a poor transliteration of ἀσύμμετρος.)

**astrolabium**, *n.*, (planispheric) astro-labe. (Gemma Frisius' **astrolabum** [ii. 185] was presumably consciously modelled on the Greek ἀστρολάβον ὄργανον of (for example) *Almagest* V. 1. The adjective **universale** (or **generale**) [iii. 162–7], almost invariably indicates incorporation of the **saphea** (**Arzachelis**), *q.v.*).

**atalasimet, altasimet** [iii. 253; iii. 255], pertaining to talismanic images (**ymagines**). (See [iii. 257] for Millás's attempt to link the word with that for azimuth.)

**auricalcum**, *n.*, brass. (*Class. Lat.* ori-chalcum.) **auricalceus, -a, -um,** of brass, brazen.

**aux** (*pl.* **auges**), *f.* [A: 278 (5, 24); E: 214 (5); iii. 170, 175–7], apogee, most distant point of epicycle or deferent circle. (Synonym of **apsis** or **absis**, *q.v.* For **aux communis, aux propria**, see [iii. 177]. **Aux** is from the Arabic *awj*, apogee.)

**azamena**, *n.*, indecl. [K: 558 (4); ii. 375–6], degree of the zodiac corresponding to human debility.

**azimuth**, *n.*, indecl. [A: 330 (11); iii. 35], azimuth (a great circle from the zenith to the nadir, or the direction of such). (From either *al-samt*, direction, or the *pl.* form *al-sumūt*. The word is much used in combination with others in Arabic astronomical vocabulary. *Cf.* **cenit, nadir.** For Millás's association of azimuth with **altasimec**, see [iii. 257].)

**billi**, *m. pl.* [S: 538 (5); i. 538 n. 1], bills, a game.

**buxus**, *m.* [ii. 159], box (-wood).

**caliga**, *f.* [H. 474 (7); H: 476 (24); i. 475 n. 5; i. 477 n. 11; ii. 338], weight barrel or drum (of the going or striking train of a clock); [4: 486 (3); H: 490 (21); ii. 353], tube or pipe (in a system of two or more coaxial arbors). (Thus **caliga diurna, caliga Solis.** See also **rota dialis.**)

**caput Draconis**, *n.*, **cauda Draconis**, *f.* [E: 224 (4–5)], the head and tail of the Dragon (i.e. the ascending and descending lunar nodes). (For some of the mythology behind the Dragon idea, see W. Hartner, *Oriens—Occidens*, pp. 349–404, 268–86. See also **geusahar** and **nodus.**)

**cardaga**—*v.* **kardaga.**

**casus, -ūs**, *m.* [A: 368 (17); i. 369 n. 1; ii. 285; iii. 215], an eclipse duration (or a half the duration of partial eclipse). (*Cf.* **mora.**)

**cata** *or* **catha** (**coniuncta/disiuncta**), *f.* [Q: 92 (2); Q: 124 (3); Q: 154 (9); ii. 63; i. 79 n. 4], cata, sector figure, figure for the theorem of Menelaus. (The disjunct case is where only one side of the triangle is cut externally.)

**cathetus**, *m.* [A: 262 (31)], a perpendicular. (In the text discussed in Appendix 16, the word denotes the *hypotenuse* of a right-angled triangle. Occasionally the word seems to be used adverbially, as a synonym of **orthogonaliter.**)

**cenit, cenith** (**capitis**), *n.* indecl. [Q: 134 (16)], zenith. (From Arabic *samt al-ra's*, direction of the head. Beware the use of **cenith** unqualified, for it occurs in Latin (rarely, however) as a synonym of **azimuth**, *q.v.*)

**centrum** (**medium/verum**), *n.* [A: 352 (5); iii. 175] (mean/true) centre, elongation of epicycle centre from line of aux.

**chilindrum**, *n.* [M: 9 (6); iii. 114], cylindrical sundial of characteristic type (sometimes known as a 'pillar dial'.)

**cifra, ciphra**, *f.* [Q: 32 (18)], zero. (From Arabic *ṣifr*, from a root meaning 'to be empty'.)

**cimba**, *f.* [H: 472 (21); i. 473 n. 6], bell (of the clock).

**circulus**, *m.*—*v.* **rota** (for **circulus** meaning gear-wheel).

**circulus declivus**, *m.* [A: 290 (5)], inclined circle (giving the Moon's apparent path, in the instance cited).

**circulus deferens**—*v.* **deferens.**

**circulus directus** (*or* **rectus**), *m.* [Q: 110 (25); i. 111 n. 5; Q: 114 (26); Q: 134 *passim*; iii. 147], right horizon. (This can be a synonym for **equi-**

**noxialis,** *q.v.*; *cf.* **sphera recta** for equinoctial. Richard of Wallingford uses **horizon rectus,** *q.v.*, for which **circulus directus** is a synonym.)

**circulus ecentricus**—*v.* **ecentricus, -a, -um.**

**circulus equans,** *m.* [A: 302 (29); iii. 173], equant circle, circle centred on the equant point (and around which the epicycle centre moves uniformly).

**circulus equinoctialis,** *m.* [Q: 98 (29)], the equinoctial, the celestial equator.

**circulus involutus,** *m.* [A: 324 (16); ii. 179–80], spiral. (There is no technical name for the curve described in II. 19.)

**circulus obliquus,** *m.* [Q: 158 (2); Q: 110 (25); i. 111 n. 5; iii. 147], oblique circle, ecliptic. (*Cf.* **circulus rectus.** Note the very different meaning of **rota obliqua,** *q.v.*, an oval wheel.)

**circulus tortuosus,** *m.* [H: 520 (27); iii. 68], the oval wheel (also called **ecentricus,** *q.v.*, and **rota obliqua,** *q.v.*).

**clausura,** *f.* [S: 537 (3)], fence; [H: 482 (1); i. 483 n. 5], locking device (for the strike of the clock).

**clava,** *f.* [H: 514 (13); i. 515 n. 4], club, mace (?).

**clima,** *n.* [A: 382 (1)], plate of an astrolabe (for a given climate, or geographical latitude).

**clok,** *f.* (?) [H: 480 (1); i. 481 n. 1; ii. 370], clock or bell for a clock (*cf.* **cloca** in *MLWL2*).

**coaptare** [H: 510 (18)], to mesh, engage (as of one wheel with another). (Generally in passive.)

**colurus,** *m.* [Q: 110 (31)], colure, a great circle passing through the poles. (It is a mistake to define the colures as though there were only two, the **colurus equinoctium** and the **colurus solsticiorum.** *Cf.* **meridianus.**)

**componere** [Q: 58 (11); i. 59 n. 4], to compound or combine (used of ratios).

**composicio,** *f.* [Q: 68 (3)], arrangement (of ratios).

**compotesta, computesta,** *m.* [E: 186 (11)], calculator (or possibly 'computist' in the sense of one who works with

the **computus** for calendar calculation). (The translator has 'countoure'.)

**conclavare** [H: 514 (11)], to nail, fix with rivets. (*Cf.* **tenaculum.**)

**conclusio,** *f.* [A: 248 (8); i. 249 n. 1], proposition, construction.

**concordare** [H: 502 (7)], to mesh, engage (as of one gear wheel with another).

**coniunct(a)**—*v.* **cata coniuncta.**

**coniungere,** to add together.

**consequenter,** consecutively.

**consideracio,** *f.* [A: 324 (12); i. 325 n. 2], (astronomical observation, investigation.

**conversus,** *m.* [ii. 312], lay brother.

**corda, corda recta,** *f.* [Q: 26 (2); R: 414 (6)], chord (of an arc of a circle. (For potential ambiguity, see **sinus,** and also [i. 137 n. 1]. For **corda versa,** see [R: 414 (7); i. 415 n. 5].)

**cordare** [M: 10 (30); A: 288 (13)], to span with a chord (as in **cordare arcum**).

**costa,** *f.* [A: 296 (1)], side (of a square, in the example cited).

**coxa,** *f.* [R: 408 (6)], (hinged) joint (of the instrument **rectangulus**). (*Cf.* **tibia.**)

**decanus,** *m.* [E: 204 (8); ii. 111], decan, interval of 10 degrees of the zodiac (this being one-third part of a sign). (Usually astrological, synonym of **facies,** *q.v.*)

**declinacio (ab ecliptica, ab equinoxiali, a zodiaco),** *f.* [Q: 158 (16); Q: 128 (1); i. 129 n. 2], declination *or* (ecliptic) latitude. (The ambiguity parallels that of **latitudo,** *q.v.* Note **declinacio tota** [Q: 124 (15)] and **declinacio maxima** [Q: 152 (20)], maximum declination of the Sun, i.e. obliquity of the ecliptic. **Declinacio** occurring alone and meaning '(ecliptic) latitude' is found, for example, at [A: 292 (2)].)

**declivus**—*v.* **circulus declivus.**

**deferens,** *m.* and *f.* A: 278 (13, 24); iii. 173], deferent, circle carrying the centre of the epicycle. (The word is often used adjectivally in the same sense, as in **ecentricus deferens** [A: 278 (22, 29)], strictly the 'carrying eccentric [circle]', rather than the 'eccentric deferent' of [A: 278 (13)].)

**demere**—to subtract.

**demonstrare** [Q: 24 (2); i. 25 n. 1], to prove, evaluate, test.

**denominacio,** *f.* [H: 450 (21)], denominator (of a fraction); [Q: 58 (9); i. 59 n. 1; ii. 56–7], denomination, integral or rational number expressing a ratio in its lowest terms.

**dens,** *m.* and *f.* [H: 444 (7)], tooth (of the wheel of a clock); [H: 476 (16)], pin (on a wheel, having the same function); [H: 480 (12)], peg, pin, stop (of the drum of the striking mechanism [iii. 67]. (Note **dens testudinatus** (H: 446 (15)], snail tooth, and **dens perforata** [H: 510 (8); i. 511 n. 2; ii. 353], slotted tooth (?), spur tooth (?); and note the random gender. See also **testudinatus, -au, -m.**)

**dialis,** *adv.* —*v.* **rota dialis.**

**dialis,** *f.* [H: 474 (4); i. 473 n. 6], dial.

**diametare** [Q: 114 (31)], to divide by, or provide with, a diameter.

**diameter, -tri,** *f.* [A: 266 (15); i. xiii], diameter.

**diametĕre** [Q: 114 (31)], 'to diameter'. (Thus, using the verb in the passive voice, 'a circle is diametered by such and such a line'. Mathematicians were ever thus.)

**differencia,** *f.*, difference; point (in common to two circles, as in the phrase **differencia communis inter duos circulos**).

**dignitas,** *f.* [E: 182 (28); ii. 108], dignity (of a planet). (Synonyms given in [E: 200 (29–31)] are: **fortitudo, potestas, testimonium.** The dignities are **domus, exaltacio, triplicitas, terminus,** and **facies,** all of which see.)

**directorium,** *n.* [ii. 259; iii. 114], an astrological instrument.

**directus, -a, -um,** direct (as opposed to retrograde, in motion). (*Cf.* **progressivus, progressus.**)

**disiunct(a)**—*v.* **cata disiuncta.**

**dispositorium,** *n.* [ii. 283] an eclipse instrument (so named by Apian).

**diversitas aspectus,** *f.* [A: 374 (2)], parallax. (Thus **diversitas aspectus in circulo altitudinis apud orizontem** [A: 286 (2)], horizontal parallax in altitude.)

**diversitas diametri,** *f.* [iii. 178], a function in the Ptolemaic theory of planetary longitude. The phrase hints at the variation (**diversitas**) in the apparent size (**diameter**) of the epicycle, as between aux and opposite aux, which variation is responsible for change in the function giving the equation of the argument.)

**doleum,** *n.* [H: 480 (9); i. 481 n. 5; i. 483 n. 6; ii. 335], barrel or drum. (Used for the barrel which counts out the strike of the clock.)

**domicilium,** *n.* [E: 202 (23); ii. 102], (planetary) house, domicile. (Richard of Wallingford usually prefers the word **domus,** *q.v.*)

**dominus anni (mensis, hore),** *m.* [E: 228 (20); E: 234 (19, 24); ii. 121–2], lord of the year (of the month, *or* of the hour).

**domus, -ūs,** *f.* [E: 200 (31); A: 382 (23); ii. 102], (planetary) house, domicile. (*Cf.* **domicilium.** Note **domus accidentalis** [E: 202 (11)], secondary house, '[signum] in quo creatus fuerat', and (in this sense, and with an almost identical definition) **domus planete** [K: 559 (12)].)

**dulkarnon,** *n.*, indecl. [A: 250 (13); Q: 38 (39); i. 251 n. 4], the theorem of Pythagoras, Euclid I. 47.

**ecentricus,** *m.*, eccentric (deferent circle). (See **deferens.**)

**ecentricus, -a, -um,** eccentric (as in [A: 278 (13)] **ecentrica deferens**). (See **deferens.** A related but far from obvious meaning is seen in the phrase **circulus ecentricus** [H: 498 (lower table, etc.)], the oval wheel, also called **obliquus** and **tortuosus,** but constructed about eccentric centres [iii. 68].)

**eclipsabilis** [A: 376 (2)], eclipsable, liable to eclipse.

**eclipsare** [A: 378 (2)], to eclipse.

**eclipsorium**, *n.* [ii. 268–9], an eclipse instrument (in particular that by Peter of Dacia).

**econverso**, conversely. (Consistently written as one word.)

**elevacio**, *f.* [Q: 120 (19)] (potential synonym of) ascension. (See **ascensio**, But also [i. 121 n. 4].)

**elongacio**, *f.* [A: 268 (25); iii. 172, 175] elongation. (of the Moon). (Used in the current technical sense; but note **longitudo (duplex)**, with this sense.)

**embolismalis** [A: 358 (27); ii. 203–4] embolismal (year), having 13 lunations.

**embolismus (annus)**, *m.* [A: 358 (21); ii. 203–4], embolism, embolismal year.

**epacta**, *f.* [A: 358 (27); i. 359 n. 3], epact, age of the moon on 1 January. (This is not the only possible use of the word in the Middle Ages. See [ii. 204–5].)

**epiciclus**, *m.* [A: 278 (1); iii. 173], epicycle, small circle carrying the planet. (The centre of the epicycle moves on the **deferens**, *q.v.*)

**epistilum**, *n.* [R: 410 (6); i. 411 n. 1], epistyle, architrave, head of a column (*archit.*).

**equacio argumenti, equacio epicicli**, *f.* [A: 258 (8); A: 274 (7); iii. 175], equation of the argument, prosthaphairesis for anomaly.

**equacio capitis (ab equatore)**, *f.* [ii. 198; iii. 155], an equation of use in the theory of trepidation.

**equacio centri**, *f.* [A: 390 (2); iii. 175], equation of the centre.

**equacio dimidii diametri (circuli parvi)**, *f.* a parameter of the theory of trepidation.

**equans**—*v.* **circulus equans.** (The phrase **punctum equans** for the equant point [iii. 173] is much rarer in the Middle Ages than **centrum equantis**, centre of the equant (circle). See, for this usage, [A: 300 (14, 19)]; and for **equans** as a substantive, [A: 300 (17)].)

**equare** [E: 196 (32)]—*v.* **adequare.**

**equatio**—*v.* **equacio.**

**equator**, *m.* [A: 344 (3)], scale (of a certain generic type, on the albion). (See **(ad)equare, equatorium, tropicus.**)

**equatorium**, *n.* [ii. 249; iii. 114; iii. 132], equatorium (an astronomical instrument for 'equating' planetary positions, etc.)

**equidistans**, *adj.* [Q: 44 (4)], parallel. (Used of almucantars [A: 332 (2–3)]. As a noun, **parallela**, *f.*, is usual.)

**equidistanter**, *adv.* [A: 272 (30)], parallel.

**equidistare** [Q: 100 (27)], to be parallel.

**equilaterus, -a, -um** [Q: 40 (30); i. 41 n. 4], isosceles (of a triangle, not necessarily equilateral in the current sense).

**equinoctialis, equinoxialis**—*v.* **circulus.**

**exafrenon**, *n.*, (a work, etc.) in six parts.

**exaltacio**, *f.* [E: 200 (31); ii. 108, 112], exaltation (of a planet).

**extremum**, *n.* [Q: 58 (18)], term in a **proporcio**, *q.v.*

**facies**, *f.* [E: 200 (32); K: 560 (22); ii. 111], face (astrological).

**falsigraphus**, *m.* [Q: 146 (1); i. 147 n. 1], pseudography. (*Cf.* [ii. 35].)

**feria**—*v.* **littera ferie.**

**fiala**, *f.* [iii. 236], wheel (of a clock). (Compare class. **phiala**, dish or saucer. Perhaps **fiala** denotes what would now be called a 'dished' wheel.)

**filigraphus**—pseudographic form of **falsigraphus**, *q.v.*

**filum**, *n.*, thread. (Note that threads of parchment instruments in medieval books often survive, and that they are almost invariably of fine coloured silk.)

**firmacio**, *f.* [R: 408 (30)], a fixing, making firm. (Note **firmamentum**, and **firmare** [R: 410 (11)], to firm, make firm.)

**firmaculum**, *n.* [H: 516 (25)], support.

**firmamentum,** *n.* [A: 338 (10, 22)], a fixing, fixing device.

**fortitudo planete,** *f.* [E: 200 (29); K: 560 (32)], dignity of a planet (*v.* **dignitas**); [E: 210 (16)], strength of a planet (in a different sense from the foregoing).

**geusahar, geuzahar, geuzar,** *n.* [A: 322 (28); A: 342 (33); i. 524], planetary node, (especially common as) lunar node. (From Arabic and Persian *al-jawzahar* or *al-jawzahr*. Owing to a faulty transcription which scribal ignorance could only preserve, the word is written **gensahar** perhaps more often than **geusahar.** See **caput/cauda Draconis,** and the article by Hartner mentioned there.)

**gradus, -ūs,** *m.,* degree. (The degrees of the zodiac are qualified in numerous ways in such astrological works as **Exafrenon.** Thus they may be **fumosi, umbrosi, lucidi, tenebrosi, vacui, pleni,** or (and these are more often descriptions of complete signs) **orientales, meridionales, occidentales, septentrionales, ignei, aeri, aquatici, terrei, masculini, feminini. Gradus putei** are added in [K: 558 (14)].)

**gradus cum quo stella mediat celum** [Q: 126 (12); i. 127 n. 2; T: 433 (2, 19); T: 436 (7–8); iii. 144–5], mediation of the star, *or* right ascension of the star (according as it is understood that the **gradus** is **zodiaci** or **equinoxialis**). (The expression is much used as a heading in star-coordinate lists, but the low level of computation makes it difficult to decide meaning without a detailed canon to the table. It is probable that the ambiguity of the bare expression led Richard of Wallingford astray when he proposed to show **quis erit punctus** *L* **in zodiaco** [Q: 130 (24–5); *cf.* Q: 126 (32–3)] and yet ended by finding a point (*K*) on the equator: **cum eodem gradu provenit stella ad medium celi** [Q: 134 (11–12)]. See [ii. 71–3; ii. 243] and *cf.* [ii. 302–

4]. *Tract III* [i. 436 (30–2)] uses the basic phrase categorically for both concepts together.)

**gradus longitudinis ex utraque parte** [iii. 159–61], marginal longitude (a coordinate for astrolabe work).

**gradus nonagesimus—***v.* **nonagesimus gradus.**

**gradus verus (in zodiaco)** [M: 17 n. 6; T: 435 (4)], ecliptic longitude.

**hallidada, halidada—***v.* **allidada.**

**hasta,** *f.* [H: 500 (29)], shaft, arbor (of clock). (Synonym [R: 418 (6); H: 514 (21)] of **regula,** *q.v.,* in connection with the rectangulus and the clock, and of **lancea,** *q.v.* Note **hasta malleoli** [H: 476 (19)], the shaft carrying the hammer (for the clock bell).)

**hore naturales,** *f. pl.* [H: 478 (8–9)], natural, canonical, unequal hours.

**horizon obliquus,** *m.* [Q: 158 (2)], horizon (in the modern sense of the word, i.e. the local horizon).

**horizon rectus,** *m.* [Q: 112 (9)], right horizon (i.e. the great circle through the poles and the eastern and western points of an observer's **horizon obliquus,** *q.v.*). (Synonym of **circulus directus,** *q.v.*)

**horologiarius,** *m.* [ii. 315–17; ii. 363], clock maker.

**horologium astronomicum,** *n.* [H: 444 (3); ii. 114], astronomical clock. (The second reference reports that Richard of Wallingford was by some regarded as the inventor.)

**ieusahar—***v.* **geuzahar.** (This version is found in MS. C.C.C.(D) 144, suggesting a consonantal pronunciation of the **i.**)

**imitari** (*dep. vb.*) [A: 286 (19)], to be approximately equal to.

**inchoare** (and derivatives)—favoured medieval spelling of class. **incohare,** to begin, commence.

**inferius** (as noun) [Q: 62 (4)], denominator (of a fraction), 'that which is below'. (Occurs in the phrase **secundum superius ad inferius.** *Cf.* **sub/supra.**)

**instrumentum,** *n.* [H: 484 (28); i. 485 n. 5], worm (gear), screw (as of a wine press). (The phrase is **ad modum instrumenti torcularis.**)

**instrumentum semidiametrale,** *n.* [ii. 283], an eclipse instrument (being Apian's version of that on the albion).

**intervallum,** *n.* [Q: 64 (4); i. 65 n. 1], simple ratio, distinct ratio.

**intrare** [H: 450 (11)], to enter (a table).

**inutilis** [Q: 66 (11); i. 67 n. 3], not needing to be further considered, to be ruled out. (Said of possibilities in an argument, that is, of 'modes' of the proportionality associated with Menelaus' theorem.)

**invicem, ab invicem** [A: 338 (25)], mutually, from one another.

**involutus**—*v.* **circulus involutus.**

**iomyn,** *n.,* indecl. [A: 312 (8); ii. 169], equation of days. (Presumably from Arabic *al-yawm bi-lailatihi,* which translates the Greek νυχθήμερον, a night and a day.)

**iudicium,** *n.* [E: 192 (11)], (astrological) judgement.

**kardaga, kardaja,** *f.* [Q: 32 (2, 22); ii. 46–7], fraction of a circle (here 15 degrees).

**kata, katha**—*v.* **cata.**

**lamen,** *f.* [H: 488 (6, 7)], plate. (Synonym of **lamina,** whose gender it retains.)

**lamina,** *f.,* plate. (Use of the word is to refer to thin plate typical of class. Latin, except perhaps in [R: 406 (15)], where the **lamina** is half a digit thick and only two wide, and in [H: 478 (7)], where the **lamine** seem to be planks out of which the clock-housing is made.)

**lancea,** *f.* [H: 484 (26); i. 485 n. 4], shaft, arbor. (Synonym of **hasta,** *q.v.*)

**lapis calculatorius,** *m.* [iii. 133], reckoning slate, exchequer board, abacus (?).

**latitudo,** *f.,* breadth, width, lateral displacement (esp. angular). (The word does not have a fully specific technical meaning equivalent to modern English 'latitude'. *Cf.* **longitudo, declinacio.** Such phrases as **latitudo stelle ab equinoxiali** [M: 17 n. 6], are a common enough prototype for modern usage. See [iii. 144–5].)

**limbus,** *m.* [A: 348 (24)], edge, rim, limb (in the same sense). (Although **limbus** is seemingly synonymous with **rota,** *q.v.,* in *Albion* Part III, this is probably only because the scales under discussion are, very loosely speaking, on the *rims* of the discs. Note the possibility that **limetur** [H: 494 (4); i. 495 n. 1] is formed from **limbus.**)

**linea numeri, linea numerorum,** *f.* [A: 390 (14); A: 388 (25)], line or column of numbers (usually down the left-hand side of a table, by which numbers the table is entered). (The first is the more usual alternative throughout the Middle Ages. The (first) phrase is a translation of the Arabic *suṭūru'l-ʿadad.*)

**littera ferie,** *f.* [A: 328 (5–6); i. 329 n. 2], day letter. (Of this the dominical letter is a special case.)

**littera numeri,** *f.* [R: 412 (29)], numeral, written number. (**Littera** is used alone with the same meaning [A: 334 (37)].)

**longitudo,** *f.,* length, distance (esp. angular). (The word does not always have the technical sense of modern English 'longitude'. Thus **longitudo duplex** [A: 268 (2)] is synonymous with **elongacio,** elongation; **longitudo longior/propior** [A: 266 (31); iii. 178] has the technical sense of 'greatest/least distance of the epicycle centre' (the comparative arising from a mistranslation of the Arabic superlative forms, which correctly represent *Almagest*); for an unusual use of **longitudo propior** see [A: 344 (9); i. 345 n. 2]; **longitudo secundum equinoxialem** [Q: 126 (29)] is right ascension, for which **gradus verus in zodiaco** (*q.v.*) is much more frequently used; but see also **gradus cum quo stella mediat celum.**)

**machina,** *f.* [ii. 315], machine. (The word is used of the clock by Leland.)

**machinacio,** *f.* [iii. 137], the art of working with a **machinamentum** (*q.v.*).

**machinamentum,** *n.* [iii. 137], instrument with moving parts (contrasted with static gnomon).

**marginare** [R: 412 (13)], to mark out a margin.

**margo,** *f.* [R: 412 (15, 30)], 'the narrow surface between two parallel straight lines, placed close together'. (This extension of the classical word's meaning seems to have no parallel, even in English. **Margo media** would normally be a contradiction in terms.)

**medium celum,** *n.* [A: 372 (6); i. 373 n. 2; ii. 102; ii. 233], mid-heaven (in the technical sense explained in the note). (Note also **imum medium celum,** *ibid.*)

**meridianus, -a, -um** [Q: 114 (31); i. 115 n. 6], of the nature of a meridian or colure. (The **circulus meridianus** cited is a colure of the equinoxes.)

**minutum,** *n.*, one sixtieth part. (Thus **minutum hore, minutum diei, minuta proporcionalia** [iii. 188].)

**modus (coniunctus/disiunctus),** *m.* [Q: 104 (1); ii. 60–1], (conjunct/ disjunct) mode (of the expression of Menelaus' theorem, etc.). (See also **cata.**)

**mora, mora media,** *f.* [A: 368 (2); i. 369 n. 1; i. 289 n. 2; iii. 25; ii. 285], duration (or half-duration) of the total phase of an eclipse. (But note the ambiguity recorded in the second reference. *Cf.* **casus.**)

**morphea,** *f.* [S: 551 (3) and n. 1], morphew.

**motus, -ūs,** *m.* [E: 184 (2); A: 326 (11–13); iii. 175–7], motus (as in **medius motus, motus verus**), movement. (The ambiguity in translating the word as 'motion' is avoided here by treating 'motus' as an English word; *cf.* Mid. English 'mote' with

this meaning. **Motus** in these technical senses is usually an (angular) *position.* Note that the combination **medius motus argumenti** [A: 326 (12)] is synonymous with **medium argumentum.** Note the non-astronomical **motus molaris** [R: 410 (12)], rotary (lit. of a mill) motion. Observe the *caveat* of [ii. 198] as regards Richard's use of **motus octave spere,** which may denote an angular movement (in ecliptic longitude) of the 'fixed' stars or the **motus accessionis et recessionis octave spere** [iii. 155], from the theory of trepidation.)

**nadir,** *n.* [E: 194 (8); A: 330 (8); A: 384 (2); i. 331 n. 2], nadir, point diametrically opposed on the celestial sphere. (The modern meanings 'lowest point' and 'point opposite the zenith' are misleading, since they are not sufficiently general. Arabic *nazīr* means 'opposite', also as substantive *al-nazīr* in an astronomical sense, especially in astrolabe theory, while *nazīr al-samt* means 'the opposite of the zenith'. *Cf.* **cenit,** zenith.)

**navicula,** *f.* [iii. 113], 'the little ship', sundial resembling a ship. (Also **navis** [iii. 113].)

**nodulus,** *m.* [M: 9 (15); i. 9 n. 3], point of intersection.

**nodus,** *m.* [A: 290 (5)], node, point of intersection of two great circles of the sphere. (Usually the ecliptic and the Moon's path are intended. *Cf.* **geusahar, caput/cauda Draconis,** which are near-synonyms in the Latin West.)

**nonagesimus gradus,** *m.* [A: 372 (2); i. 373 nn. 1, 3], mid-ecliptic, in the sense of a point 90 degrees along the (risen) ecliptic westwards from the ascendent (**ascendens,** *q.v.*).

**nux,** *f.* [H: 458 (5); H: 474 (3); H: 484 (25)], pinion, small gear (presumed solid), nut (in archaic sense). (*OED*— Nut, II. 9—lists six uses of the English word in this sense, between 1426 and 1825, and the word was still in use in Clerkenwell in the 20th century.)

**obliquus**—*v.* **circulus obliquus, rota obliqua.**

**occasus, -ūs,** *m.*—*v.* **ortus.**

**occulte** [A: 296 (35)], lightly, faintly (of inscribed or drawn lines). (Contrast with **sensibiliter** [A: 308 (11)].)

**oliveta,** *f.* [E: 242 (5); i. 243 n. 2], olive-yard (class. Lat.). (But note the Mid. English translation 'oil maker'.)

**opimetr**(—?)—*erroneous form of* **diamet**(—). (See [ii. 386].)

**opus,** *n.* [iii. 226], a train of gear wheels.

**orbis,** *m.*—probably always used here as synonymous with **circulus.** Thus **orbis signorum** [T: 436 (21); Q: 148 (33)], zodiac; and **orbis equacionis diei, orbis meridiei, orbis orizontis,** and **orbis magni** [Q: 150 (24–8); T: 436 (28)]. Perhaps the most unusual use is where **orbis** denotes a 'circle' in the sense of 'circular scale', e.g. in [A: 300 (10)].

**oriolum,** *n.* [S: 537 (40) and n. 2], porch (synonym of **porticum**).

**orizon**—*v.* **horizon.**

**ortus, -us,** *m.*, a rising (of a heavenly body). (Like **occasus,** a setting, class. Lat., but note technical uses deriving from Ptolemy: **occasus (ortus) vespertinus, occasus (ortus) matutinus** [E: 216 (24–5); ii. 118].)

**papirus,** *m. and f.* [ii. 218], paper.

**pars proporcionalis,** *f.* [iii. 178], proportional part. (This is a technical expression of the Ptolemaic theory of planetary longitude, as well as a phrase with an obvious meaning.)

**partire** [Q: 148 (16)], to divide (one quantity by another).

**paxillus,** *m.* [A: 338 (5); iii. 38; R: 408 (31); iii. 56], pin, pivot, hinge (as at the centre of the albion, or at the top of the rectangulus column, **paxillus basis** [R: 416 (9)]).

**pellvis,** *f.* [H: 520 (2)], basin (?), skin (?). (See [i. 521 n. 1; iii. 73].)

**perforatus, -a, -um**—*v.* **dens.**

**perpendiculariter** [A: 372 (9); R: 418 (12)], vertically, perpendicularly. (Both references are to the modern, and in the Middle Ages relatively rare, geometrical sense. **Perpendiculum** is the class. Lat. word for 'plumb-line'.)

**petitio** [Q: 74 (21); i. 75 n. 7], a premiss. (Used here in the theory of the eighteen modes, the word was common in logic.)

**pinnula,** *f.* [R: 408 (25); R: 426 (13)], pinnule, sighting vane (on the rectangulus, perforated as on an astrolabe; a synonym of **tabella,** *q.v.*); [H: 474 (10); i. 475 n. 8], pin, cylindrical rod (as in the phrase **pinnule strob**); [H: 488 (19); i. 489 n. 9], rod (of considerable size).

**planeta,** *m.* [E: 236 (1)], planet. (Late Lat. for class. **stellae errantes**—rare but not unknown in Richard of Wallingford [A: 278 (29)]. The known 'planets' were: **Sol,** *m.*, **Luna,** *f.*, **Saturnus,** *m.*, **Iovis,** *m.*, **Mars,** *m.*, **Venus, -eris,** *f.*, **Mercurius,** *m.* [E: 236 (7–10)]. The planets were qualified as **calidus** or **frigidus, siccus** or **humidus** [E: 236, 238], **inferior** or **superior** (below or above the Sun in the order then supposed). The allocation of the degrees of the zodiac to the planets is shown, with their symbols, at [ii. 109].

**polus mundi,** *m.* [Q: 126 (21–2); R: 424 (4)], (celestial) pole, pole of the equinoctial.

**polus zodiaci,** *m.* [Q: 126 (22)], pole of the ecliptic.

**pondus, -eris,** *n.* [H: 472 (21); i. 472 n. *c*; i. 473 n. 7], weight (for driving the clock). (Contrast with the weights on the foliot, i.e. of the **quadratura plumborum,** *q.v.*)

**posta,** *f.* [H: 478 (3); i. 479 n. 2], post (of the clock).

**progressivus, -a, -um** and **progressus** (*participial adj.* ?) [E: 212 (1, 8)], progressive, direct (of motion along the zodiac, i.e. following the direction of the Sun in its annual motion). (The translator, perhaps rightly, takes **progressus** to be the fourth declension noun, 'progression'.)

**proporcio,** *f.* [Q: 24 (6)], ratio, proportion.

**puncta ecliptica,** *n. pl.* [ii. 285], eclipse magnitude.

**quadratura plumborum,** *f.* [H: 476 (8); i. 477 n. 4], rectangular lead (weights for the foliot of the clock); *or* cross-piece (of the foliot carrying the) leads. (Contrast **pondus,** for the driving weights.)

**quadripartitum,** *n.,* (a work, etc.) in four parts.

**racio,** *f.* [Q: 78 (12); Q: 96 (24)], reasoning, computation.

**racionabilis** [H: 454 (16)], rational, uniform, regular (?).

**radius,** *m.,* ray (as in **radius visualis** [A: 388 (15)], **radius Solis**). (Note that the word is not commonly a synonym of **semidiameter** until the 17th century. For an astrological use, see **radii virtutis planete** [ii. 115 (33)], planetary rays of virtue; *cf.* iii. 128.)

**radix,** *f.* [E: 188 (4); iii. 176–7], root, value of a variable at some fundamental epoch; [A: 340 (10)], numerical parameter. (This second use is less usual. It occurs again in **radices antiquorum** [A: 260 (8)].)

**ratio**—*v.* **racio.**

**rectangulus,** *m.* [R: 406 (3); iii. 57; iii. 114], the rectangulus (Richard of Wallingford's astronomical instrument).

**regula,** *f.* [R: 416 (11); iii. 57], rule (as attached to the limbs of the rectangulus; [H: 468 (3)], (procedural) rule; [H: 488 (18); i. 515 n. 7], rule, pointer (on the face of the clock, showing the Sun's position and its nadir).

**retardativum,** *n.* [ii. 333], double pallet (resembling the **semicirculus,** *q.v.* ?).

**rete, rethe,** *n.* [E: 192 (13); iii. 132], rete (of an astrolabe). (This is the pierced disc overlaying the plate or **clima** (*q.v.*), and usually carries the star map.)

**retrogradarius, -a, -um** [E: 212 (8)] and **retrogradus, -a, -um** [E: 210 (32)], retrograde (of a planet's motion along the zodiac). (Contrasted with **progressivus** [E: 212 (8)] and **progressus** [E: 212 (1)], *q.v.* Synonymous with **contra signa** [A: 280 (24)].)

**rota,** *f.* [A: 340 (23); H: 444 (1)], wheel, gear-wheel, revolving disc, plate (of the instrument albion). (**Circulus** [H: 446 (19)] is often used synonymously, for a gear wheel, but **rota obliqua,** *q.v.*, is to be distinguished from **circulus obliquus,** *q.v.*)

**rota dialis,** *f.* [H: 472 (21–2)], wheel with a daily motion, journal wheel. (Note also **caliga dialis** [H: 474 (5–6)] and **caliga diurna** [H: 512 (8)]. For the connection with the English 'dial', see **dialis,** *f.*)

**rota obliqua,** *f.* [H: 490 (24); i. 491 n. 6; iii. 68], the oval wheel (also called **circulus ecentricus** and elsewhere **tortuosus**).

**runtellus,** *m.* [iii. 235], wheel (of a clock).

**safea, saphea,** *f.* [M: 9 (5); A: 330 (29); ii. 188; iii. 114; iii. 132], saphea (of Arzachel), plate with a universal astrolabe projection. (The Arabic word *ṣafīḥa* was used of astrolabe plates in general, for which the Latin word **lamina** was well established by the time Arzachel's universal plate reached the Latin West. *Saphea Arzachelis* is a rendering of *al-ṣafīḥat al-zarqāliyya*, Arzachel's plate.)

**sciencia communis,** *f.* [Q: 44 (12)], common notion, premiss, axiom (in modern sense).

**sector,** *m.,* sector, part of a circle between two radii and the contained arc; [*De sectore,* opening def., i. 170], figure for Menelaus' theorem on a sphere, the 'sector figure'. (See **cata.**)

**semicirculus,** *m.* [H: 480 (6); iii. 64; ii. 330–4], semicircle, double pallet (for the clock escapement).

**semissa,** *f.,* semicircle; [ii. 260], planetary equatorium (of the type designed by Peter of Dacia).

**sensibiliter**—*v.* **occulte.**

**senus, -a, -um** [A: 298 (8)], six.

**signum 60 graduum,** *n.* [H: 470 (25); i. 471 n. 2], 'physical sign', sign of 60 degrees (as opposed to the 'common sign' of 30 degrees, **signum zodiaci,** *q.v.*).

**signum (zodiaci),** *n.* [E: 188 (2); K: 562 (1)], sign (of the zodiac). (That is, one twelfth part of the zodiac, also identified with the planetary houses. See **domus, domicilium.** For the names of the signs and their symbols, see [iii. 169]. For astrological descriptions of the signs, see under **gradus**; and note other adjectives, e.g. **fixus, mobilis, communis** in [K: 562 (10–13); ii. 378]. Note also **contra signa** [A: 280 (24)], synonymous with **retrogradus,** *q.v.*)

**similibet** [i. 306 n. *e*, 307 n. 1], *for* **sibimet**?

**sinus, -ūs,** *m.* [Q: 48 (2–5); ii. 43; iii. 142–3], sine (to a standard radius rarely of unity). (Usually in combination, as **sinus rectus, sinus versus, sinus duplatus,** these phrases being clearly distinguished in *Quadripartitum.* Note the laxity of subsequent use of **sinus** and **corda,** however [i. 53 n. 3; i. 57 n. 2; i. 255 n. 2]. For Albategni's use, which is of relevance here, see *Bibl. Math.*, vol. i, p. 521.)

**sinus per se,** *m.* [Q: 34 (10); i. 35 n. 1], a difference between sines at the terminating radii of an arc.

**sinus totus,** *m.* [A: 292 (6)], the sinus rectus of 90 degrees.

**solidatura,** *f.* [A: 296 (19)], solder.

**sonitus, -ūs,** *m.* [H: 476 (13)], the strike (of the clock). (Thus **rota sonitus, pondus sonitus, caliga sonitus, mensuracio sonitus** [H: 476 (13–26)].)

**speculum generale,** *n.* [ii. 262–3], a type of equatorium.

**spera solida,** *f.* [ii. 296; iii. 132–3], an armillary sphere, with or without observing sights; a celestial globe. (There is good evidence that the phrase was ambiguous. Here **spera materialis** [iii. 133] was no doubt truly solid; but a treatise on the spera

solida in MS. Ashmole 1522, at f. 132ᵛ (*anno* 1303) is illustrated with a globe.)

**springum,** *n.* [H: 482 (1); i. 483 n. 4], spring (such as would restrain a locking catch, etc.).

**stacio,** *f.* [E: 212 (12); ii. 117, 202], station (of a planet); [A: 310 (37)], maximum or minimum value of a variable coordinate. (The second use is probably unusual at this date.)

**stacionarius, -a, -um** [E: 212 (8)], stationary. (See **stacio.**)

**strob,** *f.* (?), indecl.? [H: 474 (8); i. 475; ii. 330–4; iii. 64], escapement, or pertaining to the escapement. (There is no evidence that the word could be applied to escapements other than of the twin-wheeled variety (with **semicirculus, hasta, quadratura plumborum,** and **pinnule,** *q.v.*) characteristic of the St. Albans clock.)

**sub** [Q: 62 (2)], denominator (of a fraction), 'that which is under'. (Used with **supra** (numerator), both treated as nouns, in the phrase **secundum sub ad supra.** *Cf.* **inferius,** and [H: 444 n. *s*].)

**summitas capitis,** *f.* [Q: 150 (4)], zenith. (A synonym of **cenith capitis,** *q.v.*)

**superius** (as noun)—*v.* **inferius.**

**suppositorium,** *n.* [H: 520 (14)], a support.

**supra** (as noun)—*v.* **sub.**

**tabella,** *f.* [R: 416 (26); i. 417 n. 2], pinnule, sighting vane (on the rectangulus). (Synonym of **pinnula,** *q.v.*)

**tabulare** [A: 398 (14)], to tabulate. (Note the word **tabularius,** a maker of astronomical tables, in M.Eng. MS. Trin. Cantab. O. 5. 26, f. 130ᵛ.)

**tanquam,** equal to. (Thus *A* **tanquam est** *B*, *A* is equal to *B*.)

**tempus casus**—*v.* **casus.**

**tenaculum,** *n.* [H: 516 (21)], bolt, rivet (?). (This is conjecture. Any sort of holding device would suffice. The *pl.* form is used, and *MLWL2* gives (*anno* 1337) **tenacula** (*pl.*), pincers, tongs. *Cf.* **firmaculum.**)

**tendere ad occasum** [R: 420 (1)], to set (as the Sun).

**tenon, -is** [H: 474 (4)], **tenonum** (?) [H: 474 (14)], pillar for the bearings (for the arbors of the clock wheels). (See [i. 475 n. 1].)

**terquinque** [E: 192 (7)], fifteen (thrice five).

**terminus** [E: 200 (32); K: 560 (12); ii. 110], term (astrological).

**testimonium,** *n.* [E: 200 (31)], (planetary) dignity. (See **dignitas.**)

**testitudinatus, -a, -um** [H: 502 (7); i. 503 n. 2]—*v.* **testudinatus.**

**testudinatus, -a, -um** [H: 446 (15); iii. 354], of the nature of a worm-gear, or a gear with contrate spiral teeth. (Cognate with **testitudo,** snail, or **testudinalis,** vaulted (both in *ML-WL*) ? Probably distinguished by orthography only from **testitudinatus,** *q.v.* The analogy between the wheel of [iii. 72, lower fig.], i.e. **rota quasi testitudinata,** and a snail shell is clear enough.)

**tibia,** *f.* [R: 406 (14)], limb (or 'leg', of the instrument **rectangulus**). (*Cf.* **coxa.**)

**tollere**—to subtract, take away.

**toketum, torquetum**—*v.* **turketum.**

**tortuosus, -a, -um**—*v.* **circulus tortuosus.**

**triangulum,** *n.* [iii. 132 *bis*], an astronomical instrument (possibly connected with the tract discussed at [iii. 136–7]).

**triplicitas,** *f.* [E: 200 (31); K: 559 (32); ii. 110], triplicity (astrological).

**triquetrum,** *n.* [M: 10 (1); i. 10 nn. 1–2],

an instrument for observing the stars, 'Ptolemy's rules', the so-called 'parallactic instrument' of *Almagest* V. 12. (Not identical with the **turketum,** *q.v.*)

**tropicus,** *m.* [A: 342 (35); A: 344 (1, 9–11)], tropic (of solar eclipse or lunar eclipse). (There is no earlier or later recorded use of the word in the sense given to it in *Albion.* Bearing in mind the use of **equator** for circles on the albion tangential to 'tropics', as is the ecliptic (not the equator, but here we should look to **equare,** not **equinoctialis**) to the tropics of Cancer and Capricorn, we need not look beyond this analogy for an etymology.)

**turketum,** *n.* [M: 9 (6); R: 406 (9); ii. 296–300, esp. 298 n. 1; iii. 262], turketum, torquetum. (An instrument for astronomical observation, with equatorial and ecliptic movements.)

**umbra (recta/versa),** *f.* [M: 8 (8); i. 11; i. 13.

**uniformitans** (*f.*?) [ii. 333], double escape-wheel (resembling that of St. Albans?). (*Cf.* **rota strob.**)

**valere,** to be equal to.

**vector,** *m.* [iii. 226], arm carrying an epicycle (in an astronomical clock).

**via conbusta,** *f.* [E: 224 (1); ii. 116], 'the burnt way', the Galaxy, the Milky Way.

**zenith**—*v.* **cenit, cenith.**

# GLOSSARY OF WORDS IN THE MIDDLE ENGLISH *EXAFRENON*

THIS short glossary of words used in the Middle English translation of *Exafrenon* includes only those words likely to cause difficulty to the reader of modern English who is unfamiliar with the English of Chaucer's time. Words still in use, or related in an obvious way to words still in use, are excluded. As in the text, there is a preference for the (northern) orthography of MS. D, and the conventions explained as to transcription are also followed here. No attempt is made to record minor variations in spelling, even within MS. D. A much more complete glossary of English words of the same period will be found in Skeat's *Oxford Chaucer*.

**Aeren,** airy.

**Alther,** *gen. pl. of* **al** *or* **alle** (*adj.*, all, every). *Found with superl., as in* **alther hieste,** most of all.

**Anente,** against (*northern form*).

**Appetid,** appetite.

**Be,** by.

**Bowen,** *tr. of Lat.* INCLINARE, *which has transf. sense of* induce. *MS. T has* **draweth** *for* **bowes** *of MS. D.*

**Buxum,** tractable, obliging.

**Clepen,** call, mention, name. **Cleped** *and* **callid** *are used indifferently.*

**Comforthe,** comfort.

**Contrariuste,** contrariousness, opposition.

**Corne,** grain, seed, *esp. of* corn.

**Countoure,** calculator, *tr. of Lat.* COMPUTESTA.

**Delen,** deal, distribute.

**Demen,** deem, suppose; judge.

**Departen,** part, separate, divide.

**Differente,** deferent (circle), *tr. of* CIRCULUS DEFERENS.

**Distrablen,** disturb. *Cf. OFr.* DESTROBLER.

**Dome,** judgment, decree.

**Eke,** also, moreover.

**Eken,** eke out, increase.

**Enprenten,** impress (*as with a seal*).

**Evenfurth,** against, opposite, straight on, just forward (*tr. of Lat.* IN DIRECTO).

**Fadir,** *see n. 2 to the translation, vol. i, p. 199.*

**Faute,** deficiency, scarcity.

**Feblid,** enfeebled.

**Felen,** perceive, feel.

**Felyngly,** perceptibly (*see* **felen**).

**Foredown** (*or* **fordoon**), destroy, *pt.* fordide, *p.p.* fordoon.

**Frenchep,** friendship, association of friends or aides.

**Fructe,** fruit.

**Gedir,** gather.

**Heleful,** healthy.

**Helpe,** helper, assistant.

**Her(e), hir(e),** her (*pers. pron.*), her (*poss. pron.*), their (*poss. pron.*).

**Hyng(e),** hang. *Cf.* **hangen, hongen.** *The forms* **heng, hing,** *are northern.*

**Iche, ich,** each.

**Ilke,** each (*northern*) or same (*northern and southern form*).

**Kynde, kinde,** nature, sort, natural disposition.

**Kyndely,** natural (*adj.*), naturally (*adv.*).

**Letten** (*pt. sg.* **lettid, letted**), hinder, thwart.

**Lykynge,** lechery. *Cf.* **likerous,** lecherous.

**Manote,** may not.

**Medle,** mix, mingle. *Cf.* **mell,** meddle, interfere. *Cf. also* **meng,** *with a similar meaning.*

**Mekil,** much.

**Mell,** *see* **medle.**

**Meng,** *see* **medle.**

**Morowtide,** morning.

**Namore,** any more.

**Ne,** not (*in compounds, e.g.* **onlyne**).

**Newen,** renew.

**Notwar,** nowhere.

**Nowther,** nowhere.

**Noyen,** annoy, injure, harass.

**Nythande,** close at hand.

**Nythe,** near, nigh. *In* **wele nythe,** well nigh.

**Oley,** olive.

**On,** on, upon; one.

**Parcenel** (*or* **partenel**), partner. *Cf. Med. Lat.* PARCENARIUS.

**Particeps,** *see* **parcenel.**

**Prospicye,** prophesy.

**Pyght,** pith, pulp.

**Reven,** rob, take away.

**Roche,** roach, fish. *Cf. Med. Lat.* ROCHIA.

**Sal,** shall. **Salt,** shalt. (*Both are northern forms.*)

**Sarcle,** circle. *Cf.* **cercle.**

**Seechen,** seek.

**Sen,** since.

**Senthen,** since. *Cf.* **sithence.**

**Sewen,** pursue, follow.

**Sharpen,** sharpen, (*of cold weather*) worsen, make colder.

**Sithy** (*or* **sithe**), time, occasion. *In* Exafrenon *cap. 2, in pl., used for* times *as in a product of numbers.*

**Smert,** smart, quick.

**Sowth,** sought (*from* **seek**).

**Stede,** stead, place.

**Strang(e),** strong. *Modern* strange *is ME* straunge.

**Suande,** *northern spelling variant, from* sewen, *or* shewen, show.

**Swithter,** swifter.

**Terren,** earthy.

**That** (*conj.*), that, so that, as, because. *It is also used to repeat* of *or* when, *as in the Prologue to* Exafrenon.

**Tho,** (*pron.*) those, (*adv.*) then.

**Travel,** travail, labour.

**Trectes,** treatise.

**Trowen,** trow, believe, think possible.

**Turbation,** confusion, disorder.

**Tyedy** (*trans. of* SEROTINUS), timely, late in the season (or day?).

**Veere, vere,** *or* **ver,** the spring.

**Verray,** *or* **verrey,** true, real, exact.

**Wedeir,** weather.

**Whilke,** which (*northern form*).

**Wyne,** vine (*not* wine).

# BIBLIOGRAPHY

Apart from a few bibliographical abbreviations (for which see vol. i, p. **xx**), references made in the notes to the text and the commentaries are self-sufficient. The following list is meant for those who wish to pursue matters further. It includes printed books and articles with a general bearing on Richard of Wallingford's life and work, and on peripheral matters of importance (e.g. palaeography). Editions—unless they include much additional material—are under the original author's name, rather than the editor's. Cross-reference is made to persons only in their capacity as authors. Many of the items included were not available when this work was under preparation.

Leo Bagrow, 'The Origin of Ptolemy's *Geographia*', *Geografiska Annaler*, xxvii (1946), pp. 318–87.

J. S. Bailly, *Histoire de l'astronomie ancienne*, 2nd edn., Paris, 1781 (1st edn. 1775).

—— *Histoire de l'astronomie moderne*, vols. i, ii, iii ('nouvelle edition'), Paris, 1785 (1st edn. 1779–82).

—— *Traité de l'astronomie indienne et orientale*, Paris, 1787.

J. Bale, *Illustrium maioris Britanniae scriptorum catalogus*, Basle, 1557.

Al-Battānī, *Opus astronomicum* (ed. & tr. into Latin by C. A. Nallino, with many notes by S. V. Schiaparelli), vols. i, ii (Pubblicazioni del Reale Osservatorio di Brera in Milano, no. 40), 3 parts, Milan, 1899, 1903, 1907.

J. H. Baxter, *v. Medieval Latin Word-list*.

Guy Beaujouan, 'L'Enseignement du "Quadrivium"', *Settimane di studio del Centro italiano di studi sull'alto medioevo*, Spoleto, 1971, 'La Scuola nell' Occidente latino dell'alto medioevo', xix (1972), pp. 639–723.

S. A. Bedini & F. R. Maddison, 'Mechanical universe: the astrarium of Giovanni de' Dondi', *Transactions of the American Philosophical Society*, lvi, part 5, 1966.

C. F. C. Beeson, *English Church Clocks, 1280–1850*, Antiquarian Horological Society, London, 1971.

F. S. Benjamin & G. J. Toomer, *Campanus of Novara and Medieval Planetary Theory*, Madison, 1971.

Carl Bezold, *v. Franz Boll*.

Joseph Bidez & Franz Cumont, *Les Mages hellénisés*, 2 vols., Paris, 1938.

Al-Bīrūnī, *The Book of Instructions in the Elements of the Art of Astrology* . . . (ed. & trans. R. R. Wright), London, 1934.

A. A. Bjørnbo, 'Studien über Menelaos' Sphärik. Beiträge zur Geschichte der Sphärik und Trigonometrie der Griechen', *Abhandlungen zur Geschichte der mathematischen Wissenschaften*, xiv (1902).

A. A. Bjørnbo, 'Thabits Werk über den Transversalensatz (*liber de figura sectore*)' [mit Bemerkungen von H. Suter, herausgegeben und ergänzt durch Untersuchungen über die Entwicklung der muslimischen sphärischen Trigonometrie von H. Bürger und K. Kohl], *Abhandlungen zur Geschichte der Naturwissenschaften und der Medizin*, vii (1924).

J. Boffito & C. Melzi D'Eril, *Almanach Dantis Aligherii, sive Profacii Judaei Montipessulani almanach perpetuum*, Florence, 1908.

Franz Boll, *Studien über Claudius Ptolemäus*, Leipzig, 1894. [Reprinted from *Jahrbücher für classische Philologie*, Supp. xxi, pp. 47–244.]

—— *Sternglaube und Sterndeutung. Die Geschichte und das Wesen der Astrologie*. [with the assistance of Carl Bezold; fourth edition, edited by Wilhelm Gundel], Leipzig & Berlin, 1931. (Reprinted 1966, with a bibliography by H. G. Gundel.)

A. Bouché-Leclercq, *L'Astrologie grecque*, Paris, 1899.

A. von Braunmühl, *Vorlesungen über Geschichte der Trigonometrie*, 2 vols., Leipzig, 1900, 1903.

Carl Brockelmann, *Geschichte der arabischen Litteratur*, 1st edn. (2 vols.), 1898, 1902; supplements (3 vols.), 1937–42; 2nd edn. (2 vols.), 1943, 1949.

H. Bürger, *v.* A. A. Bjørnbo.

Campanus of Novara, *v.* F. S. Benjamin & G. J. Toomer, Euclid.

F. J. Carmody, *Arabic Astronomical and Astrological Sciences in Latin Translation*, Berkeley and Los Angeles, 1956.

Geoffrey Chaucer, *A Treatise on the Astrolabe* (ed. W. W. Skeat), Oxford, 1872.

M. Clagett, *Archimedes in the Middle Ages*, vol. i, *The Arabo-Latin Tradition*, Madison, 1964.

A. C. Crombie, *Robert Grosseteste and the Origins of Experimental Science, 1100–1700*, Oxford, 1953 (repr. 1962).

—— (ed.), *Scientific Change* (Symposium proceedings of 1961, Oxford), London, 1963.

Franz Cumont, *v.* Joseph Bidez.

M. Curtze, 'Urkunden zur Geschichte der Trigonometrie in christlichen Mittelalter', *Bibliotheca Mathematica* (3), i (1900), pp. 321–416.

J.-B. J. Delambre, *Histoire de l'astronomie ancienne*, 2 vols., Paris, 1817.

—— *Histoire de l'astronomie du moyen âge*, Paris, 1819.

A. J. Denomy, *v.* Nicole Oresme.

Victorin Doucet, *Commentaires sur les sentences, supplément au répertoire de M. Frédéric Stegmueller*, Quaracchi, Florence, 1954.

—— *v.* also F. Stegmüller.

D. C. Douglas, *English Scholars, 1660–1730*, 2nd edn., London, 1951.

A. G. Drachmann, *Ktesibios, Philon, and Heron* (*Acta Historica Scientiarum Naturalium et Medicinalium*, iv), Copenhagen, 1948.

—— 'Heron and Ptolemaios', *Centaurus*, i (1950), pp. 117–31.

—— 'The Plane Astrolabe and the Anaphoric Clock', *Centaurus*, iii (1954), pp. 183–9.

—— *Antikkens Teknik*, Copenhagen, 1963.

—— *The Mechanical Technology of Greek and Roman Antiquity* (*Acta Historica Scientiarum Naturalium et Medicinalium*, xvii), Copenhagen, 1963.

J. L. E. Dreyer, *History of the Planetary Systems from Thales to Kepler*, Cambridge, 1906. (Reprinted under the title *A History of Astronomy from Thales to Kepler*, New York, 1953.)

Charles Ducange, *Glossarium mediae et infimae latinitatis*, Paris, 1840–50.

Pierre Duhem, *Le Système du monde. Histoire des doctrines cosmologiques de Platon à Copernic*, 9 vols., Paris, 1913–59.

A. B. Emden, *A Biographical Register of the University of Oxford to A.D. 1500*, 3 vols., 1957, 1958, 1959.

Euclid, *Elementa* (ed. J. L. Heiberg, with tr. into Latin), 4 vols., Leipzig, 1883–8.

—— *The Thirteen Books of Euclid's Elements* (tr. T. L. Heath), 3 vols., 2nd edn., Cambridge, 1926. (Reprinted New York, 1947).

—— *Elementa geometriae* [version of Campanus of Novara], Erhard Ratdolt, Venice, 25 May 1482.

S. García Franco, *Catálogo crítico de astrolabios existentes en España*, Madrid, 1945.

*Gesta abbatum monasterii Sancti Albani a Thoma Walsingham . . . compilata* (ed. H. T. Riley), vol. i (793–1290), Rolls Series, London, 1867; vol. ii (1290–1349), 1867; vol. iii (1349–1411).

Owen Gingerich, *v.* Emmanuel Poulle.

F. K. Ginzel, *Handbuch der mathematischen und technischen Chronologie*, 3 vols., Leipzig, 1906–14.

B. R. Goldstein, 'The Medieval Hebrew Tradition in Astronomy', *Journal of the American Oriental Society*, lxxxv (1965), pp. 145–8.

—— *Ibn al-Muthannā's Commentary on the Astronomical Tables of al-Khwārizmī*, New Haven, 1967.

—— 'The Arabic Version of Ptolemy's Planetary Hypotheses', *Transactions of the American Philosophical Society*, lxii, part 4, 1967.

—— *Al-Biṭrūjī: On the Principles of Astronomy*, 2 vols., New Haven, 1971.

—— 'Theory and Observation in Medieval Astronomy', *Isis*, lxiii (1972), pp. 39–57.

—— *The Astronomical Tables of Levi ben Gerson* (Connecticut Academy of Arts and Sciences, Transactions, vol. xlv), New Haven, 1974.

—— 'Levi ben Gerson's Analysis of Precession', *Journal for the History of Astronomy*, vi (1975), pp. 31–41.

E. Grant (ed.), *A Source Book in Medieval Science*, Cambridge, Mass., 1974.

—— *v.* also Nicole Oresme.

Wilhelm Gundel, *v.* Franz Boll.

R. T. Gunther, *Early Science in Oxford*, vol. ii, Oxford, 1923.

Nicolas Halma, *v.* Ptolemy.

Willy Hartner, *Oriens—Occidens*, Hildesheim, 1968.

—— 'Naṣir al-Dīn al-Ṭūsī's Lunar Theory', *Physis*, xi (1969), pp. 287–304.

—— 'Trepidation and Planetary Theories. Common Features in Late Islamic and Early Renaissance Astronomy', *Accademia Nazionale dei Lincei, Atti dei Convegni*, xiii (1971), pp. 609–29.

—— 'Ptolemy. Azarquiel, Ibn al-Shāṭir, and Copernicus on Mercury. A Study of Parameters', *Archives internationales d'histoire des sciences*, xxiv (1974), pp. 5–25.

C. H. Haskins, *Studies in the History of Mediaeval Science*, Cambridge, Mass., 1924 (revised edn. 1927).

Thomas Hearne, *v.* John Leland.

T. L. Heath, *Aristarchus of Samos, the Ancient Copernicus*, Oxford, 1913.

J. L. Heiberg, *v.* Ptolemy, Euclid.

R. P. Howgrave-Graham, 'Some Clocks and Jacks, with notes on the History ✓ of Horology', *Archaeologia*, lxxvii (1928), pp. 257–312.

Henri Hugonnard-Roche, *L'Œuvre astronomique de Thémon Juif, maître parisien du XIVᵉ siècle* (Centre de Recherches d'Histoire et de Philologie, Hautes études médiévales et modernes, xvi), Paris, 1973.

Jabir ibn Aflaḥ (Gebir), *De astronomia libri IX* . . . (issued with *Petri Apiani Instrumentum primi mobiles*), Joh. Petreius, Nuremberg, 1534.

C. Johnson, *v. Medieval Latin Word-list*.

Pearl Kibre, 'Lewis of Caerleon, Doctor of Medicine, Astronomer and Mathematician', *Isis*, xliii (1952), pp. 100–8.

E. S. Kennedy, 'A survey of Islamic Astronomical Tables', *Transactions of the American Philosophical Society*, n.s. xlvi, part 2, 1956.

—— 'Parallax Theory in Islamic Astronomy', *Isis*, xlvii (1956), pp. 33–53.

—— *The Planetary Equatorium of Jamshīd Ghiyāth al-Din al-Kāshī* (*d. 1429*), Princeton, 1960.

—— 'The Arabic Heritage in the Exact Sciences', *Al-Abhath*, xxiii (1970), pp. 327–44.

E. S. Kennedy & David Pingree, *The Astrological History of Māshā'-allāh*, Cambridge, Mass., 1971.

al-Khwārizmī, *v.* B. R. Goldstein, Otto Neugebauer, H. Suter.

D. A. King, 'An Analog Computer for Solving Problems of Spherical Astronomy: The Shakkāzīya Quadrant of Jamāl al-Dīn al-Māridīnī', *Archives internationales d'histoire des sciences*, xxiv (1974), pp. 219–42.

David Knowles, *The Religious Orders in England*, vols. i and ii, Cambridge, 1948, 1955.

K. Kohl, *v.* A. A. Bjørnbo.

Max Krause, 'Die Sphärik des Menelaos aus Alexandrien in der Verbesserung von Abū Naṣr Manṣūr b. 'Ali b. 'Iraq', *Abhandlungen der Geschichte der Wissenschaften zu Göttingen*, Philol.–hist. Kl., 3. Folge, Nr. 17 (1936).

Paul Kunitzsch, *Arabische Sternnamen in Europa*, Wiesbaden, 1959.

—— *Typen von Sternverzeichnissen in astronomischen Handschriften des zehnten bis vierzehnten Jahrhunderts*, Wiesbaden, 1966.

—— *Der Almagest. Die Syntaxis Mathematica des Claudius Ptolemäus in arabisch-lateinischer Überlieferung*, Wiesbaden, 1974.

R. E. Latham, *v. Medieval Latin Word-list*.

Harriet Pratt Lattin, *Star Performance*, Philadelphia, 1969.

John Leland, *Joannis Lelandi, antiquarii, De rebus Britannicis Collectanea* (ed. T. Hearne), Oxford, 1715, esp. vol. iv.

Levi ben Gerson, *v.* B. R. Goldstein.

R. P. Lorch, 'Jābir ibn Aflaḥ and his Influence in the West' (unpublished Ph.D. thesis), University of Manchester, 1970.

Justin McCann, *The Rule of Saint Benedict*, London, 1952.

F. R. Maddison, 'Hugo Helt and the Rojas Astrolabe Projection', *Agrupamento de Estudos de Cartografia Antiga*, xii, Secção de Coimbra, Junta de Investigaçõnes do Ultramar, 1966.

—— 'Medieval Scientific Instruments and the Development of Navigational Instruments in the XVth and XVIth Centuries', ibid., xxx, 1969.

—— *v.* S. A. Bedini, Emmanuel Poulle.

Karl Manitius, *v.* Ptolemy.

*Medieval Latin Word-list from British and Irish Sources*, prepared by J. H. Baxter & C. Johnson, London, 1934. *Revised Medieval Latin Word-list from British and Irish Sources*, prepared by R. E. Latham (for the British Academy, by the Oxford University Press), London, 1965.

C. Melzi D'Eril, *v.* J. Boffito.

Karl Menninger, *Zahlwort und Ziffer*, 2nd edn., Göttingen, 1958.

A. D. Menut, *v.* Nicole Oresme.

H. Michel, *Traité de l'astrolabe*, Paris, 1947.

J. M. Millás Vallicrosa, *Assaig d'història de les idees físiques i matemàtiques a la Catalunya medieval*, Barcelona, 1931.

—— *Estudios sobre Azarquiel*, Madrid–Granada, 1943–50.

—— *Estudios sobre historia de la ciencia española*, Barcelona, 1949.

K. P. Moesgaard, 'Thābit ibn Qurra between Ptolemy and Copernicus: an Analysis of Thābit's Solar Theory', *Archive for History of Exact Sciences*, xii (1974), pp. 199–216.

Ibn al-Muthannā, *v.* B. R. Goldstein.

C. A. Nallino, *Raccolta di scritti editi e inediti*, vol. v, *Astrologia, Astronomia, Geografia*, Rome, 1944.

—— *v.* also al-Battānī.

Otto Neugebauer, 'The Alleged Babylonian Discovery of the Precession of the Equinoxes', *Journal of the American Oriental Society*, lxx (1950), pp. 1–8.

—— 'The Transmission of Planetary Theories in Ancient and Medieval Astronomy', *Scripta Mathematica*, xxii (1956), pp. 165–92.

—— *The Exact Sciences in Antiquity*, 2nd edn., Providence, R.I., 1957.

—— *The Astronomical Tables of Al-Khwārizmī* (Kgl. Danske Vidensk. Skrifter, 7. R., Hist. og filos. Afd., vol. iii, no. 1), Copenhagen, 1962.

—— *A History of Ancient Mathematical Astronomy*, 3 vols., (*Studies in the History of Mathematics and Physical Sciences*, i), Berlin, etc., 1975.

—— & A. Sachs, 'The "Dodekatemoria" in Babylonian astrology', *Archiv für Orientforschung*, xvi (1953), pp. 65–6.

—— & H. B. Van Hoesen, *Greek Horoscopes* (Memoirs of the American Philosophical Society, no, 48), Philadelphia, 1959.

P. V. Neugebauer, *Tafeln zur astronomischen Chronologie*, 3 vols., Leipzig, 1912, 1914, 1922.

J. D. North, 'Werner, Apian, Blagrave and the Meteoroscope', *British Journal for the History of Science*, iii (1966), pp. 57–65.

—— 'Opus quarundam rotarum mirabilium', *Physis*, viii (1966), pp. 337–72.

—— *Tolomeo* (in the series I Protagonisti), C.E.I., Milan, 1968.

—— 'Kalenderes enlumyned ben they. Some Astronomical Themes in

Chaucer', *Review of English Studies*, N.S., xx (1969), pp. 129–54, 257–83, 418–44.

—— 'A post-Copernican Equatorium', *Physis*, xi (1969), pp. 418–57.

—— 'The Astrolabe', *Scientific American*, January 1974, pp. 96–106.

—— 'Monasticism and the First Mechanical Clocks', in *The Study of Time*, vol. ii (ed. J. T. Fraser & N. Lawrence), New York, 1975, pp. 381–98.

—— 'The Medieval Background to Copernicus', in *Copernicus* (Vistas in Astronomy, vol. xvii, ed. A. Beer & K. Å. Strand), Oxford, Pergamon Press, 1975, pp. 3–24.

—— 'Medieval Cosmology and the Age of the World', in *Cosmology—History —Theology* (forthcoming), ed. A. D. Breck & W. Yourgrau.

—— 'Astrology and the Fortunes of Churches', in the proceedings of the Second International Colloquium in Ecclesiastical History (Oxford, September 1974, publication forthcoming), ed. D. Baker.

—— 'The Alfonsine Tables in England', in the *Festschrift* for Willy Hartner (forthcoming), ed. W. Saltzer & Y. Maeyama.

—— with Ole Østerby & Kurt Møller-Pedersen, '*Summa ratione confectum.* An Astrolabe drawn by Computer', *Archives internationales d'histoire des sciences*, xxv (1975), pp. 73–81.

Nicole Oresme, *Le Livre du ciel et du monde* (ed. A. D. Menut & A. J. Denomy), Madison, 1968.

—— *Nicole Oresme and the Kinematics of Circular Motion. Tractatus de commensurabilitate vel incommensurabilitate motuum celi* (ed. E. Grant), Madison, 1971.

W. A. Pantin, *Documents Illustrating the Activities of the General and Provincial Chapters of the English Black Monks, 1215–1540*, 3 vols. Camden Third Series, vols. xlv, xlvii, liv, London, 1931, 1933, 1937.

Pappus, *v.* A. Rome.

M. B. Parkes, *English Cursive Book Hands, 1250–1500*, Oxford, 1969.

Olaf Pedersen, 'A Fifteenth Century Glossary of Astronomical Terms', *Classica et Mediaevalia*, ix (1973), pp. 584–94.

—— 'Logistics and the Theory of Functions. An Essay in the History of Greek Mathematics', *Archives internationales d'histoire des sciences*, xxiv (1974), pp. 29–50.

—— *A Survey of the Almagest* (*Acta Historica Scientiarum Naturalium et Medicinalium*, vol. xxx), Odense, 1974.

—— & Mogens Pihl, *Early Physics and Astronomy*, London and New York, 1974.

David Pingree, 'Historical Horoscopes', *Journal of the American Oriental Society*, lxxxii (1962), pp. 487–502.

—— 'The Greek Influence on Early Islamic Mathematical Astronomy', *Journal of the American Oriental Society*, xciii (1973), pp. 32–43.

John Pits, *Relationum historicarum de rebus anglicis tomus primus*, Paris, 1619.

R. L. Poole, *Studies in Chronology and History*, Oxford, 1934.

J. B. Post & A. J. Turner, 'An Account for Repairs to the Westminster Palace Clock', *The Archaeological Journal*, cxxx (1973), pp. 217–20.

Emmanuel Poulle, 'Astrologie et tables astronomiques au XIIIᵉ siècle: Robert Le Febre et les tables de Malines', *Bulletin philologique et historique (jusqu'à 1610)*, 1964, pp. 793–831.

Emmanuel Poulle, 'Les Instruments astronomiques du moyen âge', *Le Ruban rouge*, xxxii (1967), pp. 18–29; reprinted by Museum of the History of Science, Oxford, 1969.

—— 'Un instrument astronomique dans l'occident latin, la "saphea"', in *A Giuseppe Ermini*, Centro Italiano di Studi sull'alto medioevo. Spoleto, 1970, pp. 491–510.

—— 'Oronce Fine et l'horloge planétaire de la Bibliothèque Sainte-Geneviève', *Bibliothèque d'Humanisme et Renaissance*, xxxiii (1971), pp. 311–51.

—— 'Les Mécanisations de l'astronomie des épicycles: l'horloge d'Oronce Fine', *Comptes rendus de l'Académie des Inscriptions et Belles-lettres*, Jan.–Mar. 1974, pp. 59–79.

—— 'L'Équatoire de l'empereur', *Archives internationales d'histoire des sciences*, xxv (1975), pp. 127–36.

—— E. Poulle & O. Gingerich, 'Les Positions des planètes au moyen age: applications du calcul électronique aux tables Alphonsines', *Comptes rendus de l'Académie des Inscriptions et Belles-lettres*, May 1968, pp. 531–48.

—— & F. R. Maddison, 'Un Équatoire de Franciscus Sarzosius', *Physis*, v (1963), pp. 43–64.

D. J. de Solla Price, *The Equatorie of the Planetis. Edited from Peterhouse MS. 75. I*, Cambridge, 1955.

—— 'On the Origin of Clockwork, Perpetual Motion Devices and the Compass', *Contributions from the Museum of History and Technology*, United States National Museum, Bulletin, 218 (1959), Paper 6, pp. 81–112.

—— 'Gears from the Greeks. The Antikythera Mechanism—a Calendar Computer from *ca*. 80 B.C.', *Transactions of the American Philosophical Society*, N.S. lxiv (1974), part 7.

Profatius, *v.* J. Boffito & C. Melzi D'Eril, M. Steinschneider.

Ptolemy, [The Handy Tables, ed. & tr. by the Abbé Nicolas Halma in:] *Commentaire de Théon d'Alexandrie sur les tables manuelles astronomiques de Ptolemée*, 3 parts, Paris, 1822, 1823, 1825.

—— *Claudii Ptolomaei opera*, I: *Syntaxis mathematica*, ed. J. L. Heiberg, 2 vols., Leipzig, 1898, 1905.

—— *Claudius Ptolemäus: Handbuch der Astronomie*, tr. Karl Manitius, 2 vols., Leipzig, 1912, 1913 (repr. 1963).

—— *Tetrabiblos*, ed. and tr. into English by F. E. Robbins, Cambridge, Mass., & London, 1940 (repr. 1948, 1956).

—— *v.* also B. R. Goldstein.

H. Rashdall, *The Universities of Europe in the Middle Ages*, 2nd edn., ed. by F. M. Powicke and A. B. Emden, Oxford, 1936.

H. P. J. Renaud, 'L'origine du mot "Almanach"', *Isis*, xxxvii (1947), pp. 44–6.

*Rerum Britannicarum medii aevi scriptores*, Rolls Series, London, 1858–96, 99 works in 244 volumes.

H. T. Riley, *v. Gesta Abbatum*.

F. E. Robbins, *v.* Ptolemy.

A. Rome, 'Premiers essais de trigonométrie rectiligne chez les Grecs', *L'Antiquité Classique*, ii (1933), pp. 177–92.

A. Rome, *Commentaires de Pappus et de Théon d'Alexandrie sur l'Almagest*, 3 vols., Rome, Biblioteca Apostolica Vaticana, Studi e Testi, liv (1931), lxxii (1936), cvi (1943).

G. Sarton, *Introduction to the History of Science*, vol. i (Homer to Omar Khayyam), 1927; vol. ii, part 1 (Rabbi ben Ezra to Ibn Rushd), 1931; vol. ii, part 2 (Robert Grosseteste to Roger Bacon), 1931; vol. iii, 2 parts (fourteenth century), 1947, 1948. Carnegie Institute of Washington.

A. Sayili, *The Observatory in Islam and its place in the General History of the Observatory*, Ankara, 1960.

R. Schram, *Kalendariographische und chronologische Tafeln*, Leipzig, 1908.

W. Skeat, *v.* Geoffrey Chaucer.

W. M. Smart, *Text-Book on Spherical Astronomy*, 4th edn., Cambridge, 1944.

F. Stegmüller, *Repertorium commentariorum in sententias Petri Lombardi*, 2 vols., Würzburg, 1947.

—— *v.* also Victorin Doucet.

M. Steinschneider, 'Prophatii Judaei Montepessulani Massiliensis (a. 1300) Prooemium in Almanach', *Bulletino di Bibl. e di St. d. Sc. Mat. e Fis.*, ix (1876), pp. 594–614.

—— *Die arabischen Übersetzungen aus dem Griechischen* (articles 1889–96, reprinted Graz, 1960).

—— *Die hebraeischen Übersetzungen des Mittelalters und die Juden als Dolmetscher* (1893, reprinted, Graz, 1956).

—— *Die europäischen Übersetzungen aus dem Arabischen bis Mitte des 17. Jahrhunderts* (1904, 1905, reprinted, Graz, 1956).

H. Suter, 'Die Mathematiker und Astronomen der Araber und ihre Werke', *Abhandlungen zur Geschichte der mathematischen Wissenschaften*, x (1900).

—— *Die astronomischen Tafeln des Muḥammed ibn Mūsā al-Khwārizmī in der Bearbeitung des Maslama ibn Aḥmed al-Madjrīṭī . . .*, Copenhagen, 1914. (Kgl. Danske Vidensk. Skrifter, 7. R., Hist. og filos. Afd., vol. iii, no. 1.)

Thomas Tanner, *Bibliotheca Britannico-Hibernica: sive de Scriptoribus qui in Anglia, Scotia, et Hibernia . . . floruerunt . . . Commentarius*. London, 1748.

Thābit ibn Qurra, *v.* A. A. Bjørnbo.

Theon of Alexandria, *v.* Ptolemy and A. Rome.

Lynn Thorndike, *History of Magic and Experimental Science*, vols. i, ii, New York, 1923; vols, iii, iv, 1934; vols. v, vi, 1941.

—— *The* SPHERE *of Sacrobosco and its Commentators*, Chicago, 1949.

—— & Pearl Kibre, *A Catalogue of Medieval Scientific Writings in Latin*, revised edn., Mediaeval Academy of America, London, England, 1963.

G. J. Toomer, 'A Survey of the Toledan Tables', *Osiris*, xv (1968), pp. 5–174.

—— 'The Chord Table of Hipparchus and the Early History of Greek Trigonometry', *Centaurus*, xviii (1973), pp. 6–28.

—— 'Prophatius Judaeus and the Toledan Tables', *Isis*, lxiv (1973), pp. 351–5.

—— *v.* also F. S. Benjamin.

B. Tuckerman, *Planetary, Lunar, and Solar Positions, A.D. 2 to A.D. 1649, at five-day and ten-day intervals* (Memoirs of the American Philosophical Society, vol. lix), Philadelphia, 1964.

A. J. Turner, *v.* J. B. Post.

B. L. Van der Waerden, 'Die handlichen Tafeln des Ptolemaios', *Osiris*, xiii (1958), pp. 54–78.

—— *Science Awakening*, vol. i, New York, 1961; vol. ii, Leyden and New York, 1974.

W. E. Van Wijk, *Le Nombre d'or. Étude de chronologie technique, suivie du texte de la* MASSA COMPOTI *d'Alexandre de Villedieu*, The Hague, 1936.

Lynn White, Jr., *Medieval Technology and Social Change*, Oxford, 1962.

R. R. Wright, *v.* Al-Bīrūnī.

Ernst Zinner, *Verzeichnis der astronomischen Handschriften des deutschen Kulturgebietes*, Munich, 1925.

—— 'Die Tafeln von Toledo', *Osiris*, i (1936), pp. 747–74.

—— *Deutsche und niederländische astronomische Instrumente des 11.–18. Jahrhunderts*, Munich, 1956: revised edition 1967.

# INDEX OF INCIPITS

No distinction is drawn between titles (or rubrics) and the opening words of treatises proper, and a work might thus be listed more than once.

# INDEX OF MANUSCRIPTS

REFERENCE (by volume and page number) is made to all MSS. cited. When a particular MS. has been collated for the text, or taken as a standard of reference for a subsequent edition (e.g. Simon Tunsted's of *Albion*), the fact is recorded in the appropriate place, and repeated reference to the MS. on this account is not to be expected in the index.

The letters 'f.t.' refer to the folding table between pp. 136 and 137, vol. ii.

# GENERAL INDEX

In orthography the conventions of medieval Europe are usually followed, although an attempt is made in the commentaries to distinguish between the two *personae* of Arabic writers. Thus al-Battānī is given this name when Arabic originals are being discussed, whereas in commentaries on the Latin tradition the usual spelling (Albategni) is followed. The convention is often difficult to apply, but it obviates the constant need to consider possible mistranslation, faulty ascription, and interpolation.

Most of the medieval European names included are from before the period of the surname proper, and they are here ordered by Christian name (but with much cross-reference). Where precedence has been given to an early surname, the Christian name follows in full.

This is primarily an index of proper names, but it is supplemented to some extent by the glossaries. Neither stars nor planets are included. Libraries and other owners of manuscripts are included only exceptionally. Printers' names are given for a few early books. The titles of texts are introduced into some of the longer entries, notably that on Richard of Wallingford.

# PLATES

I. British Museum, MS. Cotton Nero D.7, f. 20ʳ: Richard as Abbot of St. Albans, pointing to his astronomical clock on the abbey wall. See vol. ii, p. 18 and vol. iii, p. 115

II. Cambridge University Library, MS. Gg.VI.3, f. 68<sup>r</sup>: tables of compound ratios for *Quadripartitum*, II. 5. This is the manuscript with the *De sectore* version. See vol. i, pp. 64–5, 174

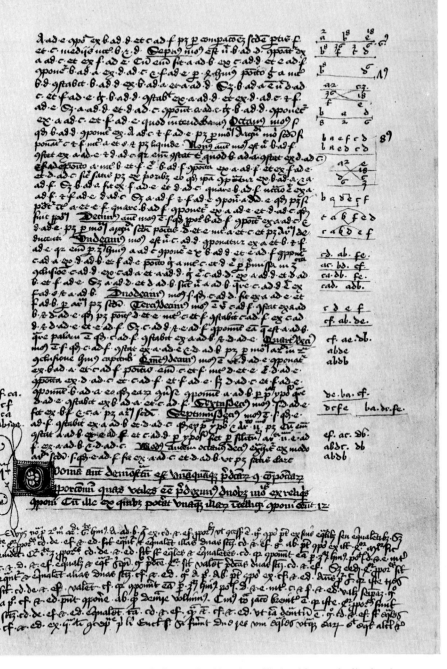

III. Bodleian Library, MS. Digby 178, f. 22ʳ: *Quadripartitum*, II. 6, with marginalia showing proportions, and also the autograph of Lewis of Caerleon. See vol. ii, p. 36

V. Corpus Christi College, Oxford, MS. (D)144, f. 59ᵛ: *Albion*, II. 19 with the spiral scale.
The marginal note concerns the Abbot's albion. See vol. ii, p. 180

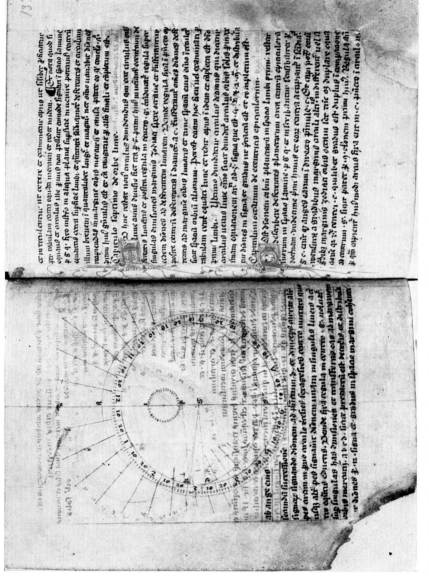

VI. Bodleian Library, MS. Ashmole 1796, ff. 132ᵛ–133ʳ: Albion, II. 6–8. This manuscript once belonged to the abbey of St. Albans, and contains elsewhere the draft of the treatise *Horologium*

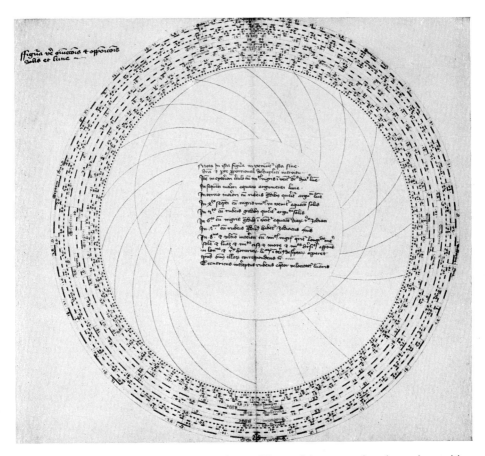

VII. Vienna, Nationalbibliothek, MS. 2332, f. 221: 'Figure of the true conjunction and opposition of the Sun and Moon'. This manuscript contains John of Gmunden's edition of *Albion*.
Cf. Plates VIII and IX

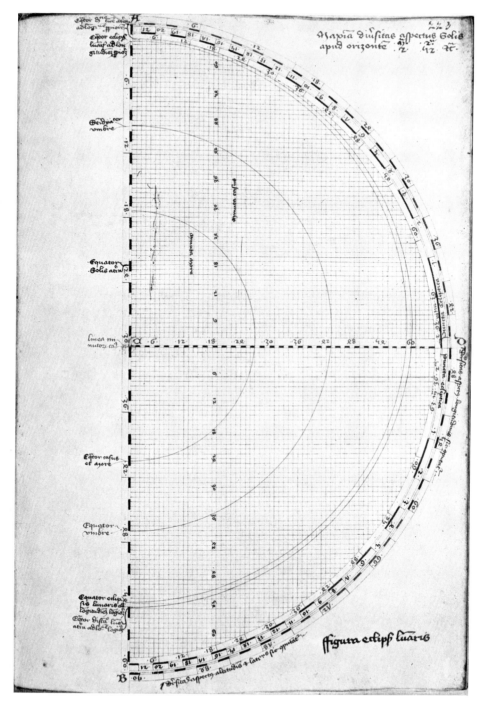

VIII. Vienna, Nationalbibliothek, MS. 2332, f. 223ʳ: 'Figure of the eclipse of the Moon'.
Cf. Plates VII and IX

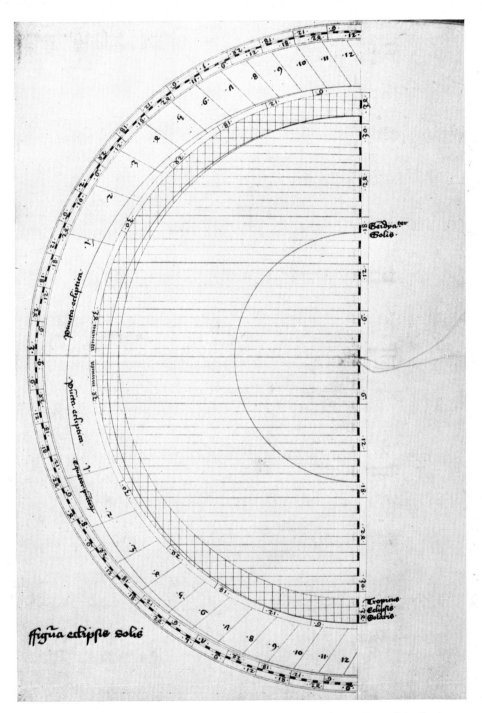

IX. Vienna, Nationalbibliothek, MS. 2332, f. 223ᵛ: 'Figure of the eclipse of the Sun'.
Cf. Plates VII and VIII

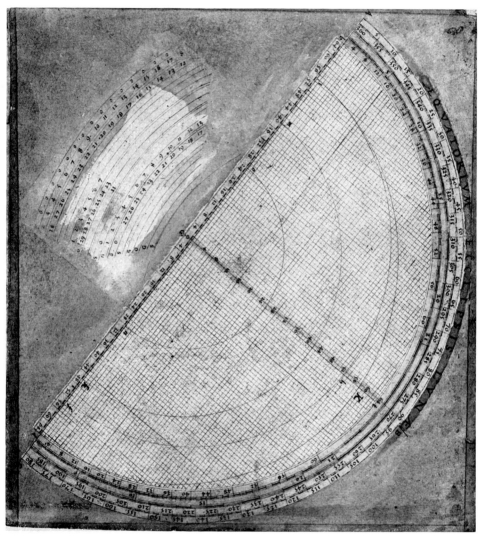

X. Bodleian Library, MS. Savile 100, f. 3ʳ: 'Equatorium of the eclipse of the Moon'. For details of this mid-sixteenth century manuscript see vol. ii, pp. 277–8

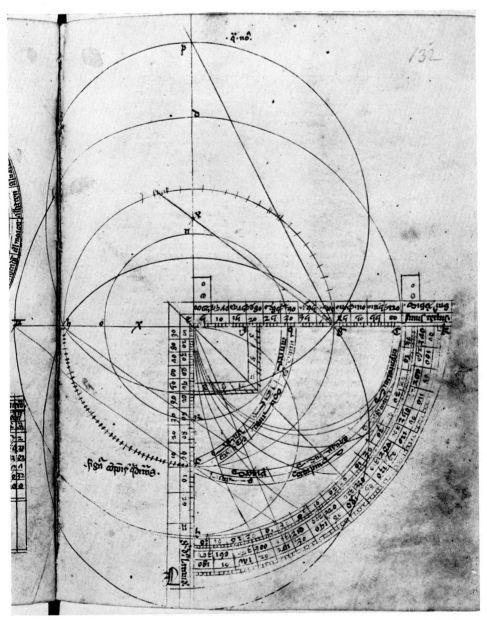

XI. Bodleian Library, MS. Ashmole 1522, f. 132ʳ: the Profatius quadrant and its geometrical construction. See vol. ii, p. 184

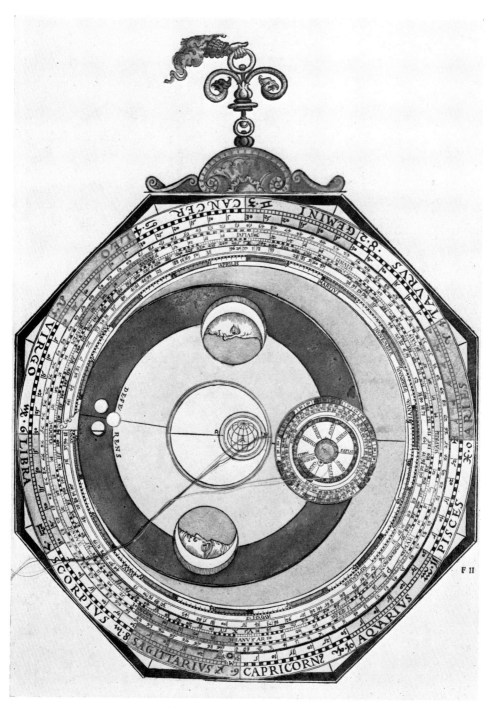

XII. Apian, *Astronomicum Caesareum* (Ingolstadt, 1540), *enunciatum* 5: the lunar equatorium. See vol. ii, p. 281

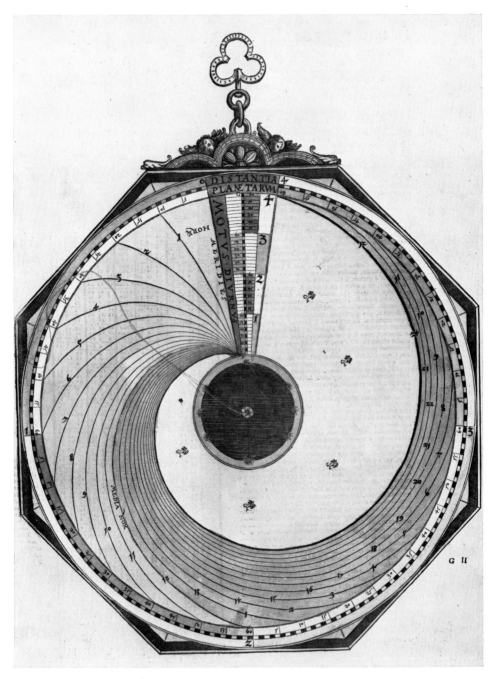

XIII. Apian, *Astronomicum Caesareum* (Ingolstadt, 1540), *enunciatum* 21: an instrument showing the influence of *Albion*. See vol. ii, p. 281

## ENVNCTIATVM VICESIMVM SEPTIM.

Perſpecta iam Lunæ defectus poſſibilitate,vt ſic loquar, id eſt, tempore quo poſſit contingere, per inſtrumentum enunctiati vic. Vlterius nunc vtrum ſcilicet nec certò fiat, quantáq; eclipſis futura ſit præſenti figura per⸗ ſpicere.

Ppoſitionem veram diei,qno futuram Lunæ obſcurationem præuideras,poſ ſibilem ſaltem, ſimulatq perſpexeris, Argumenta Solis & Lunæ ad diem eundem per enunctiata 32 & 25 comporta. Verum quoq Lunæ locū, eiuſdem latitudinem, Latitudiniſq ar⸗ gumentum,necnon capitis draconis lo cum verum per enunctiata 37 & 38 congere. His enim congeſtis, ea quæ ad eclipſes attinent, cuncta contuleris.

Per inſtrumentū q enūct: 38 eclipſis poſſibilitas, tam Solis quàm Lunæ haberi poteſt,& melius q in prio ri, hoc modo: Vbi indicem capitis draconis rite locaueris, ad tem⸗ pus ſcz veræ oppoſitionis vel coniunctionis,locum quoq Lunæ in Zodiaco contueberis, per quem filum directū ſi inter d & a ( Lu na iuxta caput draconis morante) inter g & e (Luna caudæ draconis vicina) ceciderit, fateri aude Lunam in illa oppoſitione defecturam. Viciſſim Solis defectum conſpicaberis in ſ ne quadā hac lege. Tempora ſ nis veræ cum filo ſigna, quod ſi inter b & c iuxta caput draconis, vel inter f & g iuxta caudam draconis feratur, contingens eſt Solis deliquium in hac coniunctione. Verū Solis obſcurationes propter aſpectuum diuerſitates non ita ad vn guem, ſine calculandi opera, ſicut Lunæ haberi nobis potuerunt, eas tamen, deo volente, in poſterum quoq in inſtrumenti formam redacturos nos, quo ad licebit, nō diffidimus. Poſſibilitate eclipſis Lunaris agnita, lineam in plano aliquo præſcribe, quam H D no mina, cuius medium A ſit. Huic alia perpendicularis inducta B F dicatur. Pedem nunc circini alterum in A litera fige,extenſus alter circulum occultum pro Arbitrio ſcribat, qui idem in linea B F & punctis B F ſecetur. Eo facto Lunæ latitudinem conſide⸗ ra, quæ ſi Septentrionalis fuerit, tum oportebit Argumentum eius verum eſſe 5 Signorum & 18 graduum ad minus, vſq in 6 Si gna etiam,vel Signi nullius, graduumq 12. Iam ſi Argumentum ſit Signi nullius, diuidatur quarta B D in 18 partes ſic, primo in tres, deinde ſingulas in alias tres, poſtea quamlibet in duas, & habebis 18 portiones, quarum ſigulæ 5 gradus continent. Pro ximo B literæ puncto C aſcribe, rectámq lineam à centro A per C educito. Si vero Argumentum latitudinis Signorum ſit 5, tum B H quartam diſpeſce, primæ autem poſt B literam ſe⸗ ctioni, M literam appone, perq M & A centrum lineam,vt an tea, diduc, Quod ſi meridionalem Lunam conſpexeris, Argumen tum habens 6 Signorum, & inſuper aliquot graduum, tum ab F litera D verſus proximum punctum elige, eundem cum T litera ſignans. Argumento 11 Signorum exiſtente, ab F verſus H cum puncto huiuſmodi digredere, cui aſcribe G. Hac in parte no moueatis quicquam, ſi gradus aliquando Signis non adiunctos vi⸗ deas, obſeruatis diligenter Signis tantummodo.

Hæc omniū & c.

*Quomodo eclipſis Lunæ in plano ſit for manda.*

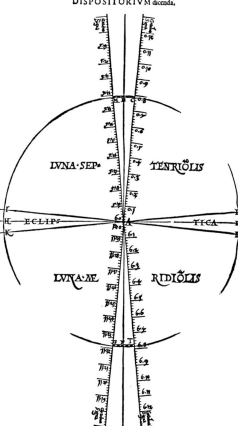

LVNA·SEP⸗       TENRIOLIS

ECLIP⸗        ⸗TICA

LVNA·ME       RIDIOLIS

Deinde in limbo exteriori ſequentis figuræ, quæ poſt hac,inſtru mentum Semidiametrale vocabitur, Solis Argumentum veſtiga, cui filum vbi ſuperpoſueris, nota eiuſdem, & limbi interioris (quæ vmbrarum varietates per ſecundam metitur) contactum, ſecunda ve ro illa à filo abſciſſa, ſeorſum excipe. Argumentum deinde Lunæ perquirens,cum filo pariter tranſmiſſo ſigna,quod filum binas tibi corporis vicz Lunaris & vmbræ terræ, ſemidiametros aperit. Fi lo ſic durante, circinum in centro fige, pede altero in punctū vſq, in quo nigrior area à filo tangitur, extenſo, illa enim extenſio ſemi diameter corporis Lunæ eſt, quam ita inuariatam reſerua : Alium poſt hac circinum ſini prioris diametri iam habiti infige, Pedem al⸗ terum eiuſdem circini vſq ad areæ fuſcæ terminum protendens, ſe⸗ midiametrum quoq vmbræ terrenæ habebis. Aream hanc circun ferentia quadam candenti circuncirca includi vides,vltra quã aliam, calami latitudine,ſuperficiem cernis,quæ in 60 ſecunda diſtribu⸗ ta eſt. Sed cum iam antea ſecunda varietatæ vmbræ didiceris, pe⸗ dea

des circini pro tot ſecundoruth diſtantia contrahi debent, & vera vmbræ ſemidiameter habebitur. Talem poſtea circinum in A pla ni prius ad hoc præparati, vbi firmaueris, circulumq deſcripſeris, veram vmbræ quantitatem cernis,vmbræ inquam terræ,quæ in Lu næ tranſitu, oppoſitionis illius tempore ſit . Mox in inſtrumento cui nomen diſpoſitorio, Lunæ Argumentum in lineis,ſiue A C ſiue A M, ſiue A G, ſiue A T ſit, contemplator. Loco obl:⸗ to pedem circini inſere, extendaſq alterum in argumenti punctum, Eandem poſtea circini extenſionem parato plano inſer,pedeq alte⸗ ro in A fixo, cum altero punctum in linea paulo antea , huic rei deputata, exprime, literaq M ſigna. Iam circinum require,exten ſum, iuxta corporis Lunaris ſemidiametrum, quem vbi fixeris in M, cumq eodem circulum deſcripſeris, corpus Lunæ te habere pu ta. Ex illis nunc demum certus eſſe potes, de contingenti huius op poſitionis eclipſi,deq quantitate eiuſdem . Si enim corpus Lunæ totum ſub vmbram conceſſerit,vniuerſalis, ſi partim, particularis,ſin omnino non,nulla eclipſis euenit.

XV. Apian, *Astronomicum Caesareum* (Ingolstadt, 1540), *enunciatum* 27: *instrumentum diametrale*, showing the influence of *Albion*. See vol. ii, pp. 283–4, and cf. Plates VIII–X

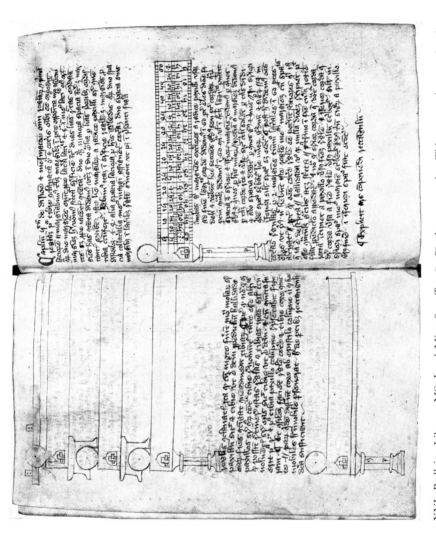

XVI. Bodleian Library, MS. Laud Misc. 657, ff. 47ᵛ–48ʳ: the rectangulus. This manuscript contains the unique Simon Tunsted version of *Albion*

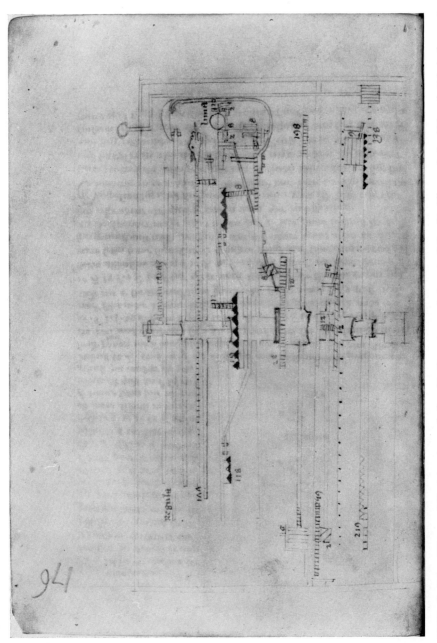

XVII. Bodleian Library, MS. Ashmole 1796, f. 176ʳ: the astronomical part of the St. Albans clock. Although this is the most elaborate of the manuscript drawings of the clock, it shows only two of the four compartments of the frame. See vol. ii, pp. 351–6

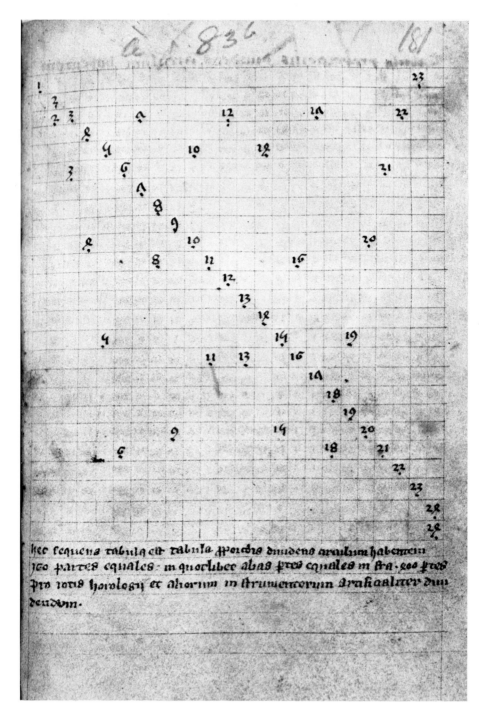

XVIII. Bodleian Library, MS. Ashmole 1796, f. 181ʳ: the 'figure' of the device for controlling the striking of the St. Albans clock. For a solution see vol. ii, p. 335

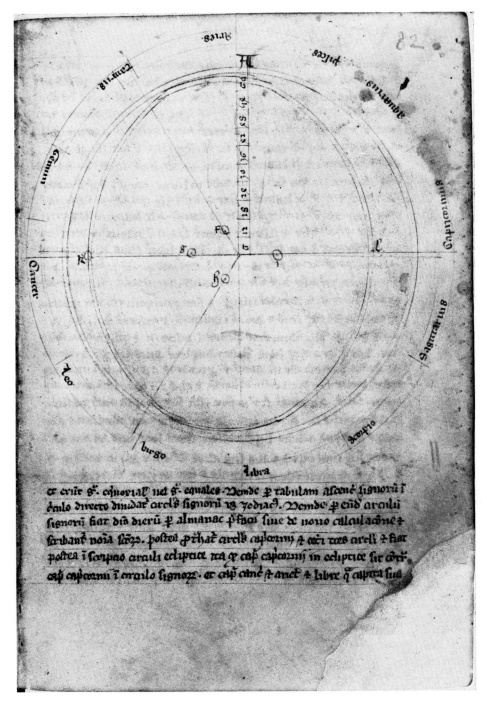

Aries

Taurus

Pisces

Aquarius

Gemini

Capricornus

Cancer

Sagittarius

Leo

Scorpio

Virgo

Libra

et erunt g̃s. eq̃novial vel g̃s equales. Deinde p̃ tabulam afcen͂e signorū in
dailo diveco dividat arel̃s signorū uɜ ʒodiac̃. Deinde p̃ eūd aroilii
signorū fiat diū dierū p̃ almanac p̃facit fiue de nouo calcul̃acõne et
fcribant noīa ſacɜɜ. poſtea p̃ t̃lat arel̃s caplorum̃ A exi͂ ceo arel̃ et fiat
poſtea ĩ fcrupao aroili ecliptice ita q̃ caṕ caplcorni͂ in ecliptice ſit c̃r̃c̃
caṕ caplcorni͂ ĩ aroilo fignoɜ. et caṕ arnet ſt arnet A libre q̃ caplcea ſuo

XIX. Bodleian Library, MS. Ashmole 1796, f. 82ʳ: the geometrical construction for the outline of the oval contrate wheel of the St. Albans clock. See vol. ii, pp. 342–8

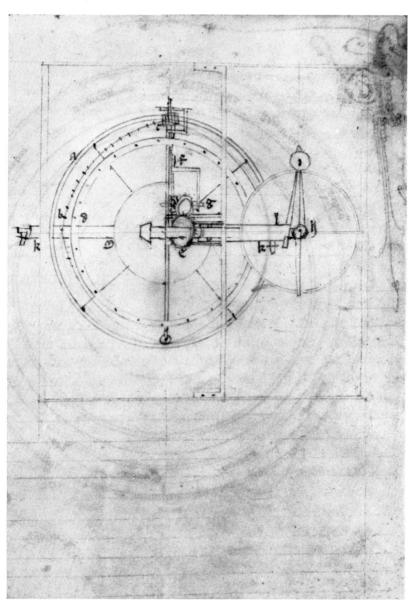

XX. Bodleian Library, MS. Ashmole 1796, f. 167ᵛ: the variable-velocity drive in the St. Albans clock. See vol. ii, p. 341. The drawing is in poor perspective

XXI. Bodleian Library, MS. Ashmole 1796, ff. 176ᵛ and 177ʳ: various wheels from compartment d/e of the St. Albans clock (see vol. iii, p. 63), and a marginal sketch showing spiral contrate teeth

XXII. Bodleian Library, MS. Digby 46, inside front cover: wooden gear-wheels. The triangular tooth-form of these fourteenth century wheels is typical of the period

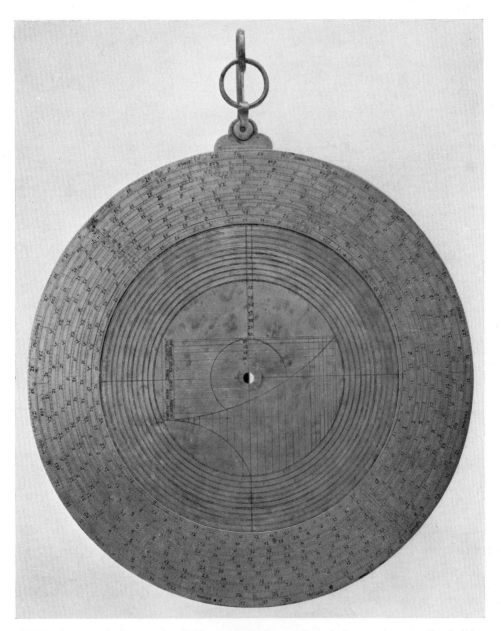

XXIII. Osservatorio Astronomico di Monte Mario, Rome: the only known extant medieval albion (mater, first face). The central disc (Plate XXIV) has been removed and the incomplete engraving of the central region is non-standard

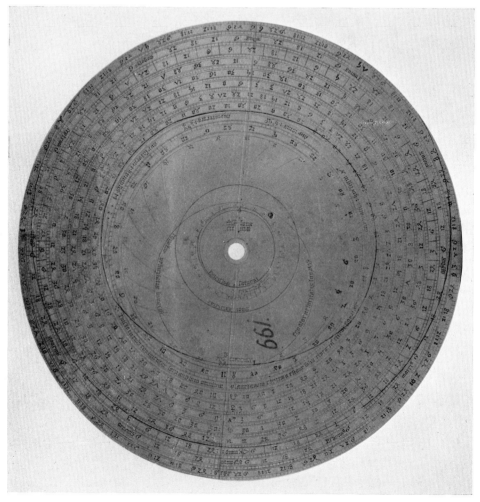

XXIV. Osservatorio Astronomico di Monte Mario, Rome: the sole surviving disc (the first face of the first disc) of the unique medieval albion (cf. Plate XXIII, which is reproduced on a smaller scale). The reverse of the disc is fully engraved with eclipse scales, etc.